Sustainable Development and the Evolution of Civil Engineering

Sérgio António Neves Lousada
University of Madeira, Portugal

IGI Global
Scientific Publishing
Publishing Tomorrow's Research Today

Vice President of Editorial Melissa Wagner
Director of Acquisitions Mikaela Felty
Director of Book Development Jocelynn Hessler
Production Manager Mike Brehm
Cover Design Jose Rosado

Published in the United States of America by
IGI Global Scientific Publishing
701 East Chocolate Avenue
Hershey, PA, 17033, USA
Tel: 717-533-8845
Fax: 717-533-7115
Website: https://www.igi-global.com E-mail: cust@igi-global.com

Library of Congress Cataloging-in-Publication Data

LCCN: 2024053581 (CIP Data Pending)
ISBN13: 9798369324486
EISBN13: 9798369324493
Isbn13Epub: 9798369324509

British Cataloguing in Publication Data
A Cataloguing in Publication record for this book is available from the British Library.

Table of Contents

Detailed Table of Contents

This chapter examines the complex housing challenges facing Machico, which is experiencing increased pressure on its housing market due to the rise of tourism. The authors explore the key issues impacting housing affordability and availability, with a particular focus on the strain placed on urban areas. The chapter identifies several strategies aimed at mitigating these challenges. These include regulating short-term rentals to curb rising prices, incentivizing the construction of affordable housing units, implementing smart urban planning and zoning practices, promoting sustainable housing solutions, and fostering community engagement in the decision-making process. The authors conclude that while the housing challenges in Machico are multifaceted, a comprehensive approach that combines regulatory measures, innovative housing solutions, and community involvement can pave the way for a more equitable and sustainable housing market. This chapter provides valuable insights for policymakers and urban planners grappling with similar issues in other tourist destinations.

This chapter evaluates the pivotal role of Life-Cycle Assessment (LCA) in civil engineering, focusing on its application in promoting sustainable infrastructure development. LCA provides a comprehensive framework to assess environmental impacts across an infrastructure project's full lifecycle—from material extraction to decommissioning. It highlights practical case studies, recent advancements in LCA methodologies, and the integration of digital technologies that increase precision. Despite challenges like data scarcity and methodological complexity, the chapter explores solutions through enhanced educational tools, technological innovation, and cross-disciplinary efforts. It concludes by advocating for LCA's broader adoption to advance sustainable, resilient, and circular infrastructure practices in the industry.

The engineering structure known as breakwaters (piers) of seaports serves an essential purpose in protecting the coasts from the damaging effects of storms and waves. Even minor harm can lead to instability, impacting the structure's elements based on incidence and frequency. At Klaipėda State Seaport, monitoring points were placed on piers for behavior analysis and stability assessment. Continuous geodetic and geotechnical measurements were conducted, including crack widths and displacements in concrete pier constructions. The analysis of monitoring data set has been done regarding determined positional changes of monitoring points and crack-gauges correlation with environmental temperature and wind speed (below 10 m/s; over 10 m/s). Due to the methodology developed, and the results obtained, no significant variations were identified in the displacements of the analysis points, relative to the breakwaters (piers)under study. It is recommended to apply this methodology as a form of monitoring to be carried out in the near future in relation to the case study.

The mining sector produces a large amount of waste and is a major contributor to environmental deterioration. However, as a sustainable strategy, new efforts have been undertaken to investigate novel techniques to repurpose mining waste in the manufacturing of bricks. The manufacturing of bricks for construction industries mostly dependent on natural resources like minerals(and clay. Huge amount of iron ore tailing waste and fly ash may be utilized for construction and building industry, which will resolve the pollution issues in sustainable way and help both mining and construction industries. Many economic and environmental advantages can be realized

Preface

INTRODUCTION

The field of Civil Engineering stands at a transformative crossroads—where tradition meets innovation, and where structural ingenuity must now align with environmental responsibility. This edited volume, *Sustainable Development and the Evolution of Civil Engineering*, is a reflection of the urgent necessity to rethink how we design, construct, and maintain the built environment in the face of pressing global challenges.

In recent decades, civil engineering has increasingly embraced a multidimensional role—not only ensuring structural integrity and functionality but also promoting economic growth, community well-being, and environmental stewardship. This book recognizes that civil engineers must now approach their work holistically, weighing technical feasibility alongside long-term social, economic, and ecological impacts. Sustainability is no longer a complementary objective; it is the foundation of future engineering practice.

The intention behind this work is to explore and portray the evolution of civil engineering through the lens of sustainability. It surveys not only the historical trajectory and achievements of the discipline but also its potential directions in the decades to come. By characterizing various academic subfields within civil engineering—from transportation and geotechnical engineering to water resources and structural systems—this volume highlights how each has responded and continues to adapt to sustainability imperatives.

The chapters herein are contributed by a diverse and accomplished array of scholars and practitioners. Each contribution enriches our understanding of how civil engineering can—and must—transform to meet the needs of future generations. This book serves as a platform for disseminating critical insights, fostering interdisciplinary collaboration, and inspiring ongoing innovation.

Designed for a wide readership—including academicians, researchers, policy-makers, environmentalists, scientists, technicians, decision-makers, practitioners, and students—this reference book aims not only to document the current state of the discipline but also to provide a lasting resource for future consultation and development. It is both a chronicle of progress and a roadmap for what lies ahead.

Let this book be a testament to the potential of civil engineering to build not only infrastructure, but also a sustainable and resilient future.

ORGANIZATION OF THE BOOK

This edited volume, *Sustainable Development and the Evolution of Civil Engineering*, brings together scholarly contributions that underscore the dynamic interplay between sustainability and the modern evolution of civil engineering. Each chapter delves into a vital area of the discipline, offering both theoretical insight and practical guidance to foster innovation and environmental responsibility in infrastructure development.

Chapter 1: Challenges of Housing in Machico: Strategies for Urban Areas Under Pressure

This opening chapter sets the stage by exploring the acute housing challenges facing Machico—a region grappling with the effects of burgeoning tourism. The authors critically examine issues surrounding housing affordability and availability, especially in urban areas under increasing strain. They propose a suite of strategic interventions, including the regulation of short-term rentals, incentivization of affordable housing projects, smart zoning practices, and sustainable urban design. A key emphasis is placed on the importance of community engagement in urban planning. This chapter provides valuable perspectives for urban planners and policymakers navigating similar pressures in other rapidly developing or tourist-driven regions.

Chapter 2: Enhancing Infrastructure Sustainability through Life-Cycle Assessment

In this chapter, the discussion centers on the critical role of Life-Cycle Assessment (LCA) in sustainable infrastructure development. The authors illustrate how LCA tools enable comprehensive evaluations of environmental impact across the entire life span of civil engineering projects—from raw material extraction through to demolition. Drawing on case studies and digital advancements, the chapter identifies both opportunities and barriers to LCA implementation. It highlights the need for interdisciplinary education, improved data access, and regulatory support to facilitate LCA's wider adoption. This exploration offers a forward-looking blueprint for embedding sustainability at the core of civil engineering practices.

Chapter 3: Heat Recovery Systems from Sewage: The Current State and Development Prospects

This chapter turns to an often-overlooked aspect of urban sustainability: heat recovery from wastewater. Highlighting the untapped potential of domestic and industrial wastewater as renewable energy sources, the authors evaluate technologies capable of reclaiming heat from sewer systems. While recent efforts have advanced domestic heat exchanger efficiency, larger-scale implementations remain underdeveloped. The chapter investigates emerging materials and rehabilitation techniques, such as plastic pipe applications, to extend sewer infrastructure longevity while maximizing energy recovery. It lays the groundwork for future research and infrastructure innovation in urban energy systems.

Chapter 4: Innovations and Sustainability in Geotechnical Engineering for Resilient Infrastructure

Addressing the subsurface dimensions of sustainability, this chapter focuses on geotechnical engineering's evolution in service of resilient infrastructure. Through an integration of theory, emerging materials, and enhanced soil stabilization techniques, the authors present a robust overview of sustainable practices within the subdiscipline. Global case studies highlight the practical challenges and solutions in real-world contexts, while the chapter emphasizes the necessity of supportive policies and professional education. It calls for a collaborative, multi-stakeholder approach to ensure infrastructure systems meet long-term sustainability goals.

Chapter 5: Innovative Fly Ash and Polypropylene Fiber Concrete: Advancing Sustainable Civil Engineering

This chapter investigates the development of geopolymer concrete using industrial by-products like fly ash and synthetic fibers to minimize the construction industry's carbon footprint. The authors explore how this alternative to traditional Portland cement reduces greenhouse gas emissions while improving material performance and durability. The chapter combines laboratory results and theoretical analysis to demonstrate the mechanical benefits of polypropylene fiber inclusion and the long-term environmental advantages of geopolymer concrete. It advocates for broader adoption of these materials to enhance the sustainability of contemporary construction methods.

Chapter 6: Integral Contributions of Civil Engineers to Transport Planning

The final chapter underscores the indispensable role of civil engineers in designing sustainable transportation networks, with a focus on India's rapidly urbanizing landscape. Through detailed analysis of the Smart Cities Mission and the integration of digital tools such as GIS, BIM, and IoT, the authors illustrate how civil engineers are addressing challenges of population density, safety, and environmental impact. The chapter explores sustainable materials and adaptive traffic solutions that align with global best practices while addressing local nuances. It highlights how civil engineering continues to evolve to meet the infrastructural demands of growing, complex societies.

Chapter 7: Long-Term Sustainability for Diverse Green Building Types: Challenges and Solutions

In this chapter, the authors delve into the intricate realities of ensuring long-term sustainability in green buildings across various typologies, including both residential and commercial developments. Addressing a spectrum of energy-efficient designs, renewable systems, and sustainable materials, the chapter brings to light the nuanced challenges of maintaining these systems over time. Through a combination of literature reviews and case analyses, the chapter introduces a practical framework categorizing sustainability challenges into technical, financial, and operational dimensions. This work is especially valuable for professionals and decision-makers striving to enhance the performance and resilience of green buildings in increasingly urbanized environments.

Chapter 8: Role of Civil Engineer in the Sustainable Water Management

This chapter outlines the indispensable role of civil engineers in addressing one of the world's most pressing challenges—sustainable water management. With more than 60% of the global population residing in water-stressed regions, the authors emphasize the need for engineering innovation to secure reliable access to clean water. Civil engineers are portrayed as the architects of water infrastructure, responsible for designing, building, and maintaining vital systems such as dams, reservoirs, pipelines, and treatment plants. The chapter underscores how civil engineering intersects with ecological sustainability, highlighting both the responsibilities and opportunities that engineers have to foster water security in a rapidly changing world.

Chapter 9: Seaport Pier Constructions, Diagnostics, and Evaluation Using Monitoring Data

Focusing on the structural integrity and longevity of seaport piers, this chapter presents a detailed case study of breakwaters at Klaipeda State Seaport. It illustrates how continuous geodetic and geotechnical monitoring—tracking elements like displacement and crack propagation—provides essential data for evaluating pier performance. The authors demonstrate how environmental factors such as temperature and wind speed influence these structural behaviors. The methodology developed here offers a replicable model for port authorities and engineers seeking data-driven approaches to infrastructure diagnostics, showcasing a proactive path to maritime infrastructure resilience and sustainability.

Chapter 10: Sustainable Mining Waste Management for Building and Construction

Turning an environmental liability into a construction asset, this chapter investigates how mining waste—particularly iron ore tailings and fly ash—can be transformed into sustainable building materials such as bricks. The authors explore the economic and ecological potential of this circular approach, presenting experimental data to validate the structural viability of mining-waste-based bricks. This research offers a compelling model for industrial symbiosis, suggesting that the integration of mining by-products into construction not only mitigates pollution but also conserves natural resources, driving both industries toward a more sustainable future.

Chapter 11: The Requirement of Sustainable Water Management and its Progress in India

This chapter presents a comprehensive analysis of India's pursuit of sustainable water management, mapping out both its urgent needs and ongoing progress. It frames sustainable practices as pivotal to resolving the country's mounting water-related challenges, which range from availability to quality. The authors review critical gaps in current research and practice while examining major national policies and infrastructure projects aimed at long-term water resilience. The inclusion of empirical data and graphical representations strengthens the discussion, providing a nuanced perspective that is both informative and action-oriented for policymakers and researchers.

Chapter 12: Understanding Mobility Patterns in Machico's Low-Density Rural Areas Through Stated Preference Surveys

Drawing on empirical data from Machico, Madeira Island, this chapter explores transportation behaviors in a low-density rural context, using stated preference surveys to analyze how residents make mobility decisions. The study identifies frequency, reliability, and travel time as key determinants, offering practical recommendations such as flexible transport services and improved public transit frequency. The insights generated are not only vital for regional planning in Madeira but also offer transferable lessons for rural mobility strategies around the globe, advocating for transportation solutions that are both user-centered and adaptable to local needs.

CONCLUSION

As editor of this volume, I am pleased to present a diverse and multidisciplinary collection of chapters that collectively underscore the critical role of engineering, planning, and innovation in advancing sustainable development. From reimagining green building longevity to transforming mining waste into valuable construction materials, and from safeguarding maritime infrastructure to optimizing rural mobility, the contributions in this book illuminate the breadth and depth of challenges and opportunities we face in the pursuit of sustainability.

A recurring theme throughout the chapters is the need for integrated, context-sensitive solutions—whether it be through policy alignment, data-driven diagnostics, or the creative repurposing of waste. Each chapter not only contributes practical insights within its domain but also invites reflection on the interconnected nature of environmental, social, and technical systems.

This compilation is intended as both a reference and a catalyst—for researchers, practitioners, policymakers, and students—to explore, adapt, and build upon the innovative approaches documented here. It is my hope that the knowledge shared within these pages will inspire continued collaboration and action toward more resilient, equitable, and sustainable futures across regions and disciplines.

Sérgio António Neves Lousada
University of Madeira, Portugal

Chapter 1
Challenges of Housing in Machico:
Strategies for Urban Areas Under Pressure

Raul Costa Alves
https://orcid.org/0000-0002-3680-6626
Universidad de Extremadura, Portugal

Sérgio António Neves Lousada
http://orcid.org/0000-0002-8429-2164
University of Madeira, Portugal

ABSTRACT

This chapter examines the complex housing challenges facing Machico, which is experiencing increased pressure on its housing market due to the rise of tourism. The authors explore the key issues impacting housing affordability and availability, with a particular focus on the strain placed on urban areas. The chapter identifies several strategies aimed at mitigating these challenges. These include regulating short-term rentals to curb rising prices, incentivizing the construction of affordable housing units, implementing smart urban planning and zoning practices, promoting sustainable housing solutions, and fostering community engagement in the decision-making process. The authors conclude that while the housing challenges in Machico are multifaceted, a comprehensive approach that combines regulatory measures, innovative housing solutions, and community involvement can pave the way for a more equitable and sustainable housing market. This chapter provides valuable insights for policymakers and urban planners grappling with similar issues in other tourist destinations.

DOI: 10.4018/979-8-3693-2448-6.ch001

INTRODUCTION

Machico, a city on the island of Madeira, Portugal, faces unique housing challenges, particularly in its most pressured urban areas (Alves, 2023a). The demand for housing has surged due to tourism growth and the region's attractiveness, while limited land availability and real estate speculation have strained the market. The median appraised value of housing in Madeira in 2022 was 1,379 euros per square meter, an 11.8% increase from 2021 (Figure 1).

Figure 1. Evolution of the median value of bank appraisal of housing in Madeira and Portugal from 2012 to 2022 (DREM, 2022a)

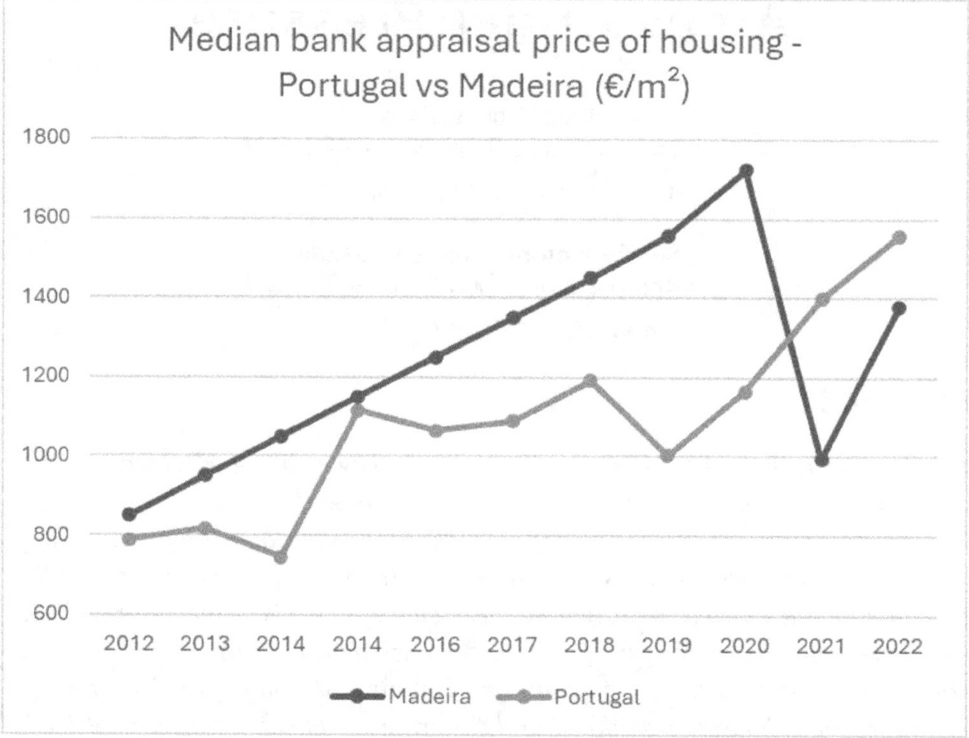

The municipality of Funchal recorded the highest price (2,109 euros/m2), while Porto Moniz had the lowest (0 euros/m²) (Figure 2).

Figure 2. Median housing appraisal values by municipality in Madeira in 2022 (DREM, 2022a)

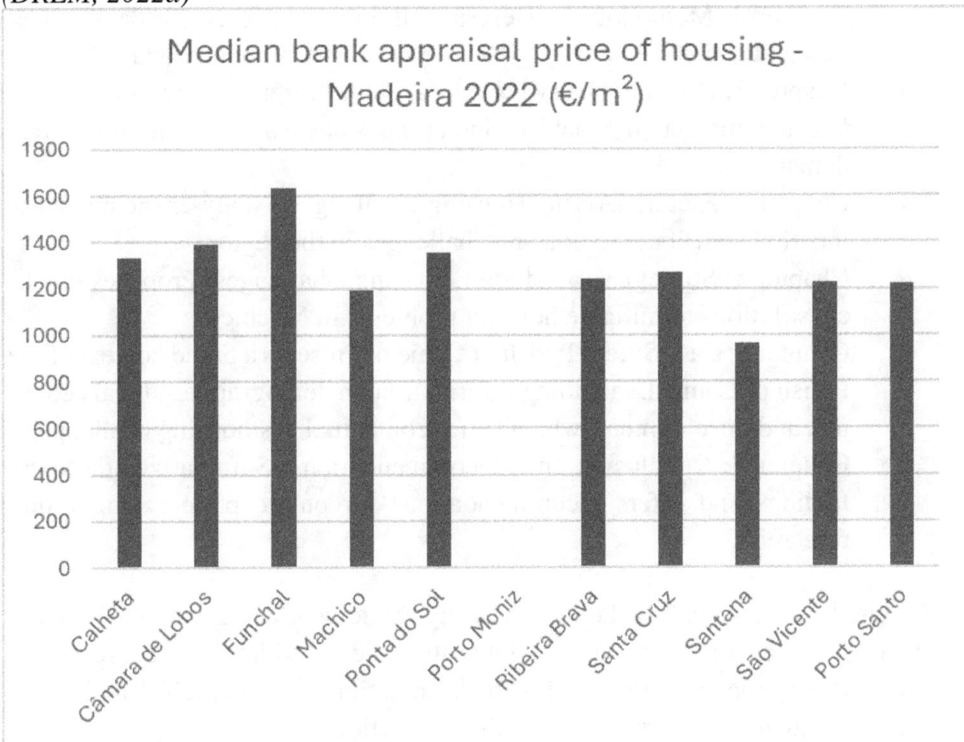

The rising pressure on the real estate market and increasing sale and rental prices make it difficult for residents, especially young adult, and low-income families, to access housing.

Analysing the housing challenges in Machico is crucial, as access to adequate housing is fundamental. In addition, understanding the dynamics of the local real estate market can provide valuable information for the formulation of effective public policies that aim to mitigate the problems faced by the population.

The main objectives of this study are:

- Analyse the main housing challenges faced by Machico.
- Identify the socioeconomic, demographic, and political factors that contribute to these challenges.
- Propose viable strategies to address the pressure on Machico's urban areas and improve access to housing.

This document is structured as follows:

- Chapter 1: Introduction: Contextualizes the problem, highlights the relevance of the study, and presents its objectives.
- Chapter 2: Methodology: Describes the methods used for data collection and analysis, including interviews and documentary analysis.
- Chapter 3: Housing Overview in Machico: Presents an overview of the housing market in Machico, including statistical data on supply and demand.
- Chapter 4: Factors Driving Housing Challenges: Analyses the main factors that contribute to housing challenges in the region.
- Chapter 5: Strategies to Address Housing Challenges: Proposes practical solutions to mitigate housing problems in Machico.
- Chapter 6: Case Study: Parish of Caniçal: Presents a broad context of the Parish of Caniçal, including information on demographics, local economy, and development, which better contextualizes housing challenges.
- Chapter 7: Conclusion and Recommendations: Summarizes the main findings and offers recommendations for public policies and future research.

This study aims to analyse the main housing challenges faced by Machico, identify contributing factors, and propose viable strategies to address these issues. The research will provide valuable insights for formulating effective public policies to mitigate the housing problems faced by the population.

METHODOLOGY

The study adopts a mixed methodological approach, combining quantitative and qualitative methods to gain a comprehensive understanding of the housing challenges in Machico. Quantitative data were collected from secondary sources, such as the Regional Directorate of Statistics of Madeira (DREM), Machico City Council, and real estate market reports. These data included property prices, occupancy rates, and demographics. Qualitative data was gathered through semi-structured interviews with residents, City Council representatives, and real estate professionals. The interviews provided direct insights into the difficulties faced by residents and their expectations regarding housing.

The quantitative data was analysed using statistical tools to identify trends and patterns in the Machico housing market. The interview transcripts were analysed using content analysis to identify emerging themes and patterns in the responses.

The study acknowledges certain limitations, including data availability, interview representativeness, and potential market changes. However, the mixed methodological approach ensures a comprehensive analysis of the housing challenges in Machico (Figure 3).

Figure 3. Diagram of the methodology used (Authors)

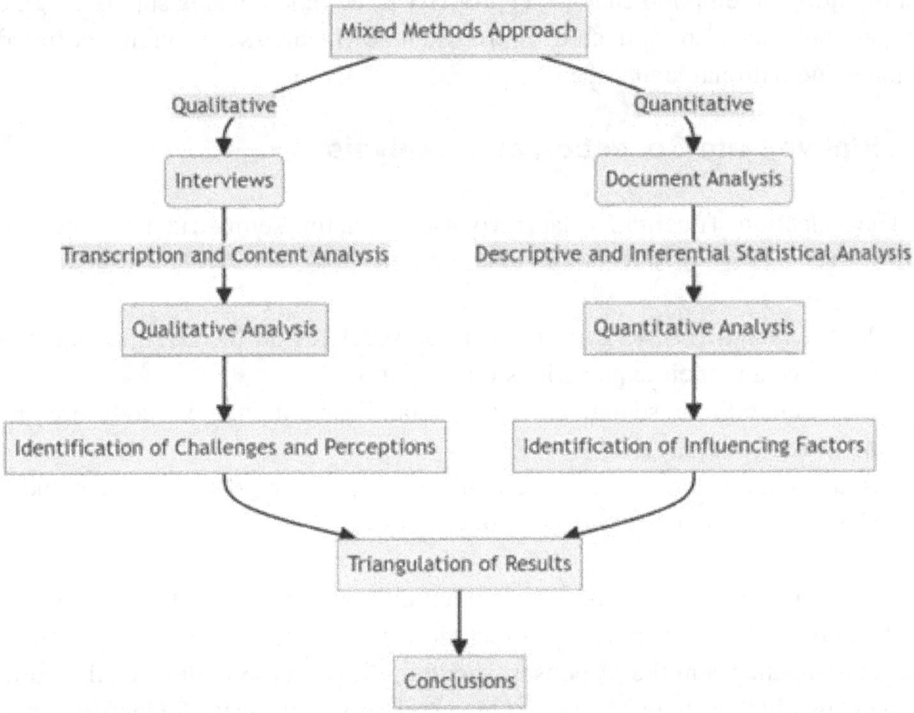

The study acknowledges certain limitations, including data availability, interview representativeness, and potential market changes. However, the mixed methodological approach ensures a comprehensive analysis of the housing challenges in Machico.

Quantitative Data Collection and Analysis

Quantitative data were sourced from reputable secondary sources, including:

- The Regional Directorate of Statistics of Madeira (DREM), which provided valuable data on property prices, occupancy rates, and demographics.
- The Machico City Council, which provided information on urban plans, housing projects, and population data.

- Real estate market reports from local agencies, offering insights into market dynamics and trends.

The quantitative data underwent rigorous analysis using statistical tools to discern trends and patterns within the Machico housing market. Descriptive analysis was employed to provide a general overview of the data, while trend analysis enabled the identification of temporal changes in property prices and housing supply. Regional comparisons were also conducted to contextualize Machico's data within the broader regional and national landscape.

Qualitative Data Collection and Analysis

The collection of qualitative data involved conducting semi-structured interviews with a diverse range of stakeholders, including:

- Machico residents, to gain firsthand perspectives on the housing challenges they face and their expectations for the future.
- City Council representatives, to understand current policies and future urban development strategies.
- Real estate professionals, to obtain insights into market dynamics and the challenges associated with providing housing.

The interviews were meticulously recorded and transcribed to facilitate in-depth analysis. The content analysis technique was employed to identify recurring themes and patterns in the responses, shedding light on the multifaceted nature of the housing challenges in Machico. Open coding was utilized to identify relevant categories and subcategories within the data, followed by axial coding to explore the relationships between these categories. This systematic approach culminated in a profound understanding of the factors that influence housing challenges in Machico.

Study Limitations

While the adopted methodology offers a comprehensive analysis, it is important to acknowledge certain limitations:

- Data Availability: Some desired information was either unavailable or incomplete, potentially impacting the depth of the analysis.
- Interview Representativeness: Although efforts were made to ensure a representative sample, the perceptions of the interviewees may not fully reflect the entirety of Machico's population.

- Market Changes: The housing market is inherently dynamic, and the data collected represents a specific point in time. Consequently, it may not capture recent or future changes in the market.

Despite these limitations, the research methodology employed in this study provides a robust foundation for understanding the housing challenges in Machico. The combination of quantitative and qualitative approaches allows for a nuanced and holistic analysis, contributing to the rigor and transparency of the research.

HOUSING OVERVIEW IN MACHICO

Machico's housing challenges are driven by growing demand, geographical limitations, and socioeconomic dynamics. The demand for housing is influenced by several factors, including tourism growth, the region's attractiveness to foreigners, and the needs of the local population. In 2022, 382 new housing units were completed in Madeira, but demand outstrips supply, leading to increased property prices (DREM, 2022a). The scarcity of flat land suitable for construction exacerbates the situation, concentrating urban pressure in already built-up areas.

The housing supply in Machico is characterized by a predominance of single-family homes, a deficit of apartments, and a shortage of social housing. In 2022, 83.1% of new constructions for family housing were single-dwelling buildings (Alves, 2022b). The construction of apartments, representing only 16.9% of new housing units in 2022, has not kept pace with demand (Alves, 2022b). The limited supply of social housing further restricts access for low-income families.

The growth of tourism and the region's attractiveness to foreigners have put additional pressure on the real estate market. In 2021, 43.9% of buildings in Machico needed some kind of repair (Alves, 2022b). The rehabilitation of old buildings can be an alternative to increase the housing supply, but the proportion of licensed rehabilitation works has decreased in recent years.

The pressure on the real estate market has driven up property prices. In 2022, the median bank appraisal value of housing in Machico was 1,192 euros per square meter. The increase in prices is especially notable in the purchase and sale market. The rental market has also been impacted, with the median value of new contract rents reaching €4.59/m2 in 2022.

Machico, located on the island of Madeira, Portugal, presents complex housing challenges, driven by a growing demand for housing, geographical limitations, and socioeconomic dynamics (CMM, 2022). The demand for housing is influenced by several factors, including the growth of tourism, the region's attractiveness to foreigners, and the need for housing for the local population (Alves, 2022a).

Distribution of Supply

The housing supply in Machico is characterized by a predominance of single-family homes, a deficit of apartments, and a shortage of social housing, reflecting the complexity of the housing challenges in the city (Figure 4).

Figure 4. City of Machico (Authors)

In 2022, 83.1% of new constructions for family housing were buildings with a single dwelling, demonstrating the preference for houses. While this housing type is traditional, it may not suit everyone's needs, especially those with lower purchasing power or who are looking for more compact and affordable alternatives (DREM, 2022b).

The construction of apartments, which represents only 16.9% of new housing units in 2022, has not kept pace with demand, especially in urban areas under greater pressure (Alves, 2023a). This gap exacerbates the shortage of supply and contributes to rising prices. Additionally, the lack of apartments makes it difficult for young adult and families to access housing who are looking for more affordable options that suit their needs.

The limited supply of social housing in Machico exacerbates housing challenges, restricting access to housing for low-income families. The lack of effective public policies to promote the construction of housing at controlled costs perpetuates social inequality and hinders the inclusion of vulnerable groups.

The growth of tourism and the region's appeal to foreigners have increased demand for both short-term and long-term accommodation (Alves, 2022a). This dynamic put additional pressure on the real estate market, competing with the demand for housing from the local population and contributing to rising prices.

In 2021, 43.9% of buildings in Machico needed some kind of repair, with 26.8% classified as medium repairs and 10.2% as deep (DREM, 2022a). The rehabilitation of old buildings can be an alternative to increase the housing supply at more affordable costs, revitalize the city centre and promote sustainability. However, the proportion of licensed rehabilitation works in Machico has decreased in recent years, reaching only 18.4% in 2022 (DREM, 2022a).

The unequal distribution of housing supply and the scarcity of affordable options have significant socioeconomic implications for Machico. Rising prices and lack of adequate housing can lead to social exclusion, exodus of residents, and difficulty attracting and retaining young professionals and families in the city (Alves, 2023a).

Property Prices

The pressure on the real estate market has driven up property prices in Machico. In 2022, the median bank appraisal value of housing in Machico was 1,192 euros per square meter, an increase of 2.1% compared to the previous year (DREM, 2022a). This value, although lower than the median for the Autonomous Region of Madeira (1,379 €/m^2), represents a significant increase in recent years and reflects the growing demand for housing in the area.

Table 1. Median housing appraisal values in Madeira in 2022

Indicator	Caniçal	Machico	RAM
Resident Population	3.565	19.508	253.259
Effective Growth Rate (‰)	2,2	-5,6	2,2
Natural Balance (people)	-8	-140	-1.345
Migratory Balance (people)	17	31	1.911
Aging Index (elderly/young)	100,5	194,9	165
Unemployment Rate (%)	12,1	10,9	7
Median Sale Price (€/m^2)	1.480	1.123,83	1.571

continued on following page

Table 1. Continued

Indicator	Caniçal	Machico	RAM
Median Rental Value (€/m^2)	4,59	4,59	7,35
New Rental Contracts	61	75	1.353

(DREM, 2022a)

This price surge is particularly evident in the purchase and sale market. In 2022, the median sale value of family housing in Madeira was €1,571 per square meter, a 9.4% increase from 2021 (DREM, 2022b). Although Machico's median sale value of €1,142.01 per square meter, was lower than the regional median, it remains significant.

The rental market has also been affected, with the median value of new contract rents in Machico reaching €4.59 per square meter in 2022. This upward trend in property prices, both for purchase and rental, is raising concerns about housing affordability for the local population, particularly for young adults and families with lower incomes.

The booming real estate market in Madeira, fuelled by increased demand, especially from foreigners, is exacerbating these affordability challenges. This surge in demand is placing upward pressure on prices, making housing increasingly out of reach for many locals, potentially leading to displacement or out-migration.

FACTORS DRIVING HOUSING CHALLENGES

Socioeconomic Factors

Although Machico's population decreased by 0.6% between 2021 and 2022, it grew by 2.9% in the last decade (2011-2021). This population increase, driven by both natural growth and internal and external migration, intensifies the demand for housing and puts pressure on the local real estate market (DREM, 2022b).

Furthermore, Machico, like the Autonomous Region of Madeira (RAM), is experiencing an aging population. In 2022, the proportion of elderly people (20.6%) was higher than that of young adult (16.4%), with an aging index of 194.9 elderly people per one hundred young adults. This population aging increases the demand for housing adapted to the needs of the elderly, such as accessibility and proximity to health services (DREM, 2022b).

In addition to population growth and aging, the structure of households has also been changing in Machico and RAM. In 2021, the average size of private households was 2.6 people, lower than the national average (2.8 people) (INE, 2021). At the same time, the proportion of single-person households, that is, people living alone, increased from 19.5% in 2011 to 23.6% in 2021 (INE, 2021). This change in household composition has direct implications for housing demand, with a greater need for smaller dwellings and specific types, such as T0 and T1 (Figure 5).

Figure 5. Machico parish (Authors)

Tourism in Machico, despite being an important economic engine, contributes to the seasonality of housing demand (Lousada, 2024). During peak tourism months, demand for temporary accommodation increases significantly, which can lead to an increase in rental prices and a shortage of supply for residents. This seasonality can create additional challenges for the local housing market, especially for those who rely on renting as a means of accessing housing (Alves, 2022a).

Political and Urban Planning Factors

Housing policies in Machico have proven insufficient to meet the growing demand for affordable housing. The lack of incentives for the construction of social housing and the slowness of licensing processes are significant obstacles. Additionally, the predominance of licensed homes for the construction of T0 and T1 (17.1%) and T2

(39.9%) types in 2022, despite representing a decrease compared to 2017, may not be meeting the needs of larger families (DREM, 2023).

Urban planning in Machico faces challenges arising from the island's geographical constraints and the need to preserve the environment. The location of the city between the sea and the mountains limits the areas available for urban expansion. Furthermore, areas with potential for construction are often protected by environmental legislation, which makes it difficult to develop new housing (Figure 6).

Figure 6. Parish of Porto da Cruz (Authors)

Machico has two approved land management instruments: the Machico Municipal Master Plan (CMM, 2005), which covers the entire municipal territory and defines the spatial structure and land use rules, and the Strategic Urban Rehabilitation Plan (CMM, 2019), which aims to define and present urban rehabilitation strategies for the Urban Rehabilitation Area of the Municipality of Machico - ARU (Alves, 2022b).

The effectiveness and articulation between these instruments are crucial to ensure sustainable and balanced urban development. However, the complexity of the territory, the pressure of the real estate market, and the need to reconcile diverse interests (economic, social, and environmental) can hinder the implementation of effective planning.

Machico's geographical location, in a coastal area with rugged terrain, makes the city vulnerable to various natural hazards, such as floods from the Machico river (Alves, 2023c) and coastal flooding (Alves, 2023b). These risks are aggravated by climate change, which can increase the frequency and intensity of extreme events.

Urban planning must take these risks into account, through the identification and mapping of danger zones, the implementation of prevention and mitigation measures, and the regulation of land use in vulnerable areas (Alves, 2023d). Lack of adequate planning and occupation of risk areas can amplify the impacts of these events, resulting in significant human and material losses (Alves, 2023e).

A study carried out in 2023 (Alves, 2023b) on the assessment of the consequences of coastal flooding in the Port of Machico, using the AHP (Analytic Hierarchy Process) methodology, demonstrated the importance of considering different spatial indicators, such as the distribution of population, infrastructure, built heritage and land use, in risk assessment and in defining prevention and mitigation measures.

Public participation in the definition of housing policies and plans is essential to ensure that the needs and aspirations of the local population are considered. However, the lack of transparency and effective participation mechanisms can lead to the exclusion of certain groups and decision-making that does not reflect the interests of the community (Alves, 2023a).

It is important that the urban and housing planning process in Machico is more participatory and transparent, involving the local population in defining priorities and finding solutions to housing challenges.

Political and urban planning factors, including insufficient housing policies, geographical constraints, natural hazards, and limited public participation, further complicate the housing challenges in Machico.

Economic Factors

Tourism, although a crucial economic engine for Machico, driving growth and employment, triggers a paradox in the housing market (Figure 7).

Figure 7. Quinta do Lorde tourist complex in the parish of Caniçal (Authors)

The growing demand for tourist accommodation, such as apartments and local accommodation, driven by tourism, reduces the availability of housing for residents, contributing to rising prices and making access to housing difficult for the local population (Lousada, 2024).

At the same time, the real estate market in Machico has attracted significant investments, fuelled by tourist demand and the interest of foreigners in residing on the island. However, this investment has focused on tourism and luxury housing projects, neglecting the urgent need for affordable housing for the local population (Alves, 2023a).

For example, in 2021, the proportion of Gross Value Added (GVA) of companies in high and medium-high technology sectors in Machico was 6.55%, and the proportion of personnel employed in these activities was 4.86% (DREM, 2022b). These data suggest that investment in high value-added sectors may not be generating enough jobs and income for the local population to compete in the real estate market, exacerbating the housing problem.

To summarize, economic factors, particularly the tourism industry and the focus on high-value-added sectors, play a role in shaping the housing challenges in Machico by influencing demand, investment patterns, and affordability.

Cultural and Social Factors

The cultural preference for single-family homes over apartments, observed in Machico and RAM, contributes to the pressure on the real estate market. The construction of houses, which in 2022 represented 83.1% of new constructions for family housing in Machico, requires larger land and increases construction costs, limiting supply and raising prices (DREM, 2022b).

Furthermore, the strong cultural appreciation of "home ownership" in Portugal, including Madeira, drives demand for home purchases over renting. In 2021, 55.7% of households in Machico lived in their own homes, with 28.3% having expenses with the purchase of housing (INE, 2021). This cultural preference intensifies competition in the real estate market, making access to housing difficult for those who cannot afford to buy a property.

The lack of affordable housing and the consequent overcrowding in some areas can have negative impacts on social cohesion and quality of life in Machico (Alves, 2023a). Inadequate housing and the financial pressure associated with high housing costs can generate stress and instability, affecting the well-being of the population.

The lack of affordable housing can lead to displacement of communities and social isolation, especially among young adult and low-income families. In 2021, 12.4% of the population residing in Machico changed residence, with the majority (8.0%) moving to another municipality (INE, 2021). This movement may reflect the difficulty in finding affordable housing in Machico, impacting the continuity of social relations and community cohesion.

Another important social challenge to be considered is the accessibility of housing for people with disabilities. The lack of housing adapted to their needs limits their options and can contribute to their social exclusion. In 2021, 5.8% of the population aged fifteen or over in RAM reported having great difficulty walking or climbing stairs, and 1.2% said they were unable to do so (INE, 2021). Although there is no specific data for Machico, it is likely that the city faces similar challenges in terms of housing accessibility.

In short, cultural, and social factors, such as the preference for single-family homes, homeownership, and the impact of housing affordability on social cohesion and accessibility, add another layer of complexity to the housing challenges in Machico.

STRATEGIES TO ADDRESS HOUSING CHALLENGES

The housing challenges faced by Machico resonate with global trends and theoretical discussions on housing affordability, social housing, and sustainable urban development. Addressing these challenges requires a comprehensive approach that recognizes the right to adequate housing and prioritizes the needs of vulnerable populations.

The concept of 'housing justice' emphasizes the fundamental right to adequate housing for everyone, regardless of their income or social status (Haffner & Hulse, 2017). In the context of Machico, this framework highlights the need to address the systemic inequalities that perpetuate housing precariousness and social exclusion. The role of the state in providing affordable housing has been widely debated, with various policy approaches proposed, including direct public investment, public-private partnerships, and regulatory interventions (Scanlon & Whitehead, 2008). The potential benefits and drawbacks of these approaches need to be carefully considered in the context of Machico's specific socioeconomic and political conditions. Furthermore, the principles of sustainable development, which emphasize the need to balance economic growth, social equity, and environmental protection, should guide housing policies and urban planning in Machico. The increasing financialization of housing markets, where housing is treated primarily as an investment asset rather than a social good, has exacerbated affordability issues in many cities worldwide (Forrest & Hirayama, 2015). The impacts of climate change, such as rising sea levels and extreme weather events, pose additional challenges to housing security and urban resilience (Alves, 2023d). Addressing these complex and interconnected challenges requires a holistic and integrated approach that recognizes the fundamental right to housing and prioritizes the needs of vulnerable populations (Alves, 2022a).

Intervention Proposals Tax Incentives

Granting tax benefits to builders and investors who dedicate themselves to the construction of social housing can be an effective tool to increase the supply of affordable housing (CMM, 2019). These incentives may include property tax exemptions (IMI) and reduction of construction fees (Figure 8), making social housing projects more attractive for private investment (Alves, 2022b).

Figure 8. Informative banner of the tax benefits of Machico City Council (Source: CMM, 2019)

Public-Private Partnerships (PPPs): Creating partnerships between local government and private companies can leverage resources and expertise to finance and build new affordable housing units. The government can make land available and regulatory support, while private companies contribute their experience and financial capacity (Alves, 2022b).

Public Funding: Direct public investment in the construction of social housing is essential to ensure that supply meets the needs of the population, especially the most vulnerable groups. This investment can be complemented by European and national funds, aimed at affordable housing projects (CMM, 2019).

Rehabilitation of the Existing Housing Stock

Urban Rehabilitation: The renovation of abandoned or underutilized buildings in the historic centre and in Urban Rehabilitation Areas (ARU) can increase the housing supply and revitalize degraded areas (CMM, 2016). Renovation programs may include subsidies for owners who rehabilitate old properties for rent at affordable prices (Alves, 2022b).

Energy Efficiency and Sustainability: Rehabilitation should focus on improving the energy efficiency and sustainability of homes, using ecological building materials, solar panels, rainwater collection systems and other technologies that reduce environmental impact and costs of use in the long term (CMM, 2019).

Support for Affordable Rentals

Rent Subsidies and Grants: The creation of rent subsidy programs for young adult and low-income families can facilitate access to housing and reduce the financial burden on families (CMM, 2019). These subsidies can be awarded directly to tenants or landlords, encouraging them to rent at lower prices (Alves, 2022b).

Regulation of Short-Term Rentals: Regulation of short-term rentals, such as local accommodation, is crucial to balance the supply of housing for residents and tourists (Alves, 2023a). Imposing limits on the number of local accommodations and creating specific zones for this type of accommodation (Figure 9) can help protect the rental market for residents (Alves, 2023a).

Figure 9. Tourist vocation areas in Terra Batista in the parish of Porto da Cruz (Alves, 2023a)

Diversification of Housing Supply

Construction of Apartments: Increasing the supply of apartments, especially larger types (T2, T3 and above), is essential to meet the needs of diverse types of households. The construction of buildings with diversified typologies can offer housing solutions for young couples, families with children and the elderly.

Cooperative Housing and Co-housing: The promotion of alternative housing models, such as cooperatives and co-housing, can offer innovative and affordable solutions for different social groups. These models can foster the creation of more cohesive and supportive communities, promoting the sharing of resources and common spaces.

Mitigation of Natural Risks

The geographical location of Machico, nestled between the sea and the mountains, renders it susceptible to a range of natural hazards, including floods, landslides, and coastal erosion. The increasing frequency and intensity of extreme weather events due to climate change further amplify these risks. Therefore, a comprehensive approach to risk mitigation must encompass both traditional natural hazards and the specific geotechnical challenges posed by the region's geological and topographical characteristics.

Geotechnical Aspects of Housing Development: The geological context of Machico, characterized by volcanic formations and steep slopes, presents unique challenges for housing development. The presence of highly weathered rocks and debris deposits can increase the risk of landslides, particularly during periods of heavy rainfall. Moreover, the complex geological structures, such as faults and dikes, can influence the stability of slopes and the suitability of various locations for construction.

Therefore, it is crucial to conduct thorough geotechnical investigations before any housing development to assess the stability of the terrain, identify potential hazards, and design appropriate foundation and slope stabilization measures. This will ensure the safety and longevity of new constructions and protect residents from geotechnical risks.

Geotechnical Conditions, Construction Costs, and Housing Affordability: The geotechnical conditions of a site can significantly impact the cost of construction. In areas with steep slopes or unstable soils, more complex and expensive foundation solutions may be required, such as deep foundations or retaining walls. These additional costs can translate into higher housing prices, affecting affordability, particularly for low- and middle-income residents. As highlighted in the case of the Machico hypermarket construction (Mateus de Brito, 2008), inadequate consideration of geotechnical conditions can lead to significant cost overruns and delays due to the need for extensive slope stabilization measures.

It is important to recognize the interplay between geotechnical conditions, construction costs, and housing affordability in Machico. Developing affordable housing in geotechnically challenging areas may require innovative construction techniques and financial incentives to offset the higher costs associated with foundation and slope stabilization measures.

Historical Instances of Geotechnical Hazards: Machico has a history of landslides and other geotechnical hazards that have impacted housing and infrastructure. The landslide at the Penha d'Águia cliff in 1992, which resulted in the formation of a new fajã (coastal plain), is a notable example (Catanho, 1997). The 2008 landslide during the construction of the Machico hypermarket, which impacted numerous homes and infrastructure, further emphasizes the potential consequences of overlooking geotechnical risks (Mateus de Brito, 2008). Such events highlight the potential risks associated with building on steep slopes and the importance of implementing effective risk mitigation measures.

Analysing historical instances of geotechnical hazards can provide valuable lessons for future housing development in Machico. It is essential to learn from past mistakes and incorporate these lessons into urban planning and building regulations to prevent similar incidents in the future.

Mitigation Works: Planning and implementation of engineering works, such as the construction of retaining walls, drainage systems and retention basins (Alves, 2023d), can reduce flood risks and protect urban areas (Alves, 2023e).

Land Planning and Management: The revision and updating of the Municipal Master Plan (PDM) and the Strategic Urban Rehabilitation Plan (PERU) must incorporate risk areas (Alves, 2023e) and mitigation measures, restricting construction in vulnerable areas and promoting sustainable urban development (Alves, 2022b).

In summary, mitigating natural risks in Machico necessitates a combination of engineering works, land planning, and community awareness to protect residents and infrastructure.

Success Stories

In Portugal, cities like Lisbon and Porto have implemented urban rehabilitation programs that can serve as a model for Machico. The "Reabilitar para Arrendar" program in Lisbon, for example, offers tax incentives and financing to owners who rehabilitate old properties for rent at affordable prices (CML, 2012). These programs have been successful in revitalizing degraded urban areas, increasing the supply of affordable housing and attracting new residents.

The "Housing First" model, implemented in Finland, offers immediate and unconditional housing to homeless people, with the support of integrated social services (Y-Foundation, 2016). This model has been effective in reducing the number of homeless people and improving their quality of life. Adapting this model to Machico could be a solution for the most vulnerable population.

These success stories from other regions demonstrate the potential effectiveness of various policy interventions and can serve as inspiration for addressing the housing challenges in Machico.

Action Plan

The implementation of the proposed strategies requires a structured action plan and the collaboration of various actors (Figure 10)

Figure 10. Vertical diagram of the action plan to address housing challenges in Machico (Author)

Detailed Diagnosis: Conduct an in-depth study of housing needs in Machico, identifying areas with the greatest deficit, the most affected population groups, and their specific needs.

Strategic Partnerships: Establish partnerships with the private sector, housing cooperatives, financial institutions, and non-governmental organizations (NGOs) to mobilize resources and expertise.

Funding: Secure the necessary funding for the construction of social housing, the rehabilitation of buildings and the implementation of support programs for affordable rentals.

Legal and Regulatory Framework: Review and update legislation and regulations related to housing, streamlining licensing processes and ensuring the safety and quality of constructions.
Monitoring and Evaluation: Implement a system for continuous monitoring and evaluation of housing policies and programs, to ensure their effectiveness and adjust strategies as needed.

CASE STUDY: PARISH OF CANIÇAL

The parish of Caniçal (Figure 11), nestled within the municipality of Machico, offers a microcosm of the housing challenges faced across Madeira. The unique characteristics of this parish, combined with broader housing trends on the island, contribute to a complex and pressing situation. The following sections delve deeper into the specific challenges faced by Caniçal and their implications for its residents, incorporating a historical perspective and a focus on the experiences of different population groups. The complexities and trade-offs of various policy options will also be critically examined.

Figure 11. Partial map of the parish of Caniçal (Author)

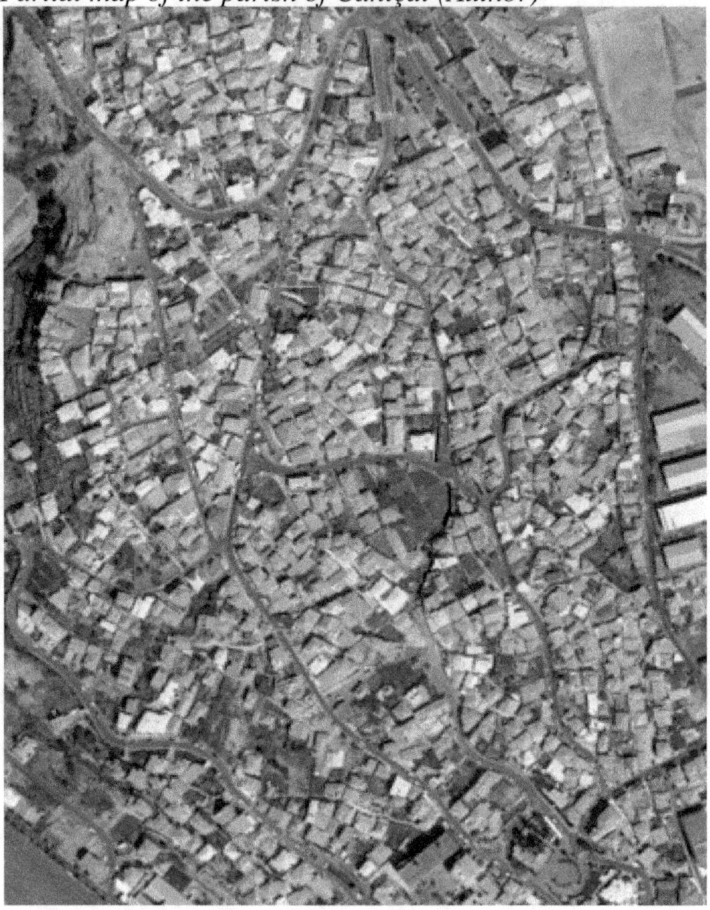

Characteristics of Caniçal

Demographics in Flux: Caniçal's population hovers around 3,565 residents, reflecting a subtle decline attributed to a negative natural balance offset by a positive migratory influx. The demographic composition reveals a diverse age distribution, with a notable aging index of 100.5 elderly people per one hundred young adults (DREM, 2022b). This aging trend, coupled with the influx of new residents, places unique demands on the housing market. The rising elderly population necessitates housing adapted to their needs, while the influx of younger residents and families creates a demand for affordable and appropriately sized homes.

Mixing the data from DREM, 2022b and INE, 2021, it was possible to build table 2, which is a comparison of indicators between the parish of Caniçal, the municipality of Machico and the Autonomous Region of Madeira.

Table 2. Comparison of indicators: Caniçal, Machico and RAM (2022)

Year	Average Sale Price (€/m²)	Median Rental Value (€/m²)
2013	1.015,27	3,8
2014	1.027,05	4,2
2015	1.020,07	4,7
2016	1.088,62	5,2
2017	1.083,61	5,9
2018	1.088,62	6,2
2019	1.198,70	7
2020	995,95	6,5
2021	1.102,40	6
2022	1.480	4,59

(Author)

Economic Transformation: Historically, rooted in fishing (Figure 12), Caniçal's economy has diversified to embrace tourism and port activities, primarily centred around the Port of Caniçal. This economic shift has spurred population growth and intensified the demand for housing, creating a dynamic interplay between economic development and housing needs. While tourism and port activities have brought economic opportunities, they have also contributed to the housing challenges by driving up demand and prices.

Figure 12. Fishing Boats "docked" at the Port of Caniçal (Author)

In summary, Caniçal's demographic shifts and economic transformation have created a complex housing landscape with diverse and often competing needs. The challenge lies in balancing the needs of an aging population, the aspirations of young residents, and the economic pressures of a tourism-driven economy.

Housing Challenges

The Weight of Geology: Machico's volcanic terrain, characterized by steep slopes and unstable soils, presents formidable challenges for construction. The need for complex foundation solutions and slope stabilization measures significantly escalates construction costs. The variability in geotechnical conditions across the region further complicates matters, as construction costs can fluctuate dramatically depending on the specific location. The 2008 landslide during the Machico hypermarket construction serves as a stark reminder of the potential financial and logistical setbacks that can arise from unforeseen geotechnical challenges (Mateus de Brito, 2008). As Catanho (1997) highlights, the variability in geotechnical characteristics across different volcanic complexes in Madeira means that construction costs can vary significantly depending on the specific location within Machico.

The Affordability Gap: The high construction costs associated with Machico's challenging geology directly impact housing affordability. Developers often pass these costs onto buyers or renters, leading to inflated housing prices that are out of reach for many residents, particularly those with low to moderate incomes. The growing disparity between income levels and housing costs is creating an affordability gap that is increasingly difficult to bridge. This phenomenon is echoed in the work of Paccoud et al. (2021), who emphasize the need to consider land prices, construction costs, and regulatory fees when addressing the housing affordability crisis.

Supply vs. Demand: A Losing Battle: The construction of new housing in Caniçal is failing to keep pace with the rising demand. In 2022, only thirty-eight buildings were completed, thirty of which were for family housing (INE, 2021). The predominance of single-dwelling buildings further exacerbates the issue, as this housing typology requires more land and resources, limiting the overall supply and driving up prices. The mismatch between housing supply and demand is a key driver of the housing crisis in Caniçal, particularly affecting young adult and low-income families who are seeking affordable options.

The Price Tag on Dreams: Property prices in Caniçal have witnessed a sharp increase, with the median bank appraisal value reaching €1,192 per square meter for apartments and €1,166 per square meter for houses in 2022 (INE, 2021). The median sale price of houses has soared even higher, reaching €1,480 per square meter in 2022. These escalating prices are fuelled by high demand, limited new construction, and the allure of the island for both residents and investors. The Table 3 illustrates the evolution of property prices in Caniçal from 2013 to 2022:

Table 3. Evolution of property prices in Caniçal (2013-2022)

Municipality	Median Bank Appraisal Value of Housing (€/m²) - 2022
Funchal	1.632
Câmara de Lobos	1.390
Santa Cruz	1.268
Machico	1.192
Calheta	1.333
Ponta do Sol	1.354
Ribeira Brava	1.242
Santana	964
São Vicente	1.227
Porto Santo	1.220
Porto Moniz	0

(Author)

The Rental Squeeze: While the median value of new rentals in Caniçal (€4.59 per square meter) is lower than the regional average of €7.35 per square meter (INE, 2021), the number of new rental contracts remains significantly below the regional average. Funchal recorded the highest value (€8.43 per square meter), while Caniçal had sixty-one new contracts, compared to the regional average of 1,353 (INE, 2021). This suggests a limited supply of rental housing, which can lead to increased competition and potentially higher rents overall. The decline of the welfare state and the increasing reliance on the private market for housing, as examined by Acolin et al. (2022), further exacerbates the affordability challenges for low-income groups. The rental market is particularly challenging for young adult and those on lower incomes, who may struggle to find affordable and secure rental options Taylor (2019).

The housing challenges in Caniçal are deeply rooted in historical trends and policy decisions, compounded by geographical constraints and economic shifts. These challenges have resulted in a housing market characterized by soaring prices, limited supply, and a growing affordability gap, with significant implications for the well-being and prospects of Caniçal's residents.

The Human Cost of the Housing Crisis: Diverse Experiences

The statistical data and trends presented thus far paint a concerning picture of the housing challenges in Machico. However, behind these numbers lie the real-life struggles of individuals and families grappling with the harsh realities of the housing crisis. The impact of these challenges is not uniform; different population groups experience the housing crisis in unique and often profound ways.

Young Adult: Dreams Deferred: For young individuals and couples in Caniçal, the dream of owning a home and establishing roots in their community is becoming increasingly elusive. Soaring property prices and limited affordable options force them to delay their plans, compromise on their living standards, or even consider leaving the island altogether in search of more affordable housing opportunities. The housing crisis is not only impacting their immediate living situations but also their long-term prospects and aspirations. The lack of affordable housing can lead to a "brain drain," where talented young adult leave the region in search of better opportunities, hindering the long-term growth and development of Caniçal.

Low-Income Families: The Struggle for Stability: Low-income families in Caniçal face a perpetual struggle to secure safe, adequate, and affordable housing. The limited availability of social housing and the rising cost of living in the private rental market leave them with few options. Overcrowding, housing insecurity, and the constant threat of displacement are common realities for many low-income families. The financial burden of housing costs can also have a ripple effect on other aspects of their lives, limiting their access to education, healthcare, and other essential services. The housing crisis can perpetuate a cycle of poverty, making it difficult for low-income families to improve their socioeconomic status and achieve upward mobility. Freemark et al. (2021) highlight the connection between income inequality and housing affordability.

The Elderly: Vulnerability and Uncertainty: The elderly population in Caniçal faces unique challenges in the housing market. Many elderly residents live on fixed incomes, making them particularly vulnerable to rising rents and the potential loss of their homes. The conversion of rental properties into tourist accommodations further exacerbates their vulnerability, as they may be forced to relocate to unfamiliar surroundings, disrupting their social networks and support systems. The housing crisis can lead to feelings of isolation, anxiety, and a diminished sense of security for the elderly, impacting their overall health and well-being. The lack of accessible and age-friendly housing options can also limit their independence and quality of life. Cox (2022) explores how these demographic shifts can shape the housing market.

Maria's Struggle to Stay: Maria, a 75-year-old pensioner, has lived in the same rented apartment in Machico's historic centre for over 30 years. The building is old and in need of repairs, but it is her home, filled with memories and a sense of belonging. Recently, her landlord informed her that he is selling the building to an investor who plans to renovate and convert the apartments into short-term tourist rentals. Maria is devastated. She cannot afford the higher rents in the newly renovated building, and the thought of leaving her familiar surroundings and community fills her with anxiety. Maria's situation highlights the vulnerability of elderly residents in the face of housing market pressures, particularly the trend towards short-term rentals driven by tourism. The displacement of long-term residents like Maria can disrupt the social fabric of the community and lead to the loss of valuable local knowledge and cultural heritage.

Sofia and Miguel, a young couple in their late twenties, have been dreaming of starting a family and settling down in their hometown of Machico. Both have stable jobs, but the soaring property prices have made their dream seem increasingly distant. They have been searching for an affordable apartment to buy for over a year, but everything within their budget is either too small, in a state of disrepair, or located far from their workplaces. The couple is now faced with the difficult choice of postponing their plans for a family, compromising on their desired living standards, or leaving Caniçal in search of more affordable housing options.

The housing crisis is not just affecting individuals and families; it is also impacting the local economy. Manuel, who owns a small bakery in Caniçal, has been struggling to retain his staff. Several of his employees have recently left their jobs because they could no longer afford to live in Caniçal. The lack of affordable housing is making it difficult for Manuel to find replacements, and he worries about the long-term viability of his business if the situation does not improve.

The pressure to build more housing in Caniçal is also taking a toll on the environment. In the outskirts of the city, a once-pristine hillside is being cleared to make way for a new housing development. The loss of this green space not only diminishes the natural beauty of the area but also increases the risk of landslides and soil erosion. The construction itself generates noise, dust, and traffic, disrupting the lives of nearby residents and impacting the local ecosystem.

In summary, the housing challenges in Caniçal are not just abstract statistics; they represent the lived experiences of real people across different generations and socioeconomic backgrounds. The inability to access affordable and adequate housing has far-reaching consequences, affecting not only individuals and families but also the social fabric and economic vitality of the community. Addressing these challenges requires a multi-faceted approach that recognizes the diverse needs of different population groups and prioritizes equitable access to housing for all.

Social Impacts

Housing Precariousness and Inequality: The lack of affordable housing options and the high proportion of overcrowded houses (10.4%) indicate a problem of housing precariousness (INE, 2021). This precariousness disproportionately affects low-income families and young adult, exacerbating social inequalities and hindering opportunities for upward mobility. The growing income inequality, as highlighted by Freemark et al. (2021), further contributes to the housing affordability crisis, particularly for those at the lower end of the income spectrum.

The Ripple Effects: The housing crisis is not just an economic issue; it also has profound social implications. The inability to access affordable and adequate housing can lead to displacement, social exclusion, and the erosion of community ties. As residents are forced to move away in search of affordable options, the social fabric of Caniçal is threatened. The work of Lee et al. (2022) underscores how housing unaffordability can lead to social exclusion and housing precarity, significantly affecting people's quality of life.

Housing Affordability Measures: To better understand the multifaceted nature of housing affordability, it is essential to expand beyond traditional metrics like the price-to-income ratio. Incorporating measures that consider housing quality, security of tenure, and the subjective well-being of residents can provide a more comprehensive picture. Barrett's (2023) article emphasizes the importance of considering housing affordability not only in economic terms but also in relation to the meanings of home and home-making practices.

Social Impacts of Housing Unaffordability: Housing unaffordability has far-reaching social consequences, including housing insecurity, financial stress, impacts on mental and physical health, and difficulty establishing a sense of belonging and community. The article by Lee et al. (2022) highlights how housing unaffordability can lead to social exclusion and housing precarity, significantly affecting people's quality of life.

In summary, the housing challenges in Caniçal are not just abstract statistics; they represent the lived experiences of real people across different generations and socioeconomic backgrounds. The inability to access affordable and adequate housing has far-reaching consequences, affecting not only individuals and families but also the social fabric and economic vitality of the community. Addressing these challenges requires a multi-faceted approach that recognizes the diverse needs of different population groups and prioritizes equitable access to housing for all.

Initiatives and Policies: A Glimmer of Hope

While the housing crisis in Caniçal is undoubtedly daunting, there are ongoing efforts and potential policy interventions that offer a glimmer of hope.

Local Policies: Room for Improvement: The current housing policies in Caniçal, while well-intentioned, are falling short of addressing the multifaceted challenges. Slow approval processes, lack of incentives for affordable housing development, and potentially restrictive zoning regulations are hindering progress. A critical analysis of the effectiveness of these policies, as suggested by Anacker (2019), is essential for future policymaking. There is a pressing need for more proactive and innovative policies that prioritize the needs of residents over short-term economic gains.

Projects in the Pipeline: Despite the shortcomings of existing policies, there are promising initiatives underway. The redevelopment of degraded areas and the construction of new housing units, such as the ongoing transformation of port areas, offer potential solutions. However, the effectiveness and impact of these projects remain to be seen. It is crucial to ensure that these initiatives prioritize affordability and cater to the diverse needs of the population, including young adult, low-income families, and the elderly.

The Role of Public Policy: Public policies play a crucial role in promoting housing affordability and guaranteeing the right to adequate housing. Anacker's (2019) article discusses various strategies to address these challenges, including construction and rental subsidies, rent control, and urban planning regulations that prioritize affordable housing development. Analysing the effectiveness of these policies in the context of Caniçal can provide valuable insights for future policymaking. A combination of supply-side and demand-side interventions, tailored to the specific needs of Caniçal, may be necessary to achieve meaningful progress.

Navigating Complexities: Trade-offs and Unintended Consequences: Addressing the housing challenges in Caniçal requires a nuanced and comprehensive approach that recognizes the complexities and potential trade-offs involved. There is no one-size-fits-all solution, and policymakers need to carefully consider the specific context of Caniçal when designing and implementing housing policies. It is crucial to strike a balance between competing interests, such as economic development, social equity, and environmental protection, to ensure a sustainable and equitable housing future for all residents. For instance, while promoting tourism can boost the local economy, it is essential to regulate short-term rentals to prevent them from displacing long-term residents and exacerbating the housing shortage. Similarly, while social housing policies are vital for ensuring access for low-income groups, it is important to avoid unintended consequences such as social segregation and disincentivizing private sector investment in affordable housing.

Future Perspectives

Urgent Need for Intervention: The housing crisis in Caniçal demands immediate and decisive action from the government. Increasing the supply of affordable housing, revitalizing degraded areas, and ensuring sustainable urban development are crucial steps towards a more equitable and liveable future for all residents. Exploring alternatives to ownership, such as housing cooperatives and shared ownership models, as highlighted by Barrett (2023), can also expand access to secure and stable housing.

Caniçal's Potential: Despite the challenges, Caniçal has the potential to become a model for sustainable and inclusive urban development in Madeira. By leveraging its unique characteristics and addressing the housing crisis head-on, Caniçal can pave the way for a brighter future. This will require strong leadership, innovative solutions, community engagement, and a long-term commitment to balancing economic growth with social equity and environmental protection. The article by Freemark and Steil (2021) emphasizes the importance of inclusive urban planning in combating socio-spatial segregation and ensuring that everyone has access to quality housing and essential amenities.

In conclusion, the housing challenges in Caniçal extend far beyond economic concerns. They have profound social ramifications, impacting the well-being, opportunities, and sense of community for residents. Addressing these challenges will necessitate a multifaceted approach that prioritizes equitable access to housing and fosters social inclusion.

CONCLUSIONS AND RECOMMENDATIONS

The study of the housing landscape in Machico reveals a complex situation, influenced by a confluence of socioeconomic, demographic, and political factors. The detailed analysis of housing challenges has shown that:

- The growing demand for housing, driven by population growth (effective growth rate of 2.2‰ in 2022) and migration (positive migratory balance of 1911 people in 2022), is not accompanied by an adequate supply of new housing units (only 30 new housing units completed in 2022) (DREM, 2022b; INE, 2021).
- The increase in property prices (median bank appraisal value of €1,192 per square meter for apartments and €1,166 per square meter for houses in 2022), exacerbated by tourism development and the lack of effective housing policies, has led to housing precariousness and social inequality (INE, 2021).
- Population aging (20.6% of the population aged sixty-five or over in 2022) and the difficulties faced by young adult in acquiring or renting affordable properties (median value of new rentals of €4.59 per square meter in 2022) further aggravate the situation (DREM, 2022b; INE, 2021).
- Public policies and urban planning have not been sufficient to mitigate these challenges, resulting in a significant deficit of affordable housing.

To address the housing challenges in Machico, it is essential to implement the proposed strategies:

- Development of Social Housing: Tax incentives and public-private partnerships are crucial to increase the supply of affordable housing to meet the demand of the population with lower purchasing power.
- Redevelopment of Degraded Areas: Urban renewal, such as the revitalization of former port areas, and the incorporation of sustainable solutions can not only increase the housing supply but also improve the quality of life of residents.
- Support for Purchase and Rental: The creation of support programs, such as subsidies and tax exemptions for young adult who want to buy a house, and the establishment of an affordable income program, can facilitate access to housing for low-income families and young adult, in addition to encouraging the rental market.
- The housing challenges in Machico require a multidimensional and integrated approach, combining efforts from local government, the private sector, and the community. The proposed interventions, based on successful examples from other regions and countries, such as Porto's affordable housing program and the redevelopment of degraded neighbourhoods in Lisbon, provide a viable path to mitigate current problems and promote sustainable and inclusive urban development.

The relationship between geotechnical conditions, construction costs, and housing affordability is a critical consideration for housing development in Machico. Addressing this issue requires a multi-faceted approach:

- Thorough Geotechnical Investigations: Conducting comprehensive geotechnical studies before any development is crucial to identify potential hazards and design appropriate mitigation measures, minimizing the risk of costly surprises during construction.
- Innovative Construction Techniques: Exploring and promoting the use of innovative and cost-effective construction techniques that can address geotechnical challenges without significantly increasing costs.
- Financial Incentives: Providing financial incentives, such as subsidies or tax breaks, to developers who incorporate geotechnical risk mitigation measures into their projects, particularly for affordable housing developments.
- Land-Use Planning: Incorporating geotechnical considerations into land-use planning and zoning regulations to guide development towards areas with more favourable ground conditions and lower construction costs.

To ensure the effectiveness of the proposed strategies, it is necessary:

Political Commitment: The engagement of local and regional leaders is essential to implement long-term housing policies, such as the creation of a municipal housing fund, and to ensure the allocation of necessary resources.
Community Participation: Involving the community in the process of planning and implementing housing policies, through public consultations and hearings, is crucial to meet the real needs of residents and promote social cohesion.
Continuous Monitoring: Establish monitoring and evaluation mechanisms to track the progress of housing initiatives, such as the creation of a housing observatory, and adjust strategies as needed.

It is expected that, with the adoption of the suggested strategies, Machico will be able to significantly reduce housing precariousness (currently at 10.4%), provide affordable housing and improve the quality of life of its citizens. The success of these interventions could serve as a model for other regions facing similar housing challenges, demonstrating that it is possible to build a fairer and more equitable future for all.

REFERENCES

Acolin, A., & Reina, V. (2022). Housing cost burden and life satisfaction. *Journal of Housing and the Built Environment, 37*(4), 1–27. DOI: 10.1007/s10901-021-09921-1 PMID: 36776144

Alves, R., Lousada, S., Cabezas, J., & Gómez, J. M. N. (2022). The importance of the strategic urban rehabilitation plan in the sustainable development of the municipality of Machico. *Sustainability (Basel), 14*(24), 16816. DOI: 10.3390/su142416816

Alves, R., Lousada, S., Cabezas, J., & Gómez, J. M. N. (2023). Local housing strategy: Analysis of importance and implementation in Machico municipality, Madeira. *Land (Basel), 12*(9), 1778. DOI: 10.3390/land12091778

Alves, R., Lousada, S., Cabezas, J., & Gómez, J. M. N. (2023b). Port of Machico, Madeira: Assessment of the Consequences of Overtopping Using AHP Methodology. In S. Lousada (Ed.), Geoinformatics in Support of Urban Politics and the Development of Civil Engineering (pp. 77-103). IGI Global. https://doi.org/DOI: 10.4018/978-1-6684-6449-6.ch004

Alves, R., Lousada, S., Cabezas, J., & Gómez, J. M. N. (2023c). Modeling of Flood Risk: Vulnerability and Adaptation of Machico City Center. In Lousada, S. (Ed.), *Geoinformatics in Support of Urban Politics and the Development of Civil Engineering* (pp. 25–49). IGI Global., DOI: 10.4018/978-1-6684-6449-6.ch002

Alves, R., Lousada, S., Cabezas, J., & Gómez, J. M. N. (2023d). Urban Flooding Risk in Machico and Planning Its Territory as a Form of Prevention. In Lousada, S. (Ed.), *Geoinformatics in Support of Urban Politics and the Development of Civil Engineering* (pp. 50–76). IGI Global., DOI: 10.4018/978-1-6684-6449-6.ch003

Alves, R., Lousada, S., Cabezas, J., & Gómez, J. M. N. (2023e). Flood Risk and Vulnerability: An Approach to Urban Floods in Ribeira de Machico in the Autonomous Region of Madeira. In Lousada, S. (Ed.), *Geoinformatics in Support of Urban Politics and the Development of Civil Engineering* (pp. 104–131). IGI Global., DOI: 10.4018/978-1-6684-6449-6.ch005

Alves, R., Silva, A., & Lousada, S. (2022a). The impact of local accommodation in the sustainable development of Machico. In Aranzadi, T. R. (Ed.), *Planeamiento Ecológico En Las Iniciativas de Desarrollo* (pp. 305–322). Academic Press.

Anacker, K. B. (2019). Introduction: Housing affordability and affordable housing. *International Journal of Housing Policy, 19*(1), 1–16. DOI: 10.1080/19491247.2018.1560544

Barrett, P. (2023). Intersections between housing affordability and meanings of home: A review. *Kotuitui*, *18*(1), 27–44. DOI: 10.1080/1177083X.2022.2090969

Catanho, A. (1997). Contribution to the Geological and Geotechnical Characterization of the Different Volcanic Complexes of Madeira Island.

CML. (2012). Área de Reabilitação Urbana de Lisboa. Available online: https://www.lisboa.pt/temas/urbanismo/reabilitacao-urbana/area-de-reabilitacao. (accessed on 17 Maio 2024).

CMM. (2005). Plano Diretor Municipal de Machico. Available online: https://www.cm-machico.pt/municipal/machico-atua/urbanismo/pdm-plano-diretor-municipal. (accessed on 10 Janeiro 2024).

CMM. (2016). Área de Reabilitação Urbana de Machico. Available online: https://www.cm-machico.pt/municipal/machico-atua/urbanismo/aru-area-de-reabilitacao-urbana. (accessed on 10 Janeiro 2024).

CMM. (2019). Plano Estratégico de Reabilitação Urbana de Machico e do Porto da Cruz. Available online: https://www.cm-machico.pt/municipal/machico-atua/urbanismo/peru-plano-estrategico-de-reabilitacao-urbana. (accessed on 10 Janeiro 2024).

CMM. (2022). *Relatório da Estratégia Local de Habitação*. Câmara Municipal de Machico.

Cox, W. (2022). *Demographia International Housing Affordability*. Urban Reform Institute & Frontier Centre for Public Policy.

Direção Regional de Estatística da Madeira (DREM). (2022a). Estatísticas da Construção e Habitação da Região Autónoma da Madeira 2022. Funchal: DREM. Available online: https://estatistica.madeira.gov.pt/ (accessed on 27 Maio 2023).

Direção Regional de Estatística da Madeira (DREM). (2022b). Madeira em Números da Região Autónoma da Madeira 2022. Funchal: DREM. Available online: https://estatistica.madeira.gov.pt/ (accessed on 27 Maio 2024).

Forrest, R., & Hirayama, Y. (2015). The financialisation of the social project: Embedded liberalism, neoliberalism and home ownership. *Urban Studies (Edinburgh, Scotland)*, *52*(2), 233–244. DOI: 10.1177/0042098014528394

Freemark, Y., & Steil, J. (2021). Local power and the location of subsidized renters in comparative perspective: Public support for low- and moderate-income households in the United States, France, and the United Kingdom. *Housing Studies*, *0*, 1–29. DOI: 10.1080/02673037.2021.1910628

Haffner, M. E. A., & Hulse, K. (2019). A fresh look at contemporary perspectives on urban housing affordability. *International Journal of Urban Sciences, 25*(sup1), 59–79. https://doi.org/.DOI: 10.1080/12265934.2019.1687320

Instituto Nacional de Estatística (INE). (2021). Censos 2021. Available online: https://censos.ine.pt/xportal/xmain?xpid=CENSOS&xpgid=censos_quadro&QUADRO=1&selTab=tab1 (accessed on 7 Junho 2024).

Lee, Y., Kemp, P. A., & Reina, V. J. (2022). Drivers of housing (un)affordability in the advanced economies: A review and new evidence. *Housing Studies*, *37*(10), 1739–1752. DOI: 10.1080/02673037.2022.2123623

Lousada, S., Alves, R., & Fernandes, M. (2024). The impact of the Madeira Airport expansion on tourism and sustainable development in the city of Machico. *XV International Tourism Congress*. 2024. DOI: https://doi.org/DOI: 10.2478/9788367405287-019

Mateus de Brito, J. (2008). Landslide of the slope adjacent to the peripheral retaining wall of the Machico hypermarket on Madeira Island.

Paccoud, A., Hesse, M., Becker, T., & Gorczyńska, M. (2021). Land and the housing affordability crisis: Landowner and developer strategies in Luxembourg's facilitative planning context. *Housing Studies*, *0*, 1–18. DOI: 10.1080/02673037.2021.1950647

Scanlon, K., Lunde, J., & Whitehead, C. (2008). Mortgage Product Innovation in Advanced Economies: More Choice, More Risk. *European Journal of Housing Policy*, *8*(2), 109–131. DOI: 10.1080/14616710802037359

Taylor, K.-Y. (2019). *Race for Profit: How Banks and the Real Estate Industry Undermined Black Homeownership*. University of North Carolina Press. DOI: 10.5149/northcarolina/9781469653662.001.0001

KEY TERMS AND DEFINITIONS

Machico: A city located on the island of Madeira, Portugal, facing significant housing challenges due to increased demand and geographical limitations.

Tourism Growth: A key factor driving up housing demand in Machico, making it an attractive residential destination for foreigners.

Median Value of Bank Appraisal: A statistical measure used to indicate the central tendency of property appraisal values, which has risen significantly in Madeira.

Urban Planning: The process of designing and regulating the use of space within a city, which in Machico is challenged by geographical constraints and environmental legislation.

Housing Policies: Public policies aimed at addressing housing issues, which in Machico are currently insufficient to meet the demand for affordable housing.

Property Prices: The cost of purchasing or renting properties, which has been increasing in Machico, impacting housing affordability.

Environmental Legislation: Laws aimed at protecting the environment, which can limit areas available for urban development in Machico.

Geographical Limitations: Natural constraints, such as the mountainous terrain of Madeira, which limit the availability of land for construction in Machico.

Sustainability: Practices aimed at meeting current needs without compromising the ability of future generations to meet theirs, relevant in the context of housing development and urban planning.

Affordable Housing: Housing that is priced, allowing low- and moderate-income individuals to afford it, which is scarce in Machico.

Quantitative Methods: Research methods involving the collection and analysis of numerical data to identify trends and patterns in the housing market.

Qualitative Methods: Research methods involving the collection of non-numerical data, such as interviews, to gain insights into residents' experiences and perceptions.

Socioeconomic Factors: Factors related to the economic and social characteristics of the population, such as population growth and aging, affecting housing demand.

Chapter 2
Enhancing Infrastructure Sustainability Through Life-Cycle Assessment:
Sustainable Development and the Evolution of Civil Engineering

Ali Akbar Firoozi
University of Botswana, Botswana

Ali Asghar Firoozi
University of Botswana, Botswana

ABSTRACT

This chapter evaluates the pivotal role of Life-Cycle Assessment (LCA) in civil engineering, focusing on its application in promoting sustainable infrastructure development. LCA provides a comprehensive framework to assess environmental impacts across an infrastructure project's full lifecycle—from material extraction to decommissioning. It highlights practical case studies, recent advancements in LCA methodologies, and the integration of digital technologies that increase precision. Despite challenges like data scarcity and methodological complexity, the chapter explores solutions through enhanced educational tools, technological innovation, and cross-disciplinary efforts. It concludes by advocating for LCA's broader adoption to advance sustainable, resilient, and circular infrastructure practices in the industry. This analysis offers key insights into LCA's role in driving sustainability within civil engineering.

DOI: 10.4018/979-8-3693-2448-6.ch002

INTRODUCTION

The quest for sustainability within the realm of civil engineering has evolved from a peripheral concern to a central directive for the global construction industry. Infrastructure projects, from skyscrapers to bridges, significantly impact on the natural environment, economies, and societies. These projects consume vast quantities of resources, generate considerable waste, and contribute significantly to greenhouse gas emissions. Against this backdrop, the civil engineering sector faces increasing pressure to mitigate environmental impacts and contribute to the achievement of sustainability goals. A pivotal tool in this endeavor is the Life-Cycle Assessment (LCA), a methodological framework that evaluates the environmental burdens associated with the entire life cycle of a project, from material extraction through construction, use, and eventual decommissioning, (Nwodo & Anumba, 2019) (Toniolo *et al.,* 2020).

The relevance of LCA in civil engineering cannot be overstated. It offers a comprehensive perspective on sustainability by quantifying the environmental impacts of infrastructure projects over their entire lifecycle. This holistic approach enables engineers and project managers to make informed decisions that reduce negative environmental impacts while also potentially enhancing the economic and social aspects of sustainability. The application of LCA in civil engineering is aligned with the broader sustainability goals outlined in international frameworks such as the United Nations Sustainable Development Goals (SDGs), particularly those related to responsible consumption and production (Goal 12) and sustainable cities and communities (Goal 11), (Zhang *et al.,* 2020).

However, the integration of LCA into civil engineering practices is not without challenges. The complexity of LCA methodologies, the availability and quality of data, and the need for specialized knowledge to interpret LCA results are significant hurdles. Moreover, the construction industry's traditional focus on cost, time, and quality objectives often overshadows environmental considerations, making the widespread adoption of LCA practices a gradual process, (Bjørn *et al.,* 2020) (Mannan & Al-Ghamdi, 2020).

Despite these challenges, the potential benefits of LCA for infrastructure sustainability are immense. By systematically assessing environmental impacts, LCA facilitates the identification of opportunities for reducing resource consumption, minimizing waste and emissions, and selecting materials and processes with lower environmental footprints. Furthermore, LCA can enhance project value by improving environmental performance, which is increasingly recognized as a competitive advantage in the construction industry, (Corominas *et al.,* 2020).

The evolution of LCA methodologies and their application in civil engineering is a testament to the growing commitment to sustainability in the field. Innovations in LCA tools and software, coupled with increasing availability of environmental impact data, are making it more feasible for engineers to incorporate sustainability considerations into project planning and design, (Weiland *et al.,* 2021). As LCA continues to mature and become more integrated into civil engineering practices, it holds the promise of driving the industry towards more sustainable infrastructure development.

In this introductory section, we have outlined the critical role of LCA in advancing infrastructure sustainability within civil engineering. The following sections will delve deeper into the LCA methodology, its application through illustrative case studies, recent advancements, and the outlook for LCA in civil engineering, underscoring its significance as a cornerstone of sustainable infrastructure development.

LIFE-CYCLE ASSESSMENT (LCA): AN OVERVIEW

Life-Cycle Assessment (LCA) stands as a critical analytical methodology designed to assess the environmental impacts associated with all the stages of a product's life from raw material extraction through materials processing, manufacture, distribution, use, repair and maintenance, and disposal or recycling. By extending this framework to civil engineering projects, the LCA methodology illuminates the environmental footprint of infrastructure across its entire lifecycle, offering a holistic perspective that is essential for advancing sustainable practices in the field, (Moreno-Sader, Meramo-Hurtado, & González-Delgado, 2020).

Definition and Scope of LCA

LCA is defined by the International Organization for Standardization (ISO) in the ISO 14040 series as a technique for assessing the environmental aspects and potential impacts associated with a product, by:

- Compiling an inventory of relevant inputs and outputs,
- Evaluating the potential environmental impacts associated with those inputs and outputs,
- Interpreting the results of the inventory and impact phases concerning the objectives of the study, (Bahramian & Yetilmezsoy, 2020).

This comprehensive approach enables the quantification and evaluation of the environmental impacts of infrastructure projects, guiding decision-makers in identifying strategies for minimizing negative environmental effects.

Table 1 provides a succinct overview of fundamental terms and definitions related to Life-Cycle Assessment (LCA) as established by the International Organization for Standardization (ISO) standards ISO 14040:2006 and ISO 14044:2006. These definitions form the basis for understanding the LCA methodology, emphasizing the comprehensive approach LCA takes in evaluating the environmental impacts associated with all stages of a product or project's life cycle. The table aims to clarify key LCA concepts, ensuring a shared understanding of terms such as "environmental aspects," "inventory of inputs and outputs," and "impact assessment," which are crucial for conducting effective LCA studies in civil engineering and beyond.

Table 1. Key definitions and terms in Life-Cycle Assessment (LCA)

Term	Definition
Life-Cycle Assessment (LCA)	A methodology for assessing environmental aspects and potential impacts associated with a product, from raw material extraction through to disposal or recycling.
Environmental Aspects	Elements of a product or project's life cycle that can interact with the environment, such as emissions to air or water, and consumption of resources.
Inventory of Inputs and Outputs	The compilation of quantitative data on the relevant inputs (energy, materials) and outputs (emissions, waste) of a system under study.
Impact Assessment	The process of evaluating the significance of potential environmental impacts based on the inventory of inputs and outputs.

Historical Development of LCA in Civil Engineering

The application of LCA in civil engineering has evolved significantly over the past decades. Initially conceptualized for product manufacturing, LCA's relevance to infrastructure projects gained recognition as the environmental movement emphasized the importance of sustainable development within the construction industry, (Tokede & Traverso, 2020). The adoption of LCA in civil engineering practices reflects a paradigm shift towards incorporating environmental considerations in the design and implementation of infrastructure projects.

Key Components of LCA

The LCA methodology is structured around four main phases, as established by ISO standards:

1. Goal and Scope Definition: This initial phase involves outlining the study's purpose, scope, and boundaries. For civil engineering projects, this includes specifying the infrastructure's lifecycle to be assessed, the functional unit (Igos *et al.,* 2019), and the system boundaries that determine which processes will be included in the analysis (Guinée et al., 2011).
2. Inventory Analysis (LCI): The LCI phase involves collecting data on all relevant inputs (e.g., materials, energy) and outputs (e.g., emissions, waste) associated with the infrastructure project. This step is critical in identifying key areas where environmental improvements can be made, (Castellani, Beylot, & Sala, 2019).
3. Impact Assessment (LCIA): In this phase, the environmental impacts of the inputs and outputs identified in the LCI are assessed. This involves selecting and applying specific impact categories (e.g., global warming potential, water use) and indicators to quantify the project's environmental burdens, (van Der Werft, Knudsen, & Cederberg, 2020)
4. Interpretation: The final phase involves analyzing the results of the inventory and impact assessment to draw conclusions and make informed decisions. For civil engineering projects, this might include comparing the environmental impacts of different design options or identifying strategies for reducing the project's environmental footprint, (Khandelwal *et al.,* 2019).

Equation 1 represents a foundational formula used in the Life-Cycle Impact Assessment (LCIA) phase of LCA, illustrating the method by which environmental impact scores are calculated for different categories, such as global warming potential. In this equation, the “Impact Score (IS)” for a given impact category (e.g., CO_2 emissions) is determined by summing the products of the quantity of each input or output (e.g., kilograms of CO_2 emitted) and its corresponding impact factor (a coefficient that quantifies the relative environmental impact per unit of input or output). This simplified formula underscores the process of quantifying the potential environmental impacts associated with the lifecycle of civil engineering projects, facilitating the identification of areas where interventions can significantly reduce overall environmental burdens.

$$Impact\ Score\ (IS) = \sum\left(Quantity\ of \frac{Inpit}{Output} \times ImpactFactor\right) \quad (1)$$

Table 2 delineates typical inputs and outputs associated with the Inventory Analysis (LCI) phase of Life-Cycle Assessment (LCA) in civil engineering projects, providing tangible examples of what is quantified during this critical stage. The table categorizes inputs and outputs into materials, energy, water, emissions, and land use, illustrating the breadth of data collected to assess the environmental impacts of infrastructure projects. For inputs, it lists common materials like steel and concrete, energy sources like electricity and diesel, and water usage in construction processes. Outputs include construction waste, emissions such as CO_2 and CH_4, wastewater, and impacts on land use, such as habitat disruption. This table underscores the importance of comprehensively documenting both the resources consumed and the environmental burdens generated throughout the construction process, enabling the identification of key areas where environmental improvements can be made.

Table 2. Examples of inventory inputs and outputs in civil engineering projects

Category	Inputs (Examples)	Outputs (Examples)
Materials	Steel, concrete, timber, asphalt	Construction waste, surplus materials
Energy	Electricity (for machinery), diesel (for vehicles)	Heat emissions, energy consumption data
Water	Water used in construction processes	Wastewater from construction sites
Emissions	-	CO_2, CH_4, N_2O emissions from construction activities
Land Use	Land area disturbed by construction	Habitat disruption, soil erosion

Figure 1 illustrates the sequential phases of the Life Cycle Assessment (LCA) process as applied to civil engineering projects. The diagram depicts the progression from Goal and Scope Definition, through Inventory Analysis (LCI) and Impact Assessment (LCIA), culminating in the Interpretation phase. This visual representation aids in understanding the comprehensive approach to assessing environmental impacts within civil engineering projects.

Figure 1. The four phases of Life Cycle Assessment (LCA) in civil engineering

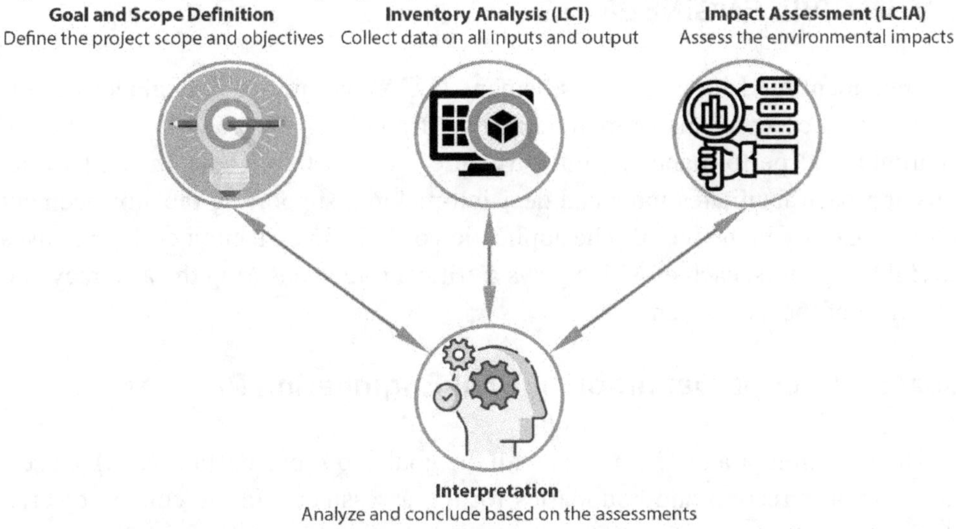

Significance of Each Component in Civil Engineering

Each component of the LCA methodology plays a crucial role in advancing the sustainability of civil engineering projects. The goal and scope definition sets the foundation for the assessment, ensuring that the study is aligned with the project's sustainability objectives. The inventory analysis provides a detailed account of the project's environmental inputs and outputs, highlighting areas for improvement. The impact assessment offers insights into the potential environmental impacts, guiding decision-making towards more sustainable options. Finally, the interpretation phase facilitates the application of LCA findings to enhance the project's environmental performance.

LCA's comprehensive approach to assessing environmental impacts makes it an invaluable tool for civil engineering, enabling the integration of sustainability into the planning, design, and management of infrastructure projects. By applying the LCA methodology, civil engineers can identify opportunities to reduce environmental impacts, optimize resource use, and contribute to the broader goal of sustainable development.

METHODOLOGICAL FRAMEWORK FOR LCA IN CIVIL ENGINEERING

Implementing Life-Cycle Assessment (LCA) within civil engineering encompasses a comprehensive methodological framework designed to evaluate the environmental performance of infrastructure projects throughout their lifecycle. This approach facilitates informed decision-making, supporting the advancement of sustainability in the sector. The application of LCA in civil engineering involves several key phases, each of which plays a critical role in ensuring the accuracy and relevance of the assessment.

Goal and Scope Definition in Civil Engineering Projects

The initiation of an LCA begins with the goal and scope definition, which sets the strategic direction and boundaries for the assessment. In the context of civil engineering, this phase is critical for establishing a clear understanding of the project's objectives, the functional unit (for example, square meters of built area for buildings or kilometers for roads), and the system boundaries that delineate which processes will be included in the LCA, (Temizel-Sekeryan, Wu, & Hicks, 2021). Determining the system boundaries is particularly challenging in civil engineering due to the sector's diverse range of activities and materials, making it essential to carefully consider which stages of the infrastructure lifecycle—such as material production, construction, operation, and end-of-life—will be assessed.

Inventory Analysis: Gathering Data on Inputs and Outputs

Following the goal and scope definition, the Inventory Analysis (LCI) phase involves the detailed collection of data on all inputs (e.g., raw materials, energy) and outputs (e.g., emissions to air, water, and soil) associated with the infrastructure project within the defined system boundaries. This stage is crucial for identifying the environmental hotspots of a project. For example, in building construction, the LCI would quantify the amounts of steel, concrete, glass, and other materials used, along with the energy consumed and the waste generated, (Lu *et al.,* 2021). The accuracy and comprehensiveness of the LCI are vital for ensuring the reliability of the LCA results, making this phase one of the most data-intensive aspects of the LCA process.

Figure 2 presents the key methods for data collection used during the Inventory Analysis phase of Life Cycle Assessment (LCA) in civil engineering projects. It highlights three main techniques: Site Surveys, which involve direct data gathering on-site; Databases, which provide secondary, standardized data crucial for LCA;

and Manufacturer Information, offering specific details about material compositions and emissions. This variety ensures comprehensive data collection, enhancing the accuracy and reliability of the LCA results.

Figure 2. Data collection techniques in LCA for civil engineering

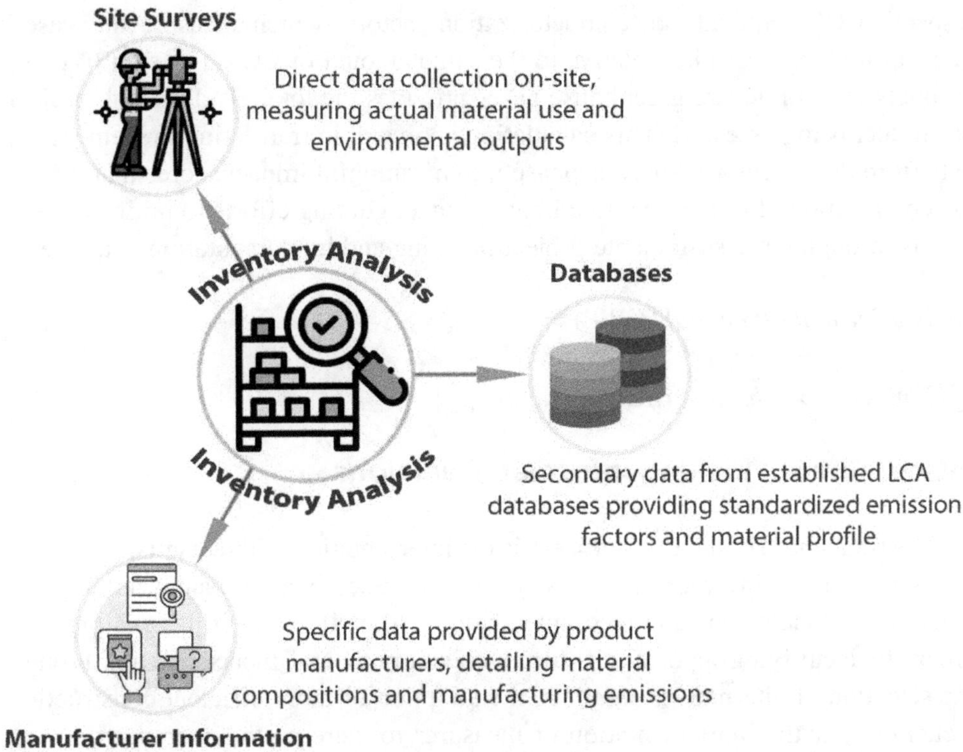

Impact Assessment: Evaluating Environmental Impacts

The Impact Assessment (LCIA) phase assesses the potential environmental impacts identified during the inventory analysis. This involves applying specific impact categories, such as global warming potential, ozone layer depletion, eutrophication, and acidification, to quantify the environmental impacts associated with the inputs and outputs cataloged in the LCI, (Kaab *et al.*, 2019). The LCIA phase is where the data collected becomes insightful, enabling the identification of the most significant environmental impacts of the project. For civil engineering projects, this might mean focusing on the contributions to climate change from greenhouse gas emissions or assessing the impact on water quality from runoff during construction activities.

Equation 2 provides a specific example of how environmental impacts, particularly Global Warming Potential (GWP), are calculated within the Life-Cycle Impact Assessment (LCIA) phase of LCA. This equation multiplies the quantity of greenhouse gas emissions (measured, for instance, in kilograms or tons) by their respective CO_2 equivalence characterization factors, which measure the gases' impact on global warming relative to the same amount of CO_2. The sum of these products for all relevant greenhouse gases provides the total GWP for the project or product being assessed. This calculation is pivotal in translating raw emissions data from the Inventory Analysis phase into meaningful impact scores that reflect the contribution of the project to climate change, guiding efforts to mitigate these effects through more sustainable project planning and implementation strategies.

$$Global\ Warming\ Potential\ (GWP) =$$

$$\sum \left(Emissions \times Characterization\ Factor_{CO2eq} \right) \quad (2)$$

Interpretation: Making Informed Decisions

The final phase of the LCA process is the interpretation of the results, where the findings from the inventory analysis and impact assessment are analyzed to draw meaningful conclusions and recommendations. In civil engineering, this phase is crucial for identifying opportunities to reduce environmental impacts, such as through the selection of alternative materials, the adoption of more efficient construction techniques, or the implementation of measures to increase the energy efficiency of buildings during their use phase, (Meglin, Kytzia, & Habert, 2022). The interpretation phase is also where the limitations of the study are acknowledged, and the reliability of the findings is evaluated, ensuring that the conclusions drawn are robust and actionable.

The implementation of LCA in civil engineering is not without challenges. Data availability and quality can vary significantly, affecting the accuracy of the assessment. Additionally, the complexity of infrastructure projects and the need for interdisciplinary expertise can make the LCA process more demanding. Despite these challenges, the application of LCA in civil engineering offers a powerful means of identifying strategies for reducing environmental impacts and advancing sustainability within the sector.

Figure 3 delineates the decision-making framework utilized to translate LCA results into actionable sustainability strategies in civil engineering. This flowchart guides the reader through the process from analyzing LCA results to monitoring the implementation of strategies, emphasizing the iterative nature of making informed, sustainable decisions. It highlights how each step, from identifying opportunities for improvement to implementing and reassessing strategies, plays a crucial role in enhancing environmental sustainability in civil engineering projects.

Figure 3. LCA decision-making framework in civil engineering

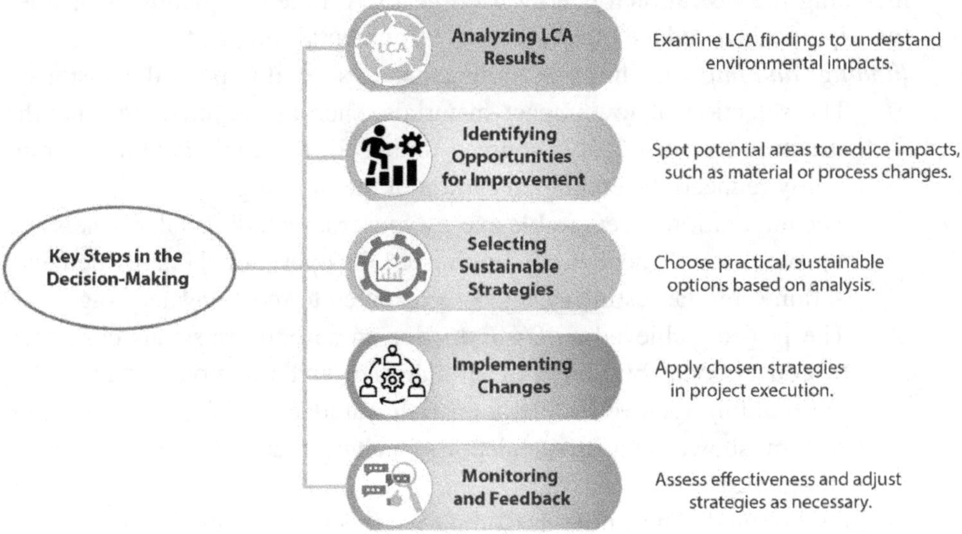

CASE STUDIES: LCA IN ACTION

The practical applications of Life-Cycle Assessment (LCA) in civil engineering projects provide insightful demonstrations of how this methodology can guide sustainable infrastructure development. This section delves into detailed case studies across different sectors within civil engineering, showcasing the implementation and outcomes of LCA. These examples highlight the versatility of LCA in identifying environmental impact reductions across various project types, offering valuable lessons and best practices for future projects.

Sustainable Building Construction

- *Project Overview:* The construction of the EcoTech Complex, a multi-use development aimed at achieving high sustainability standards, employed LCA to guide material selection, energy systems, and waste management practices. The project, located in an urban European setting, aimed to serve as a model for sustainable development in the region.

- *LCA Application:* The LCA for the EcoTech Complex focused on comprehensive environmental impact categories, including global warming potential, embodied energy, water usage, and material resource efficiency. By evaluating the entire lifecycle of the building, from construction to demolition, including the operational energy use, the LCA aimed to identify strategies that significantly reduce the project's environmental footprint.
- *Findings and Impact:* The LCA findings led to several impactful decisions:
    - o The selection of low-impact materials, such as recycled steel for the structure and sustainably sourced timber for external cladding, significantly reduced the embodied carbon of the building.
    - o The integration of renewable energy sources, including solar panels and a ground-source heat pump, minimized the operational energy demand.
    - o A rainwater harvesting system was installed to reduce water usage.
    - o The project achieved a 40% reduction in carbon emissions compared to conventional buildings of similar size and function. Additionally, the building received a Platinum rating under the LEED certification system, showcasing its sustainability achievements, (Janjua, Sarker, & Biswas, 2019).
- *Lessons Learned:* The EcoTech Complex project illustrates the critical role of LCA in driving sustainable decision-making in building construction. Key lessons include the importance of early integration of LCA in the design process and the potential of innovative materials and energy solutions in reducing environmental impacts.

Table 3 delineates the significant environmental impacts and benefits derived from the application of Life-Cycle Assessment (LCA) to the design and construction of the EcoTech Complex, a model for sustainable development. The table compares the baseline scenario, reflective of standard construction practices, with the LCA-informed design choices, highlighting the substantial reductions achieved in key environmental aspects such as CO2 emissions, energy use, water consumption, and construction waste. Specifically, the implementation of sustainable materials and energy-efficient systems led to a 40% reduction in both CO2 emissions and energy use, alongside notable decreases in water usage and waste generation. The LCA-informed design not only achieved these reductions during the construction phase but also ensured lower operational energy requirements, underscoring the comprehensive environmental benefits of integrating LCA insights into civil engineering projects. This table exemplifies how LCA results can guide more sustainable design decisions, ultimately contributing to significant environmental improvements and advancing the field toward more sustainable infrastructure solutions.

Table 3. Ecotech complex: LCA findings and impact

Environmental Aspect	Baseline Scenario	LCA-Informed Design	Reduction Achieved
CO2 Emissions	5,000 tons	3,000 tons	40%
Energy Use	10,000 MWh	6,000 MWh	40%
Water Consumption	500,000 liters	350,000 liters	30%
Construction Waste	200 tons	100 tons	50%
Operational Energy	1,000 MWh/year	600 MWh/year	40%

Road and Highway Construction

- *Project Overview:* The upgrade of a 50-kilometer section of the A45 Highway in Germany aimed to improve traffic flow and safety while minimizing environmental impacts. An LCA was conducted to evaluate the sustainability of different construction materials and methods.
- *LCA Application:* This LCA focused on analyzing the environmental impacts of asphalt production, transportation, and laying processes. It compared traditional hot-mix asphalt with newer, more sustainable options like warm-mix asphalt and recycled asphalt pavements.
- *Findings and Impact:* The analysis revealed significant benefits in opting for warm-mix asphalt, which requires lower production temperatures, thus reducing energy consumption and greenhouse gas emissions by approximately 15%, (Vega-Araujo, Martinez-Aruguelles, & Santos, 2020). The use of recycled asphalt further contributed to material savings and waste reduction, embodying the principles of circular economy in infrastructure projects.
- *Lessons Learned:* The A45 Highway project highlighted the environmental and economic benefits of adopting innovative materials and techniques in road construction. It demonstrated that sustainable practices could be integrated into public infrastructure projects without compromising on quality or safety, emphasizing the need for lifecycle thinking in civil engineering.

Figure 4 highlights the key sustainable features implemented in the EcoTech Complex. It shows the use of recycled materials such as steel and sustainably sourced timber to reduce embodied carbon, integration of renewable energy systems like solar panels and ground-source heat pumps to lower operational energy demand, and water-saving measures including a rainwater harvesting system to conserve water usage. These initiatives exemplify the complex's commitment to sustainability and the effective application of LCA to optimize environmental performance.

Figure 4. Sustainable features of the Ecotech complex

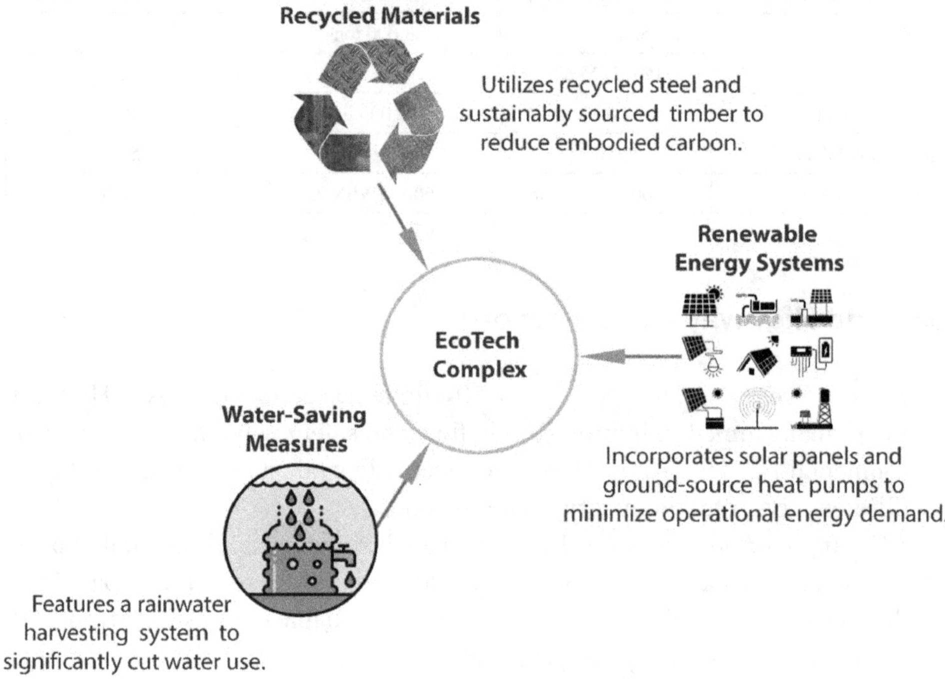

Water and Wastewater Treatment Facilities

- *Project Overview:* The GreenFlow Water Recovery Facility in California represents a state-of-the-art approach to water and wastewater treatment, designed to enhance water efficiency and reduce environmental impacts in one of the most drought-prone regions of the United States.
- *LCA Application:* An LCA was carried out to compare the environmental performance of conventional wastewater treatment technologies with the GreenFlow system, which incorporates advanced membrane bioreactors, solar-powered desalination, and water recycling processes.
- *Findings and Impact:* The LCA results indicated that the GreenFlow system significantly lowers energy use, reduces greenhouse gas emissions by 30%, and enhances water recovery rates compared to traditional treatment

methods, (Hsien *et al.,* 2019). The facility not only contributes to local water sustainability but also serves as a blueprint for future water treatment plants aiming at minimal environmental impact.

- *Lessons Learned:* This case study underscores the potential of integrating advanced technologies and renewable energy in reducing the environmental footprint of water treatment facilities. It highlights the importance of LCA in identifying and quantifying the benefits of innovative approaches to essential public services.

Table 4 provides a comprehensive comparison between traditional wastewater treatment methods and the innovative GreenFlow system, based on a Life-Cycle Assessment (LCA) focused on various environmental and operational metrics. The comparison reveals significant improvements achieved by the GreenFlow system, including a 33% reduction in energy use and a 30% reduction in greenhouse gas emissions, illustrating the system's enhanced efficiency and environmental performance. Furthermore, the GreenFlow system demonstrates a notable 21% increase in water recovery rates, highlighting its contribution to sustainable water management practices. Additionally, the reduction in chemical usage by 40% underscores the system's more environmentally friendly approach to wastewater treatment. Operational cost savings of 20% with the GreenFlow system compared to traditional methods indicate not only environmental but also economic benefits. This table underscores the value of adopting advanced and sustainable wastewater treatment technologies, as evidenced by the LCA outcomes, pointing towards significant opportunities for improving the environmental footprint of wastewater management processes in civil engineering projects.

Table 4. Comparative environmental performance of greenflow vs. traditional wastewater treatment

Environmental Metric	Traditional Treatment	GreenFlow System	Improvement with GreenFlow
Energy Use	1,500 MWh/year	1,000 MWh/year	33% Reduction
Greenhouse Gas Emissions	800 tons CO_2e/year	560 tons CO_2e/year	30% Reduction
Water Recovery Rates	70%	85%	21% Increase
Chemical Usage	10 tons/year	6 tons/year	40% Reduction
Operational Costs	$500,000/year	$400,000/year	20% Reduction

These case studies collectively illustrate the versatility and impact of LCA in guiding the sustainable development of diverse infrastructure projects. By providing a systematic framework for assessing and mitigating environmental impacts, LCA enables civil engineers and project managers to make informed decisions that promote sustainability across the project lifecycle. As such, LCA not only contributes to the advancement of sustainable infrastructure but also aligns with broader environmental and societal goals.

ADVANCEMENTS IN LCA METHODOLOGIES

The evolution of Life-Cycle Assessment (LCA) methodologies has been pivotal in addressing the complex sustainability challenges faced by the civil engineering sector. Continuous improvements and innovations in LCA tools and approaches have significantly enhanced the accuracy, efficiency, and applicability of LCAs in infrastructure projects. This section explores recent advancements in LCA methodologies, highlighting their impact on sustainable civil engineering practices.

Integration of Digital Technologies

The integration of digital technologies into LCA methodologies represents a significant leap forward. Building Information Modeling (BIM), Geographic Information Systems (GIS), and other digital tools have been increasingly utilized to collect, manage, and analyze data more effectively. For instance, the use of BIM in conjunction with LCA enables the detailed modeling of buildings and infrastructure projects, facilitating a more comprehensive assessment of materials, energy use, and environmental impacts throughout the project lifecycle, (Vega-Arujo, Martinez-Arguelles, & Santos, 2020). GIS, on the other hand, offers spatial analysis capabilities that are particularly useful for large-scale infrastructure projects, allowing for the consideration of geographical factors in LCA studies, (Laurent *et al.,* 2020).

Figure 5 illustrates the synergistic integration of Building Information Modeling (BIM) and Geographic Information Systems (GIS) within the Life Cycle Assessment (LCA) process in civil engineering. BIM provides detailed data on building materials and energy use, while GIS enhances the process with its spatial analysis capabilities. This diagram shows how data from these digital technologies flows into and supports the LCA process, enabling a more comprehensive environmental impact assessment and facilitating informed decision-making in sustainable infrastructure development.

Figure 5. Integration of BIM and GIS in LCA for civil engineering

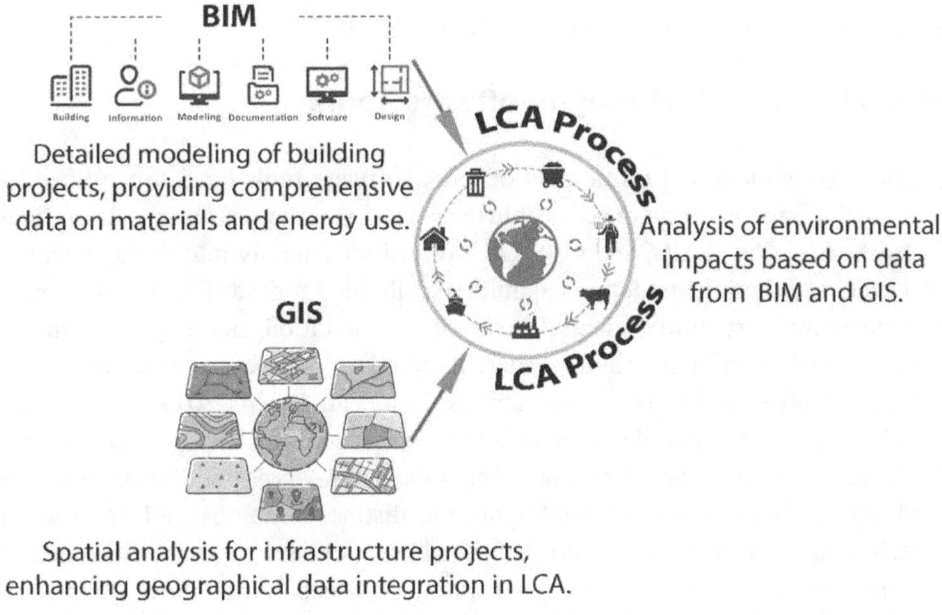

Development of Sector-Specific Databases

The development of extensive, sector specific LCA databases has greatly improved the availability and quality of data, which is crucial for conducting accurate and reliable LCAs. These databases provide detailed information on the environmental impacts of materials, processes, and technologies specific to civil engineering and construction. The International Reference Life Cycle Data System (ILCD) and the Environmental Product Declarations (EPDs) are examples of initiatives that have contributed to the standardization and transparency of LCA data, facilitating more informed decision-making in the design and construction of sustainable infrastructure, (Lü *et al.*, 2019).

Enhanced Impact Assessment Methods

Advancements in impact assessment methods have expanded the scope of LCA to include a wider range of environmental, social, and economic impact categories. New methodologies, such as life cycle sustainability assessment (LCSA), which integrates LCA with life cycle costing (LCC) and social life cycle assessment (SLCA), offer a more holistic view of a project's sustainability performance. These developments enable civil engineers to not only assess environmental impacts but also consider social and economic dimensions, leading to more balanced and sustainable outcomes, (Ternel, Bouter, & Melgar, 2021)

Streamlining LCA Through Software Tools

The development and refinement of LCA software tools have streamlined the LCA process, making it more accessible to professionals across various disciplines. Tools like SimaPro, GaBi, and OpenLCA offer user-friendly interfaces, extensive databases, and flexible modeling capabilities, allowing for the efficient assessment of complex infrastructure projects. These software solutions not only save time but also enhance the consistency and comparability of LCA studies, contributing to their growing adoption in the civil engineering sector, (Fauzi *et al.*, 2019).

Table 5 presents a detailed comparison of three leading Life-Cycle Assessment (LCA) software tools: SimaPro, GaBi, and OpenLCA, across various features and capabilities. This comparison highlights the distinct strengths and applications of each tool, offering insights into their database inclusion, user interface design, supported impact assessment methods, integration capabilities with other tools, sector suitability, cost, and available training and support services. SimaPro is noted for its comprehensive databases and broad industry applicability, making it a versatile choice for diverse LCA projects. GaBi stands out for its industry-specific databases and strong integration with product design software, catering especially to the manufacturing sector. OpenLCA offers a unique advantage as an open-source platform, making it accessible for academic research and projects that benefit from customizable features and a community-driven support structure. This table is designed to assist readers in understanding the comparative advantages of each software tool, facilitating informed decisions based on the specific needs and goals of their LCA analyses.

Table 5. Comparison of leading LCA software tools

Feature/Capability	SimaPro	GaBi	OpenLCA
Database Inclusion	Comprehensive proprietary and third-party databases	Extensive industry-specific databases	Open-source and third-party databases
User Interface	User-friendly, with advanced analytical capabilities	Intuitive design with streamlined workflow	Open-source platform with customizable features
Impact Assessment Methods	Multiple methods supported, including ReCiPe and CML	Supports a wide range of methods, tailored for specific industries	Flexible support for various methodologies, including custom methods

continued on following page

Table 5. Continued

Feature/Capability	SimaPro	GaBi	OpenLCA
Integration with Other Tools	Good integration with CAD and BIM tools	Seamless integration with engineering and product design software	Compatible with various data formats for easy integration
Sector Suitability	Broad applicability across various industries	Particularly strong in manufacturing and product design	Ideal for academic research and projects with open-source preferences
Cost	Subscription-based, with tiered pricing for different user levels	Subscription model with options for enterprise solutions	Free, with optional paid support and services
Training and Support	Comprehensive training programs and customer support	Detailed documentation and dedicated support services	Community-driven support with access to training materials

Future Directions

Looking forward, the LCA methodology is expected to continue evolving in response to emerging sustainability challenges and technological advancements. Key areas of focus include improving data quality and availability, further integrating LCA with digital design tools, and expanding the scope of impact assessments to encompass emerging environmental issues such as biodiversity loss and circular economy principles. Additionally, the development of standardized methodologies for incorporating LCA into regulatory frameworks and certification schemes is anticipated to drive broader adoption and implementation in civil engineering projects.

CHALLENGES AND OPPORTUNITIES IN IMPLEMENTING LCA IN CIVIL ENGINEERING

Implementing Life-Cycle Assessment (LCA) in civil engineering encounters a unique set of challenges and opportunities. Overcoming these hurdles is crucial for maximizing the benefits of LCA in guiding sustainable infrastructure development. This section delves into the practical aspects of LCA application, focusing on data and methodological challenges, and the bridging of theory and practice.

Overcoming Data and Methodological Challenges

One of the most significant barriers to the effective application of LCA in civil engineering is the challenge associated with data: its availability, quality, and the applicability of LCA methodologies. Accurate and region-specific data are essential for conducting meaningful LCAs, yet such data can often be scarce or unreliable. Moreover, the selection of appropriate methodologies that can accommodate the complexities of large-scale infrastructure projects adds another layer of difficulty, (Bamber *et al.,* 2020).

Strategies for overcoming these challenges include the development of more comprehensive and accessible LCA databases tailored to the civil engineering sector. Initiatives such as the European Platform on Life Cycle Assessment (EPLCA) and the U.S. Life Cycle Inventory Database contribute towards this goal by providing standardized, quality-assured datasets that can support more reliable LCA studies, (Peña *et al.*, 2021) (Safari & Azarijafari, 2021).

Figure 6 illustrates the cyclical process of addressing data and methodological challenges in Life Cycle Assessment (LCA) for civil engineering. Starting with the identification of data challenges, the flowchart progresses to problem identification, development of potential solutions such as sector-specific databases and standardization efforts, and implementation of these solutions. The feedback loop at the end emphasizes continuous refinement based on practical outcomes, ensuring the evolution of more effective and reliable LCA practices. This framework highlights the dynamic process of improving data quality and methodology in LCA studies.

Figure 6. LCA data challenge and solution framework

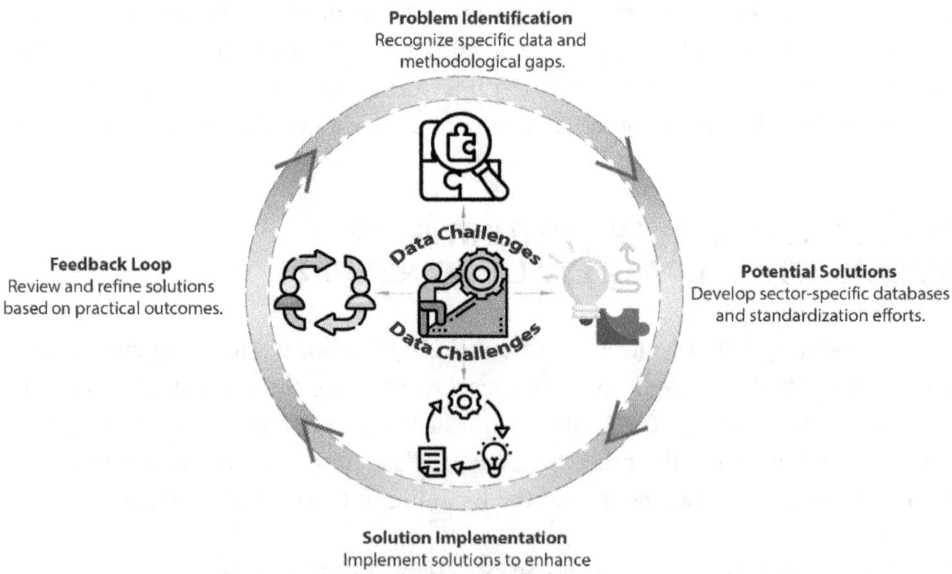

Bridging the Gap Between Theory and Practice

The gap between LCA theory and its practical application in civil engineering projs represents another significant challenge. While LCA principles are well-established in academic and research contexts, translating these principles into actionable strategies for real-world projects can be daunting, particularly given the scale and complexity of infrastructure development, (Rezaei, Bulle, & Lesage, 2019).

Bridging this gap requires focused efforts to integrate LCA training into the education of civil engineers, ensuring that future professionals are equipped with the knowledge and skills needed to apply LCA methodologies effectively. Additionally, the development of user-friendly LCA software tools designed specifically for civil engineering applications can facilitate more widespread adoption by making the process more accessible to practitioners, (Hosseinzadeh-Bandbafha *et al.*, 2022).

Moreover, fostering stronger collaboration between academia, industry, and government can drive the advancement of practical LCA applications. Joint projects and partnerships can facilitate knowledge exchange, co-create innovative solutions, and demonstrate the value of LCA in achieving sustainable outcomes in civil engineering, (Massaro *et al.*, 2021).

Embracing Technological Innovations

The rapid advancement of digital technologies offers significant opportunities for enhancing LCA implementation in civil engineering. Technologies such as artificial intelligence (AI), machine learning, and blockchain can streamline data collection and analysis, improve the accuracy of LCAs, and enhance transparency and traceability throughout the project lifecycle, (Joensuu, Edelman, & Saari, 2020).

Table 6 outlines the transformative impact of emerging technologies such as Artificial Intelligence (AI), Machine Learning, and Blockchain on the application and effectiveness of Life-Cycle Assessment (LCA) in civil engineering. Each technology brings specific applications to LCA processes, offering significant benefits that address traditional challenges in the field. AI's automation capabilities streamline the labor-intensive processes of data collection and analysis, while its predictive modeling aids in foreseeing lifecycle impacts more accurately. Machine Learning enhances LCA by mining large datasets for impactful patterns, thereby refining impact assessments and aiding in the selection of materials and processes that minimize environmental footprints. Blockchain technology contributes to LCA by safeguarding the integrity and transparency of data across the project's supply chain, ensuring verifiable traceability of materials from source to end use. Together, these technologies present a new frontier in LCA methodology, promising greater efficiency, accuracy, and transparency in assessing and mitigating the environmental impacts of civil engineering projects.

Table 6. Impact of emerging technologies on LCA in civil engineering

Technology	Applications in LCA	Benefits for LCA
Artificial Intelligence (AI)	Automating data collection and analysis; Predictive modeling for lifecycle impacts	Enhances accuracy and efficiency; Reduces time for analysis; Predicts future environmental impacts
Machine Learning	Identifying patterns in large datasets; Optimizing material selection and use	Improves the precision of impact assessments; Facilitates smarter material and process choices
Blockchain	Ensuring data integrity and transparency; Tracking material sources and uses through the supply chain	Increases trust in LCA data; Promotes transparency and accountability in material sourcing and usage

THE FUTURE OF LCA IN CIVIL ENGINEERING

As we navigate toward a future where sustainability is increasingly paramount, the role of Life-Cycle Assessment (LCA) in civil engineering is poised for significant evolution and expansion. The integration of LCA into the fabric of civil engineering practices offers a promising pathway to achieving more sustainable infrastructure projects. This section outlines potential advancements, challenges, and strategic directions for LCA in the context of civil engineering, focusing on how it can continue to contribute to sustainable development goals.

Broader Adoption and Integration

The future will likely see broader adoption of LCA in civil engineering, driven by growing regulatory pressures, stakeholder awareness, and the escalating urgency of environmental challenges. As the benefits of LCA in reducing environmental footprints become more widely recognized, we can anticipate its integration into standard practice, not only at the project level but also within strategic planning and policy-making processes. Enhanced collaboration between stakeholders, including governments, industry, and academia, is crucial to developing frameworks that facilitate this integration, ensuring that LCA becomes an indispensable part of the decision-making process in infrastructure projects.

Advancements in Methodologies and Tools

Continued advancements in LCA methodologies and tools are expected, with a focus on increasing accuracy, usability, and comprehensiveness. Innovations may include the development of more sophisticated impact assessment models, better alignment with emerging sustainability criteria (such as resilience and adaptability), and the integration of LCA with other sustainability assessment tools. The proliferation of digital technologies, such as artificial intelligence (AI) and machine learning, could significantly enhance the efficiency and depth of LCAs, enabling more dynamic and complex analyses.

Focus on Circular Economy and Resource Efficiency

The principles of the circular economy are becoming increasingly central to discussions of sustainability in civil engineering. Future LCA applications will likely emphasize the optimization of resource use and the minimization of waste through the design of infrastructure projects that facilitate material reuse and recycling. This shift towards circular economic principles will require adaptations in LCA methodologies to better account for the benefits of circular strategies, potentially transforming how materials and processes are evaluated in the context of infrastructure development.

Enhancing LCA Education and Capacity Building

To realize the full potential of LCA in civil engineering, there is a need for enhanced education and capacity building among professionals and students in the field. Universities, professional associations, and industry groups will play a pivotal role in integrating LCA principles into educational curricula and training programs. By equipping the next generation of civil engineers with the knowledge and skills to apply LCA effectively, we can ensure that sustainability considerations are embedded in the design and execution of infrastructure projects from the outset.

Addressing Emerging Environmental and Social Challenges

As global environmental and social challenges evolve, LCA in civil engineering must adapt to address new priorities. This includes expanding the scope of assessments to include emerging issues such as biodiversity loss, social equity, and the impacts of climate change on infrastructure resilience. By aligning LCA methodologies with these broader sustainability concerns, civil engineering can contribute more effectively to global efforts to address climate change, protect natural ecosystems, and promote social well-being.

The future of LCA in civil engineering is one of both opportunity and challenge. As the field continues to evolve, the adoption and refinement of LCA methodologies will be critical in steering the industry towards more sustainable and resilient infrastructure solutions. By embracing innovation, fostering collaboration, and expanding the scope of sustainability assessments, the civil engineering sector can play a pivotal role in building a more sustainable future.

CONCLUSION

As we conclude this exploration of the role of Life-Cycle Assessment (LCA) in enhancing infrastructure sustainability within civil engineering, it's clear that LCA stands as a pivotal tool in the quest for sustainability. This chapter has delved into the comprehensive methodology of LCA, demonstrating its critical importance in making informed decisions that minimize environmental impacts across the lifecycle of infrastructure projects. From the foundational principles and methodologies to real-world applications and future directions, the potential of LCA to transform civil engineering practices and contribute to sustainable development goals is immense.

Key Findings

- Comprehensive Methodology: LCA provides a systematic approach to assessing environmental impacts associated with all stages of a project's life cycle, from material extraction through to end-of-life disposal or recycling. This holistic perspective is essential for identifying opportunities to reduce negative environmental impacts.
- Real-World Applications: Through detailed case studies, we've seen how LCA can guide the selection of materials, construction methods, and operational practices that significantly lower the environmental footprint of infrastructure projects. These examples underscore the versatility and practical value of LCA in driving sustainability across different sectors within civil engineering.
- Advancements in Methodologies: The continuous improvement of LCA methodologies and tools, especially with the integration of digital technologies and sector-specific databases, enhances the accuracy, efficiency, and applicability of LCA studies. These advancements make LCA more accessible and relevant to civil engineering professionals.

- Challenges and Opportunities: Despite its potential, the implementation of LCA faces challenges, including data availability and methodological complexities. However, opportunities abound for overcoming these hurdles through education, collaboration, and technological innovation, paving the way for the broader adoption of LCA in civil engineering.

Future Directions

Looking ahead, the integration of LCA into civil engineering practices is expected to deepen, driven by the growing urgency of sustainability challenges and the recognition of LCA's value in addressing these issues. The future of LCA in civil engineering will likely be characterized by:

- Broader Adoption: As awareness of the importance of sustainability continues to grow, LCA is set to become an integral part of the civil engineering landscape, embedded in planning, design, and policy-making processes.
- Innovation and Technological Advancements: Ongoing advancements in LCA methodologies and digital technologies will further enhance the precision and ease of conducting LCAs, facilitating innovative and sustainable infrastructure solutions.
- Focus on Circular Economy: Aligning LCA with circular economy principles will be crucial in optimizing resource use and minimizing waste, contributing to the resilience and sustainability of infrastructure projects.

In conclusion, LCA represents a powerful instrument in the civil engineering toolkit for sustainability. By embracing the opportunities and addressing the challenges associated with its implementation, civil engineering professionals can make significant strides toward realizing sustainable infrastructure that meets the needs of the present without compromising the ability of future generations to meet their own needs. The journey towards sustainability is complex and multifaceted, but through the judicious application of LCA, we can navigate this path with greater clarity and purpose, ultimately contributing to a more sustainable and resilient world.

CONFLICT OF INTEREST STATEMENT

The authors declare that there is no conflict of interest regarding the publication of this chapter. The content has been composed to advance academic and practical knowledge on the application of Life-Cycle Assessment (LCA) in civil engineering to promote sustainability. The insights and conclusions drawn within this chapter are based on a comprehensive review of existing literature, case studies, and advancements in the field, and do not serve the interests of any specific organization or entity.

REFERENCES

Bahramian, M., & Yetilmezsoy, K. (2020). Life cycle assessment of the building industry: An overview of two decades of research (1995–2018). *Energy and Building, 219*, 109917. DOI: 10.1016/j.enbuild.2020.109917

Bamber, N., Turner, I., Arulnathan, V., Li, Y., Zargar Ershadi, S., Smart, A., & Pelletier, N. (2020). Comparing sources and analysis of uncertainty in consequential and attributional life cycle assessment: Review of current practice and recommendations. *The International Journal of Life Cycle Assessment, 25*(1), 168–180. DOI: 10.1007/s11367-019-01663-1

Bjørn, A., Chandrakumar, C., Boulay, A. M., Doka, G., Fang, K., Gondran, N., Hauschild, M. Z., Kerkhof, A., King, H., Margni, M., McLaren, S., Mueller, C., Owsianiak, M., Peters, G., Roos, S., Sala, S., Sandin, G., Sim, S., Vargas-Gonzalez, M., & Ryberg, M. (2020). Review of life-cycle based methods for absolute environmental sustainability assessment and their applications. *Environmental Research Letters, 15*(8), 083001. DOI: 10.1088/1748-9326/ab89d7

Castellani, V., Beylot, A., & Sala, S. (2019). Environmental impacts of household consumption in Europe: Comparing process-based LCA and environmentally extended input-output analysis. *Journal of Cleaner Production, 240*, 117966. DOI: 10.1016/j.jclepro.2019.117966 PMID: 31839696

Corominas, L., Byrne, D. M., Guest, J. S., Hospido, A., Roux, P., Shaw, A., & Short, M. D. (2020). The application of life cycle assessment (LCA) to wastewater treatment: A best practice guide and critical review. *Water Research, 184*, 116058. DOI: 10.1016/j.watres.2020.116058 PMID: 32771688

Fauzi, R. T., Lavoie, P., Sorelli, L., Heidari, M. D., & Amor, B. (2019). Exploring the current challenges and opportunities of life cycle sustainability assessment. *Sustainability (Basel), 11*(3), 636. DOI: 10.3390/su11030636

Hosseinzadeh-Bandbafha, H., Nizami, A. S., Kalogirou, S. A., Gupta, V. K., Park, Y. K., Fallahi, A., Sulaiman, A., Ranjbari, M., Rahnama, H., Aghbashlo, M., Peng, W., & Tabatabaei, M. (2022). Environmental life cycle assessment of biodiesel production from waste cooking oil: A systematic review. *Renewable & Sustainable Energy Reviews, 161*, 112411. DOI: 10.1016/j.rser.2022.112411

Hsien, C., Low, J. S. C., Fuchen, S. C., & Han, T. W. (2019). Life cycle assessment of water supply in Singapore—A water-scarce urban city with multiple water sources. *Resources, Conservation and Recycling, 151*, 104476. DOI: 10.1016/j.resconrec.2019.104476

Igos, E., Benetto, E., Meyer, R., Baustert, P., & Othoniel, B. (2019). How to treat uncertainties in life cycle assessment studies? *The International Journal of Life Cycle Assessment*, *24*(4), 794–807. DOI: 10.1007/s11367-018-1477-1

Janjua, S., Sarker, P., & Biswas, W. (2019). Sustainability assessment of a residential building using a life cycle assessment approach. *Chemical Engineering Transactions*, *72*, 19–24. DOI: 10.3303/CET1972004

Joensuu, T., Edelman, H., & Saari, A. (2020). Circular economy practices in the built environment. *Journal of Cleaner Production*, *276*, 124215. DOI: 10.1016/j.jclepro.2020.124215

Kaab, A., Sharifi, M., Mobli, H., Nabavi-Pelesaraei, A., & Chau, K. W. (2019). Combined life cycle assessment and artificial intelligence for prediction of output energy and environmental impacts of sugarcane production. *The Science of the Total Environment*, *664*, 1005–1019. DOI: 10.1016/j.scitotenv.2019.02.004 PMID: 30769303

Khandelwal, H., Dhar, H., Thalla, A. K., & Kumar, S. (2019). Application of life cycle assessment in municipal solid waste management: A worldwide critical review. *Journal of Cleaner Production*, *209*, 630–654. DOI: 10.1016/j.jclepro.2018.10.233

Laurent, A., Weidema, B. P., Bare, J., Liao, X., Maia de Souza, D., Pizzol, M., Sala, S., Schreiber, H., Thonemann, N., & Verones, F. (2020). Methodological review and detailed guidance for the life cycle interpretation phase. *Journal of Industrial Ecology*, *24*(5), 986–1003. DOI: 10.1111/jiec.13012 PMID: 33746505

Lü, G., Batty, M., Strobl, J., Lin, H., Zhu, A. X., & Chen, M. (2019). Reflections and speculations on the progress in Geographic Information Systems (GIS): A geographic perspective. *International Journal of Geographical Information Science*, *33*(2), 346–367. DOI: 10.1080/13658816.2018.1533136

Lu, K., Jiang, X., Yu, J., Tam, V. W., & Skitmore, M. (2021). Integration of life cycle assessment and life cycle cost using building information modeling: A critical review. *Journal of Cleaner Production*, *285*, 125438. DOI: 10.1016/j.jclepro.2020.125438

Mannan, M., & Al-Ghamdi, S. G. (2020). Environmental impact of water-use in buildings: Latest developments from a life-cycle assessment perspective. *Journal of Environmental Management*, *261*, 110198. DOI: 10.1016/j.jenvman.2020.110198 PMID: 32148271

Massaro, M., Secinaro, S., Dal Mas, F., Brescia, V., & Calandra, D. (2021). Industry 4.0 and circular economy: An exploratory analysis of academic and practitioners' perspectives. *Business Strategy and the Environment, 30*(2), 1213–1231. DOI: 10.1002/bse.2680

Meglin, R., Kytzia, S., & Habert, G. (2022). Regional circular economy of building materials: Environmental and economic assessment combining material flow analysis, input-output analyses, and life cycle assessment. *Journal of Industrial Ecology, 26*(2), 562–576. DOI: 10.1111/jiec.13205

Moreno-Sader, K., Meramo-Hurtado, S. I., & González-Delgado, A. D. (2020). Environmental sustainability analysis of chitosan microbeads production for pharmaceutical applications via computer-aided simulation, WAR and TRACI assessments. *Sustainable Chemistry and Pharmacy, 15*, 100212. DOI: 10.1016/j.scp.2020.100212

Nwodo, M. N., & Anumba, C. J. (2019). A review of life cycle assessment of buildings using a systematic approach. *Building and Environment, 162*, 106290. DOI: 10.1016/j.buildenv.2019.106290

Peña, C., Civit, B., Gallego-Schmid, A., Druckman, A., Pires, A. C., Weidema, B., Mieras, E., Wang, F., Fava, J., Canals, L. M., Cordella, M., Arbuckle, P., Valdivia, S., Fallaha, S., & Motta, W. (2021). Using life cycle assessment to achieve a circular economy. *The International Journal of Life Cycle Assessment, 26*(2), 215–220. DOI: 10.1007/s11367-020-01856-z

Rezaei, F., Bulle, C., & Lesage, P. (2019). Integrating building information modeling and life cycle assessment in the early and detailed building design stages. *Building and Environment, 153*, 158–167. DOI: 10.1016/j.buildenv.2019.01.034

Safari, K., & AzariJafari, H. (2021). Challenges and opportunities for integrating BIM and LCA: Methodological choices and framework development. *Sustainable Cities and Society, 67*, 102728. DOI: 10.1016/j.scs.2021.102728

Seyis, S. (2019). Pros and cons of using building information modeling in the AEC industry. *Journal of Construction Engineering and Management, 145*(8), 04019046. DOI: 10.1061/(ASCE)CO.1943-7862.0001681

Temizel-Sekeryan, S., Wu, F., & Hicks, A. L. (2021). Global scale life cycle environmental impacts of single-and multi-walled carbon nanotube synthesis processes. *The International Journal of Life Cycle Assessment, 26*(4), 656–672. DOI: 10.1007/s11367-020-01862-1

Ternel, C., Bouter, A., & Melgar, J. (2021). Life cycle assessment of mid-range passenger cars powered by liquid and gaseous biofuels: Comparison with greenhouse gas emissions of electric vehicles and forecast to 2030. *Transportation Research Part D, Transport and Environment*, *97*, 102897. DOI: 10.1016/j.trd.2021.102897

Tokede, O., & Traverso, M. (2020). Implementing the guidelines for social life cycle assessment: Past, present, and future. *The International Journal of Life Cycle Assessment*, *25*(10), 1910–1929. DOI: 10.1007/s11367-020-01814-9

Toniolo, S., Tosato, R. C., Gambaro, F., & Ren, J. (2020). Life cycle thinking tools: Life cycle assessment, life cycle costing and social life cycle assessment. In *Life Cycle Sustainability Assessment for Decision-Making* (pp. 39–56). Elsevier., DOI: 10.1016/B978-0-12-818355-7.00003-8

van Der Werf, H. M., Knudsen, M. T., & Cederberg, C. (2020). Towards better representation of organic agriculture in life cycle assessment. *Nature Sustainability*, *3*(6), 419–425. DOI: 10.1038/s41893-020-0489-6

Vega-Araujo, D., Martinez-Arguelles, G., & Santos, J. (2020). Comparative life cycle assessment of warm mix asphalt with recycled concrete aggregates: A Colombian case study. *Procedia CIRP*, *90*, 285–290. DOI: 10.1016/j.procir.2020.02.126

Weiland, S., Hickmann, T., Lederer, M., Marquardt, J., & Schwindenhammer, S. (2021). The 2030 agenda for sustainable development: Transformative change through the sustainable development goals? *Politics and Governance*, *9*(1), 90–95. DOI: 10.17645/pag.v9i1.4191

Zhang, A., Zhong, R. Y., Farooque, M., Kang, K., & Venkatesh, V. G. (2020). Blockchain-based life cycle assessment: An implementation framework and system architecture. *Resources, Conservation and Recycling*, *152*, 104512. DOI: 10.1016/j.resconrec.2019.104512

Chapter 3
Heat Recovery Systems From Sewage:
The Current State and Development Prospects

Olga Arsenyeva
O.M. Beketov National University of Urban Economy in Kharkiv, Ukraine

Ihor Biletskyi
O.M. Beketov National University of Urban Economy in Kharkiv, Ukraine

Sergiy Plankovskyy
O.M. Beketov National University of Urban Economy in Kharkiv, Ukraine

Petro Kapustenko
Sustainable Process Integration Laboratory (SPIL), Brno University of Technology, Czech Republic

Yevgen Tsegelnyk
https://orcid.org/0000-0003-1261-9890
O.M. Beketov National University of Urban Economy in Kharkiv, Ukraine

Natalia Teliura
https://orcid.org/0000-0003-0732-7789
O.M. Beketov National University of Urban Economy in Kharkiv, Ukraine

ABSTRACT

Renewable energy, crucial amid global warming, emphasizes waste heat utilization in cities. Industrial waste heat and domestic wastewater offer significant potential. In the UK, up to 6.1% of heat needs could be met by wastewater recovery. Domestic systems can save up to 50% of hot water energy. Recent research focuses on improving heat exchanger efficiency for household applications. However, larger-scale systems lack attention. Solutions for extracting heat from sewer collectors are limited. New technologies aim to restore collectors and recover heat efficiently. Plastic pipes offer promising rehabilitation methods, ensuring prolonged collector life. The proposed chapter will detail efficient heat exchangers and innovative sewer heat recovery

DOI: 10.4018/979-8-3693-2448-6.ch003

methods, addressing existing limitations and future research prospects.

INTRODUCTION

Researchers are intensifying their exploration of new renewable energy sources, focusing notably on waste heat recovery technologies in residential and industrial sectors, given the pressing concerns of global warming and environmental degradation. Industrial waste heat sources are especially promising due to their significant capacity and potential for efficient hybrid recovery schemes, which incorporate heat pumps alongside electricity generation methods like the Rankine cycle and thermoelectric generators (Farhat et al., 2022). Wastewater from domestic sources holds considerable promise for heat reclamation, potentially fulfilling up to 6.1% of the UK's heat demand, as estimated solely for England (Wilson and Worrall, 2021). Leveraging wastewater heat offers significant advantages as it is consistently generated and will persist indefinitely. Moreover, wastewater is typically produced near areas with the highest demand for heat, enhancing its accessibility and efficiency (Papagiannis et al., 2018; Vystavna et al., 2019).

Most of the research in waste heat recovery primarily targets household applications, as they often offer straightforward technical solutions. Wehbi et al. (2022) indicates that these systems can potentially save up to 50% of the energy typically consumed for hot water provision, given specific conditions. Pomianowski et al. (2020), Wehbi et al. (2022) offers a comprehensive analysis of recent advancements in household heat recovery techniques. Meanwhile, Pokhrel et al. (2022) explores the utilization of sewage effluent heat to warm multi-apartment residential complexes, employing heat pumps to deliver low-grade thermal energy to buildings (Babaev et al., 2016). An analysis of the effectiveness and potential applications of these technologies reveals a possibility of reducing CO_2 emissions by 1.1 million kilograms annually. Recent research has been concentrating on enhancing the efficiency and compactness of heat exchangers, which are the primary component of waste heat recovery systems in households. This focus has been particularly directed towards plate heat exchangers (Tovazhnyanskyy et al., 2020; Arsenyeva et al., 2022).

Research on heat recovery conditions for larger-scale systems, such as urban agglomerations and industrial operations, remains inadequate to date. Piotrowska et al. (2020) proposes a study to investigate the potential of consolidating flows from multiple buildings into collectors. This aims to augment wastewater volume and enhance heat recovery, thereby increasing profitability. Furthermore, according to Piotrowska et al. (2020), examining wastewater with varying temperatures and pollution levels is imperative for understanding heat recovery systems in sewage collectors.

Managing pollution is a critical concern in the context of harnessing heat recovery from sewage systems, with multiple strategies available for mitigation. One established approach entail utilizing air circulation to extract heat from contaminated sewage, as detailed in Shen et al. (2020). This system employs four circuits: a sewage loop, an air loop, a heat pump loop, and a hot water loop. Sewage is pumped and sprayed from the top of the sewage tower, cascading down along a guide plate, while circulating air is introduced from the bottom, rising to the top. Wastewater and air intersect in counterflow, where a portion of the wastewater evaporates into a gas state, mingling with the circulating air to transfer latent heat from the wastewater to the air.

The high-enthalpy air entering the evaporator of the heat pump absorbs heat from circulating air, warming the water in the hot water tank via a plate heat exchanger. As a byproduct, water vapor from drains condenses on the heat exchanger plates. This condensed water, deemed suitable for city landscaping irrigation according to Shen et al. (2020), poses advantages but also introduces complexities. Disadvantages include the need for sewage pumps requiring preliminary treatment, as well as the likelihood of eventual clogging in the spray system.

The alternative being considered involves heat recovery within a two-circuit system, outlined in Cecconet et al. (2019). This particular system is presently employed in the HUBERTM ThermWin commercial system, developed by HUBER SE in Berching, Germany. The two-circuit system comprises a shaft positioned at the lower section of the sewer, facilitating gravity-based connection to both an overhead heat exchanger and a heat pump. A sump serves as a reservoir for the pump supplying the heat exchanger and includes a coarse screen to prevent debris carried by the wastewater from entering the heat exchanger. The solids are subsequently reintroduced into the main system via a vertical auger. As stated in Cecconet et al. (2019), this approach consumes 58% less energy compared to a conventional system of similar capacity, encompassing cooling, heating, and associated energy needs. Moreover, an assessment of the influence of recovering heat from wastewater on subsequent biological processes indicates that this effect is minimal. The wastewater inlet temperature experiences a slight decrease of 0.045°C, while during winter and summer, there is a respective increase of 0.056°C. Consequently, these fluctuations in temperature do not significantly impact the efficiency of wastewater treatment.

A proven technique for extracting heat from sanitary sewage collectors utilizes a specialized sewer collector pipe known as Source-Energy. This pipe consists of a concrete cylinder integrated with a spiral polyethylene pipe (Dacquay et al., 2020). This pipe efficiently harnesses heat from sewers, storm drains, and the surrounding ground, making it suitable for use in a heat pump system. With inlet and outlet pipes at both ends, operating it at a depth of 4 meters ensures consistent performance. Doubling the capacity of the sewer pipe from 25% to 50% and 75% results in an 8.7% and 10.5% increase in thermal flow respectively, while still meeting sustainability goals (Dacquay et al., 2020). Notably, the system remains untouched by contaminated drains, offering a significant advantage. However, extending the length of the collector requires additional piping to complete the water circuit, which could pose logistical challenges.

Repairing sewage collectors is a pressing concern (Zhu et al., 2021), particularly in the United States, where more than 2.0 million kilometers of sewer lines were constructed during the 1950s and 1960s. Improving current water supply and sanitation systems would necessitate a minimum investment of $1 trillion (Zhu et al., 2021; Yesina et al., 2023). More than half of the sewage system in the United States relies on concrete pipelines, especially those with a diameter exceeding one meter (Selvakumar and Tafuri, 2012). A promising rehabilitation method involves utilizing plastic pipes, which can be restored using a trench-less technique in Krogness et al. (2020). The restored collectors can have a service life of up to 50 years. The Spiral Wound PVC Liner Pipe (Kivirenko et al., 2023; Kombarov et al., 2021) is a viable choice for repairing such collectors. Inspection data gathered after five years of operation reveal only structural deformations occurring in consecutive rings along the pipe. These deformations are likely attributed to utility construction activities above the pipe, possibly due to the presence of an underground electric line in that specific location (Krogness et al., 2020).

Thus, the effectiveness of recovering heat from wastewater heavily relies on the quality of heat exchangers. To make informed decisions during the design of recovery systems, it's essential to have up-to-date information on the advancements in heat exchanger technology for extracting heat from wastewater, whether it's at the household level or from sewer collectors. An innovative approach aims to address both wastewater heat recovery and the restoration of large sewer collectors simultaneously. A patented solution suggests the utilization of a spirally wound plastic pipe with a hollow box profile (Usatenko and Biletskyi, 2023). This pipe is designed to facilitate the transfer of cold water for heating purposes extracted from sewage. While this solution remains untested in practice, assessing its efficacy holds substantial theoretical and practical importance. This review endeavors to furnish researchers engaged in developing wastewater heat recovery systems with vital insights into these challenges.

EFFICIENT PLATE HEAT EXCHANGERS FOR WASTEWATER HEAT RECOVERY

The improvement of the utilization of sewage water heat potential is possible by implementing heat pump technologies (Averfalk et al., 2017). Here, the four scenarios for incorporating heat pumps into district heating systems are possible in accordance with Sayegh et al. (2018) identified four scenarios for integrating heat pumps in district heating systems: (a) placing heat pumps in the existing network without major changes, (b) placing heat pumps in an expanded network, (c) deep refurbishment of the existing district heating system, and (d) designing a new district heating system supplied by a heat pump. (1) integrating heat pumps into the current network with minimal modifications, (2) incorporating heat pumps into an extended network, (3) extensively renovating the existing district heating system, and (4) creating a new district heating system powered by a heat pump. By harnessing excess heat that would have otherwise been wasted, metropolitan areas can meet their heat requirements by selecting one of the existing alternatives. This will result in energy conservation, increased use of renewable energy sources, decreased emissions of greenhouse gases and air pollution in urban areas, providing a potential approach for moving towards environmentally sustainable, resource-independent cities (Lu et al., 2020).

The selection of an appropriate heat pump is significantly impacted by the characteristics of the available heat source. The heat source affects the positioning, connection, operating modes, thermal capacity, and operating temperature range of a heat pump unit, as well as the choice of the appropriate heat pump technology (Sayegh et al., 2018; Ruzaikin and Lukashov, 2023). The investigation concentrated on scenarios where heat pumps were successfully integrated into district heating systems. One of the good examples is the implementation of large heat pumps in the Swedish district heating system, discussed in Averfalk and Werner (2017). Such parameters as capacity utilization, competitiveness, refrigerant management, and refrigerant leakage, are examined. The investigation of the spatial and temporal dimensions of wastewater heat recovery system efficiency was presented by Spriet et al. (2020).

Heat Pumps for Residential Heating Systems

To utilise the heat from the wastewater heat from the sewage collector in the residential heating systems, the system with heat pumps is proposed. As the wastewater is aggressive fluid, the direct implementation of heat transfer equipment for heat collection with direct contact with sewage water, require expensive material and complex design. Moreover, the sewage water temperature is low, around 12–15°C, what makes it problematic to utilise directly, but can successfully be used as a low-grade heat for the heat pumping system.

The withdrawal of heat from the wastewater collector can be done from special plastic heat exchanger, described in next chapter. The proposed design makes it possible to collect lo-grade heat in a closed loop avoiding mixing, and then to direct eat, using the water as heat carrier, to the heat pump for preheating cold water for radiator heating or for tap water supply (Delil et al., 2002; Morishita et al., 2017). Additionally, such systems can be integrated with renewable energy sources or centralised district heating supply circuits to increase the supplied heat in cold periods. Such heat pumps are compatible with including low-temperature district heating systems of the 5th generation and can be implemented for residential heating of individual multi-stored buildings or on a larger scale having the potential to supply low-grade heat to a district heating networks (DHN). The potential of the accessible low-grade heat source of energy when implementing such systems for large-scale regions is estimated around 5% of the heat demand of cities (Kugler et al., 2022).

In the proposed technology the heat pump increases the temperature and potential of low-grade heat gained from the sewage collector (Figure 1). A heat pump transfers energy from a fluid at a low temperature to one at a high temperature by utilising mechanical or heat energy. This extra energy might be seen as a cost for reversing the normal heat transfer, which, as per the second law of thermodynamics, typically goes from a higher to a lower temperature.

Figure 1. Application of heat pump system for heat recovery from the sewage water

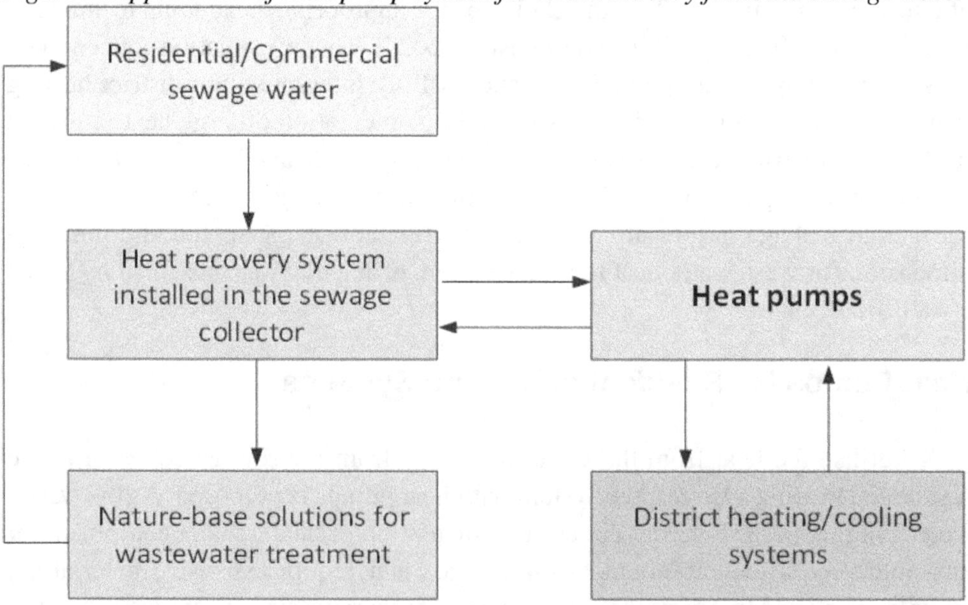

The efficiency of a heat pump is often assessed using its coefficient of performance (COP), which indicates the amount of thermal energy transferred to the medium being heated per unit of additional energy needed to operate the heat pump. The exact estimation of heat pump COP affects the design efficiency of heat pump application (Jesper et al., 2021). Generally, COP is calculated according to the following relation:

$$COP = \frac{Q_s}{P_{ed}}$$

, (1)

where Q_s is the supplied heat energy, W; P_{ed} is the driving energy, W. The energy needed to operate a heat pump can be supplied by several means such as an electric motor, petrol engine, electricity, or thermal energy.

The primary energy (PER) ratio is utilised to assess the efficiency of a heat pump system by indicating the amount of thermal energy provided to the material being heated per unit of primary energy consumed to operate the heat pump system.

$$PER = \frac{Q_{total\_s}}{P_{es}}$$

, (2)

where Q_{total_s} is the total thermal energy supplied, W; P_{es} is the energy spent, W.

According to the natural physical phenomena, different types of heat pumps (HP) exist. The widely used types include mechanical compression HP, absorption HP, ejector HP, chemical HP, thermoelectric HP, and others. The mentioned types of HP have different design principles, sources of energy and delivery methods. Correspondingly, their different performance and cost characteristics vary.

Vapour-compression heat pumps function based on the same principle as vapour-compression refrigeration systems (Green and Perry, 2008). The perfect Carnot cycle is unachievable, it is reversible, and its temperature-entropy diagram is depicted in Figure 2. The process involves adiabatic compression (1–2), isothermal heat rejection during condensation (2–3), adiabatic expansion (3–4), and isothermal heat addition during evaporation (4–1). The perfect Carnot cycle can be used estimation the temperatures that define the efficiency of the heat transfer process from low to high temperature levels.

Figure 2. Temperature-entropy diagram of the Carnot cycle

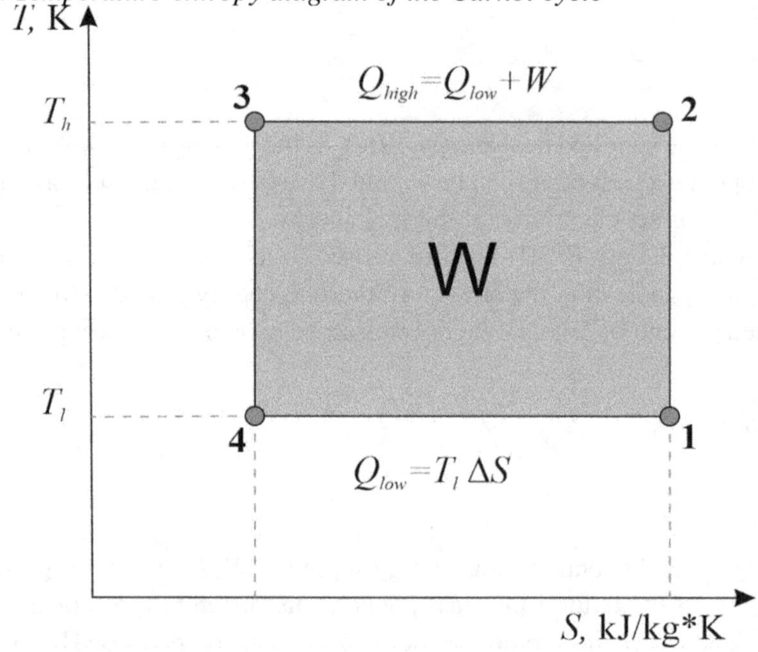

The heat transported to the hot media at the high temperature Q_{high} is the primary goal of heat pump operation. The heat transported in process Q is equal to the heat extracted from the low-temperature level. The additional energy input W is added to this quantity and then supplied to a high temperature level with entropy change is given by $\Delta S = \Delta Q/T$ according to the third postulate of thermodynamics.

The COP of a heat pump working with a Carnot cycle can be determined using the following equation:

$$COP = \frac{Q_{high}}{W} = \frac{Q_{low} + W}{W} = \frac{T_h}{T_h - T_l}$$

, (3)

where T_h and T_l are the high and low temperature levels of Carnot cycle, K.

The coefficient of performance, when extracting heat at a lower temperature, is calculated as the ratio of heat removed from the low-temperature level Q_{low} to the energy input W:

$$COP_{Tl} = \frac{Q_{low}}{W} = \frac{Q}{W} = COP - 1$$

. (4)

Vapour-compression heat pumps transmit heat to higher temperature levels by the next processes: (1) evaporation of refrigerant in the evaporator, (2) condensation in the condenser where heat is transferred to a hot medium, (3) expansion of refrigerant in an expansion device. The compression can be physically utilised in different ways, as illustrated in Figure 3. The possible processes include (1) utilising a mechanical compressor powered by an electric motor, (2) a petrol engine driving a mechanical compressor, (3) utilising pressure difference through an ejector, (4) using thermal energy for absorption and desorption. Regarding heat exchangers, a minimum of two recuperative heat exchangers are necessary to function as condenser and evaporator in all these approaches. The number of implemented heat exchangers can be greatly expanded through intricate arrangements, particularly when indirect heat exchange is necessary.

Figure 3. Basic methods to transform low-pressure vapor into high-pressure vapor

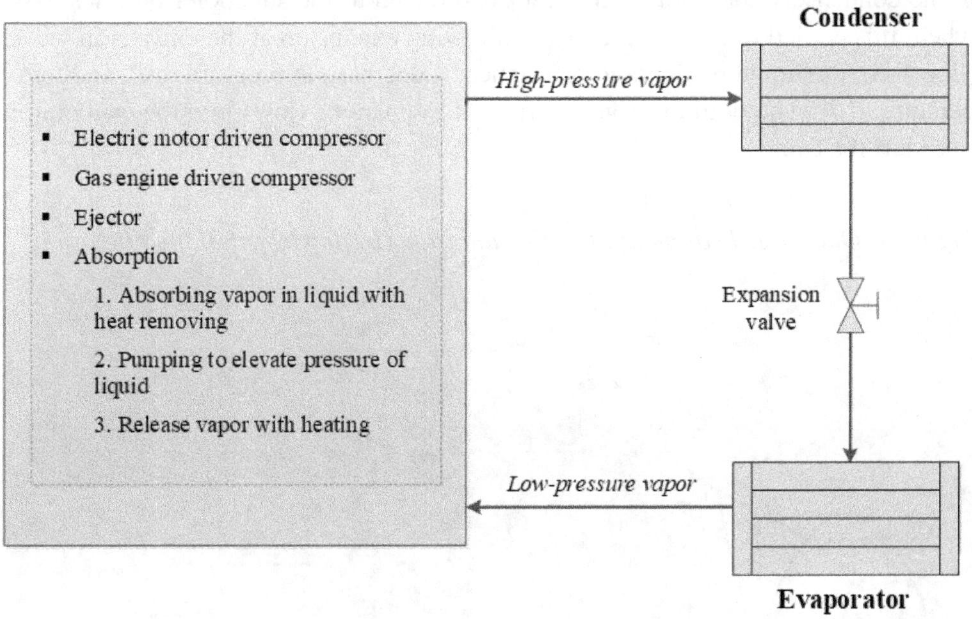

Vapor-compression cycles are the predominant principle utilised in heat pumps. The *T–S* diagram of the ideal Carnot cycle is straightforward and useful for thermodynamic study, but the process is more complex. Figure 4, *a* illustrates a vapour compression cycle (Rankine cycle) in *T–S* coordinates, closely resembling reality. An isothermal process necessitates isobaric evaporation and condensation under constant pressures. However, in practice, the refrigerant flow in the condenser and evaporator experiences a pressure decrease. Sub cooling is consistently detected in the condenser, while vapour superheating occurs at the compressor suction line due to ongoing processes in the heat exchangers and environmental factors. Subcooling and superheating are important to ensure that only liquid reaches the expansion mechanism. Superheating is advised to prevent liquid droplets from entering the compressor.

The heat pump's functioning depicted in Figure 4, *b* is carried out as follows. The low temperature medium transfers heat to the refrigerant in the evaporator. The refrigerant is boiling at pressure P_0 and saturation temperature T_0. The vapour, separated from liquid droplets in the separator, is directed to the compressor where its pressure is raised to P_c. On the *T–S* diagram, it corresponds to the line 1–2. The dashed line 1–2* represents isentropic compression, which is unattainable in practice because to friction and irreversible heat losses. The temperature of the superheated vapour increases to T_2 following compression, above the saturation temperature at pressure P_c. The vapour is cooling and condensing at the saturation temperature T_c in the condenser. The liquid refrigerant is directed to the subcooler on lines 3–4, where it is cooled to temperature T_{CO}. Following expansion at the expansion valve (line 4–5), the liquid refrigerant experiences a decrease in pressure to P_0 and temperature to T_0. The liquid refrigerant from the separator flows into the evaporator to restart the cycle.

Figure 4. One stage heat pump: (a) T–S diagram (b) flow chart of the operation

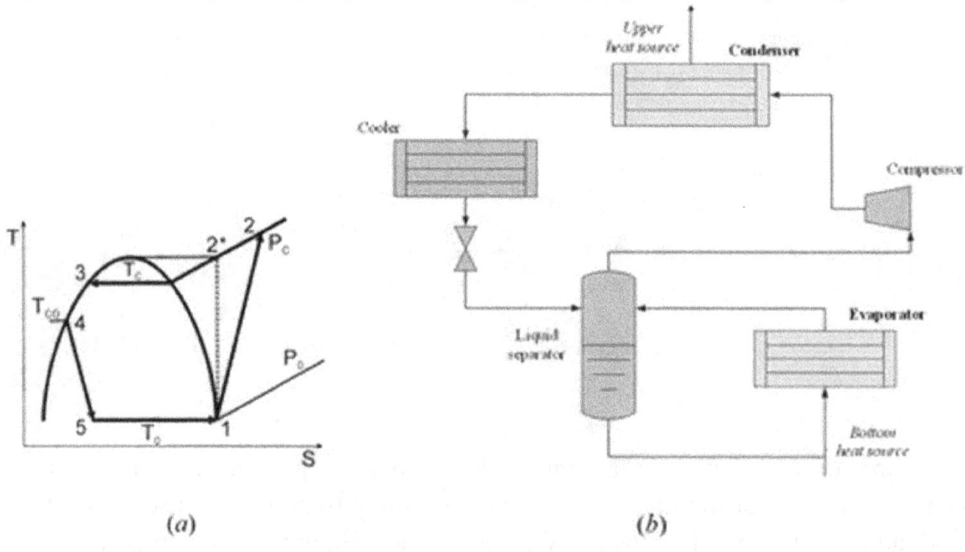

(*a*) (*b*)

Various methods can enhance the COP of a cycle, contingent upon the refrigerant and operating temperatures of the heat pump. Compression can be divided into two stages for significant differences in condensing and evaporation temperature and pressure, utilising two compressors with the same refrigerant. Two different temperature levels can be achieved by utilising condensers on each stage. A cascade system involves two systems with different refrigerants and compressors exchanging heat through a mutual heat exchanger. This heat exchanger serves as a condenser for the system with low temperatures and as an evaporator for the system with higher temperatures. Various thermodynamic cycles, such as the double Rankine cycle, Stirling cycle, Brayton cycle, and their combinations like Otto-Rankine, Stirling-Rankine, Rankine-Brayton, and Stirling-Brayton, can be utilised in the design of vapor-compression heat pumps. These systems vary in efficiency depending on the refrigerant utilised and are typically employed in certain applications. Heat pumps necessitate heat exchangers to function as evaporators and condensers.

Various types of compressors are utilised in mechanical vapor-compression heat pumps, including reciprocating, rotary such as screw, vane, scroll, and centrifugal or turbocompressors. The selection of the compressor is based on the heat pump's capacity and application type. The essential characteristic to consider when selecting heat exchangers to work with a given compressor is the level of vibrations generated in the system. The heat pump's compressor should be powered by an engine. Various types of propulsion systems can be used for this purpose. Electric motors are the most utilised because to their great efficiency, reaching up to 95% at full loads, reliability, ease of maintenance, and eco-friendliness (Pliuhin et al., 2023). When considering electricity production efficiency, which typically does not surpass 35–40% for most present power generation facilities, the primary energy usage of a heat pump with an electric motor drive can be significant, with a Performance Efficiency Ratio slightly above 1.

The crucial element determining construction and efficiency of a heat pump is working media which is taking heat at low temperature and supplying it to upper temperature level, or refrigerant. Natural refrigerants do not disrupt the ozone layer and have a low global warming potential, but many of them have other negative characteristics. Ammonia is harmful and corrosive to copper. Propane and butane are combustible. Carbon dioxide has a low COP and necessitates a condenser pressure of 100 bar or greater. Heat exchangers have certain criteria, with compactness and small internal volume being crucial for reducing the refrigerant load in a heat pump system. Qin and Tian (2022) highlighted that the economic features and operational effectiveness of an Organic Rankine Cycles (ORC) are significantly influenced by the choice of working fluid. A wide range of chemicals have been identified for ORCs, including hydrocarbons, hydrofluorocarbons, hydrochlorofluorocarbons, chlorofluorocarbons, perfluorocarbons, siloxanes, alcohols, aldehydes, ethers, hydrofluoroethers, amines, fluid mixes (zeotropic and azeotropic), and inorganic fluids. The natural refrigerants not causing harmful effect on ozone layer and has not big global warming potential, but most of them possesses other undesirable qualities. Ammonia is toxic and aggressive to copper. Propane and butane are flammable. CO_2 has low COP and requires pressure in condenser 100 bar and higher. It puts forward special requirements for heat exchangers, among which the compactness and small inside volume are the most important as it allows to decrease the charge of refrigerant in heat pump system.

Figure 5. Simplified flow diagram of Rankine cycle

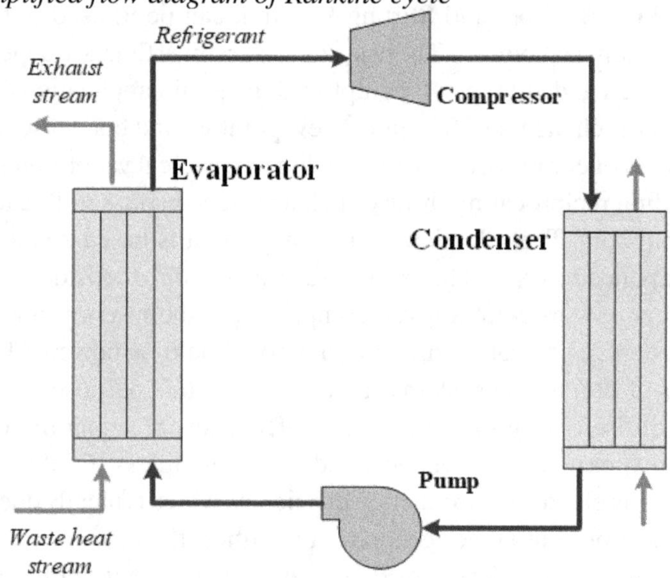

The operational principles of ORC are outlined in Figure 5. The refrigerant is transferred from a low-pressure state in the condenser to a high-pressure state in the evaporator as a liquid with minimal energy input. An optional recuperative heat exchanger can be incorporated to enhance the efficiency of the ORC. In this setup, the fluid at a higher temperature following the turbine is utilised to preheat the fluid at a lower temperature before it reaches the evaporator. The liquid in the evaporator is heated at a constant pressure by an external low-grade heat source, causing it to evaporate and turn into a dry saturated vapour. The vapour expands in the turbine and generates electric power through a generator. The vapor's temperature and pressure decrease, causing it to condense at a constant temperature in a condenser, transforming into saturated liquid before being pumped to the evaporator. Fluids used in ORC must have critical temperatures that exceed the temperatures of the operating cycle.

Heat Transfer Equipment for Heat Pumps

The ORC heat pump system utilises recuperative heat exchangers to facilitate heat transfer processes such as gas-gas, gas-liquid, liquid-liquid, evaporation, and condensation to convert low-grade heat from wastewater into power. The conditions of these processes vary significantly in terms of temperatures, pressures, pressure drops, thermal and physical qualities of streams, and inclinations for fouling and corrosion of heat transfer surfaces. Not all process circumstances may be effectively met by a single type of heat exchanger, such as the commonly used shell and tube heat exchangers, even from a technical perspective. Another factor to consider is the economic aspect, since heat transfer equipment can make up a significant portion of the initial cost for many systems that utilise low-grade heat. The advancement of heat transfer equipment to meet the needs of diverse industries has led to the construction of several types of heat exchangers for applications in industry, mining, transport, aerospace, and other sectors. Many of these varieties can be utilised in various low-grade heat applications that have unique needs.

One primary constraint of using heat transfer equipment is the narrow temperature range required for utilising low-grade heat. When the heat pumps are upgraded with heat, any extra temperature differential in the evaporator or condenser causes a fall in the COP for the entire system and a decline in system performance. The minimal temperature differential in the heat recuperation system is a key parameter in Process Integration. A reduced Mean Temperature Difference (MTD) leads to improved heat recovery and enhanced energy efficiency, but it also results in higher expenses due to the larger heat transfer area required for recuperative heat exchangers. The minimum temperature differential when employing conventional shell and tube heat exchangers typically ranges from 10 K to as low as 5 K, depending on their design characteristics. Using lower MTD values without significantly raising capital expenditures would enhance the energy efficiency of recuperative heat exchanger systems. Other types of heat exchangers can achieve this by heat transfer augmentation technology.

In a heat exchanger, the heated fluid absorbs all the thermal energy transferred from the cooling fluid, with minimal heat losses to the environment that can be reduced by insulation in certain heat exchangers. The exergy, which is the useful component of thermal energy, is lost since it is not possible to heat a cold fluid to the temperature of the hot fluid used for heating. To minimise exergy losses, heat the cold fluid as close as feasible to the supply temperature of the hot fluid and aim to minimise the temperature difference.

To choose the appropriate type and design of heat exchangers, it is needed to analyse the full range of existing heat transfer equipment to identify the optimal option on the market or enhance it through further development for the specific scenario.

The most promising type of heat exchangers to be used in heat pumps for sewage water utilisation are plate heat exchangers (PHEs). PHEs are a type of compact heat exchanger that provide increased heat transfer. Heat transmission occurs within channels with intricate geometries created by corrugated plates. Plate-and-frame, brazed, and welded PHEs are highly efficient compact heat exchangers commonly utilised in various industries (Klemes et al., 2015). The operating parameters of available commercial PHEs for liquid-vapour and gas-liquid operations, which are suitable for application as evaporators or condensers, are presented in Table 1 (Wang et al., 2022). The listed brazed PHEs and Compabloc welded PHE are produced by Alfa Laval. The last Pillow-plate heat exchanger presents the innovative design, promising for the two-phase application.

Table 1. PHEs for liquid-vapor and gas-vapor application

PHE type	Max working temperature, °C	Max working pressure, bar	Illustration
Brazed PHEs			
AC	225	54	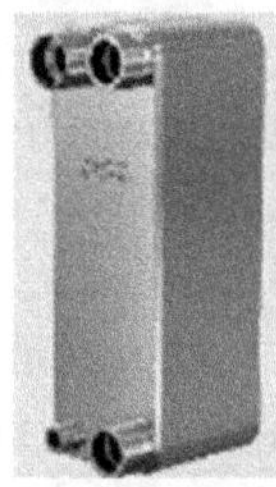
CB	204	32	
Welded PHEs			

continued on following page

Table 1. Continued

PHE type	Max working temperature, °C	Max working pressure, bar	Illustration
Compabloc	343	42	
AlfaVap	180	6	
Pillow-Plate HEs*	under investigation, specifically designed for each unit		

* LOB. WTP-heating system. https://lob-gmbh.de/en/produkt/wtp-plate-heat-exchanger/ (accessed 09 March 2024)

The problem of the PHEs design was presented in Klemes et al. (2015). The work presents the approach for achieving the minimal cost of the unit, enabling to determine the minimal needed heat transfer area and is based on estimation of the ideal geometric characteristics of the unit which will suit the specific process conditions. The approach is applicable for brazed and welded PHEs produced by Alfa Laval given in Table 1.

For optimal PHEs design, all the construction parameters, which will suit the working conditions, should be established. In general, the stream flow rates, their inlet and outlet temperatures are defined by the process as well as the thermos-physical parameters of the working fluids. In the case of two-phase media, the following parameters should be specified: flowrates of incoming two-phase flow $G_{1.in}$, kg/s; the pressure of two-phase flow $P_{1.in}$, Pa; the inlet temperature of two-phase flow $t_{1.in}$, °C; required outlet temperature of two-phase stream $t^0_{1.out}$, °C; flowrate of one-phase stream G_2, kg/s; inlet and outlet temperatures of the one-phase stream, $t_{2.in}$ and $t_{2.out}$, °C; allowable pressure drops for streams ΔP^0_1 and ΔP^0_2 in Pa.

Based on these parameters, the optimisation problem of PHE design is the minimum of the heat transfer area, which is estimated based on the geometrical parameters of PHEs:

$$F = F\left(\beta, b, \gamma, L_p, Fpl, TP_{specification}\right) \quad , \quad (5)$$

where L_p is the length of the plate, m; Fpl is heat transfer area of a single plate, m^2, b is corrugations height, m; β is the angle of corrugations inclination to the main direction of the flow, degrees; $\gamma = 2b/S$ is corrugation aspect ratio; $TP_{specification}$ is the temperatures and pressures combinations according to application requirements. The discussed geometrical parameters of heat transfer plates, which form the channels in PHEs are illustrated in Figure 6.

Figure 6. The geometric parameters of the PHEs plates corrugations: (a) plate corrugations crossing; (b) PHE channels cross-sections; (c) three-dimensional view

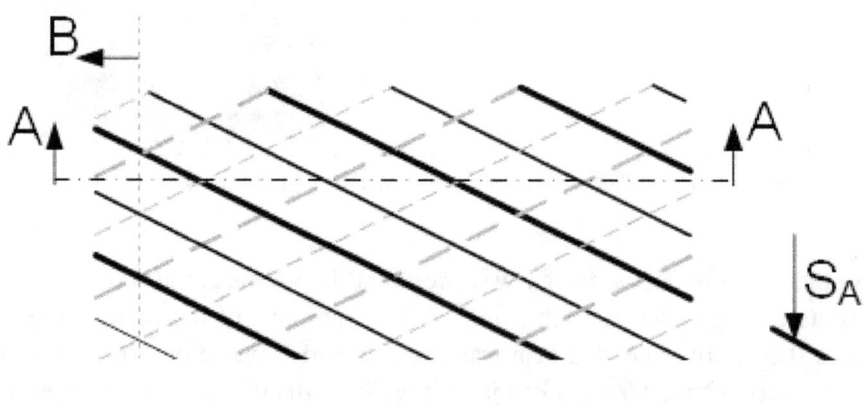

According to Equation (5), the geometrical parameters of PHEs plates that ensure the minimal value of function F should be found to satisfy the known operating conditions with additional limitations for two-phase flows:

$$t_{1.out} \leq t^0_{1.out}$$

, (6)

$$\Delta P_1 \leq \Delta P_1^0$$

. (7)

At PHE design, the available plate parameters, specified by manufacturers, should be kept. The available ranges for commercially produced plates are presented in Table 2.

Table 2. The range of parameters for commercially produced plates

Parameter	Values
Corrugation aspect ratio, γ	0.56 ± 0.04
Corrugations inclination angle β, degree	30°, 60°
Corrugations height b, mm	from 2.0 mm to 5.0 mm

The PHE design for evaporation and condensation duties should estimate the heat transfer coefficient and pressure drop of the unit. To do it, the proper correlations for single phase flow from the cold side of condensed and hot side of evaporator, as well as for two-phase flows of condensed and evaporated fluid should be used.

For the single liquid-liquid flow without the phase change, a generalised correlation for the Darcy friction factor in PHE channels with varying corrugation geometries on their walls, proposed in Arsenyeva et al. (2011), can be used. The equation is expressed as follows:

$$\zeta = 8 \times \left[\left(\frac{12 + \frac{\pi \cdot \beta \cdot \gamma^2}{3}}{Re} \right)^{12} \left(A + \left(\left(\frac{37.530 \cdot \exp(-0.157 \cdot \beta)}{Re} \right)^{16} \right)^{-\frac{3}{2}} \right) \right]^{\frac{1}{12}}$$

, (8)

$$A = \left[p \cdot \ln \left(\frac{1 + 0.1 \cdot \beta}{\left(\frac{7}{Re} \exp\left(-\pi \cdot \frac{\beta}{180} \cdot \frac{1}{\gamma^2} \right) \right)^{0.9} + 0.27 \times 10^{-5}} \right) \right]^{16},$$

$$p = \left(0.061 + \left(0.69 + tg\left(\beta \cdot \frac{\pi}{180} \right) \right)^{-2.63} \right) \cdot (1 + (1 - \gamma) \cdot 0.9 \cdot \beta^{0.01}),$$

where $\gamma = 2b/S$ is the ratio of corrugation double height to corrugations pitch S; β is the angle of corrugations inclination to the main direction of the flow, degrees.

The heat transfer coefficients in single-phase flow of PHEs can be calculated according to Kapustenko et al. (2022) as follows:

$$Nu_0 = 0.065 \cdot \mathrm{Re}^{\frac{6}{7}} \cdot \left(\frac{\psi \cdot \zeta}{F_x}\right)^{\frac{3}{7}} \cdot \mathrm{Pr}^{0.43} \cdot \left(\frac{\mu_b}{\mu_w}\right)^{0.14}, \qquad (9)$$

where *Re* and *Pr* are the Reynolds and Prandtl numbers; F_x is the enlargement factor of plate area due to corrugations; μ_b and μ_w are dynamic viscosities at the flowing bulk and at the wall, Pa·s; ζ is total Darcy friction factor; ψ is the share of pressure losses caused by friction ($\psi = \zeta\tau/\zeta$) in total pressure loss including also form drug losses and calculated as follows:

$$\psi = \left(\frac{Re}{380}\left[tg(\beta)\right]^{1.75}\right)^{-0.15\cdot\sin(\beta)} \leq 1. \qquad (10)$$

When observing the two-phase operation of PHE, the approach, which suppose the change of process parameters along the plate length was applied. The gaseous, vapour and liquid part of the channel was considered with the counter-current flow of heat carriers.

The heat transfer coefficient for the liquid part of the two-phase stream is estimated as:

$$h_f = \frac{\lambda_f}{d_e} Nu^* \cdot \left[1 + x_{tp} \cdot \left(\frac{\rho_f}{\rho_1} - 1\right)\right]^{0.48}, \qquad (11)$$

where λ_f is a thermal conductivity of the liquid part, W/(m K); ρ_f is the density of the liquid part, kg/m^3; x_{tp} is the mass quality; Nu^* is Nusselt number calculated according to Eqiation (9) for the total value of flowrate in the channel.

The heat transfer coefficient for the gaseous part of the flow is as follows:

$$h_{cnv} = \left(\lambda_g/d_e\right) \cdot Nu_0 \cdot \left(1 + 0.85 \cdot b_H\right) \cdot 4 \cdot \left(1 + \sqrt{\frac{\rho_1}{\rho_{1f}}}\right)^{-2}, \qquad (12)$$

where b_H is permeability parameter for heat:

$$b_H = \frac{c_{ps}}{c_{p1}} \cdot \frac{j_s \cdot \mathrm{Re}_1 \cdot Pr_1}{\rho_1 \cdot W_1 \cdot Nu_0}$$

, (13)

where index 1 represents the two-phase flow; *s* index is stated for steam; W_1 is the velocity of two-phase stream, m/s.

Mass transfer coefficient from flow bulk to liquid film:

$$\beta_D = (D_D/d_e) \cdot Sh_0 \cdot \frac{\ln(1+B_m)}{B_m} \cdot 4 \cdot \left(1+\sqrt{\frac{\rho_1}{\rho_{1f}}}\right)^{-2}$$

, (14)

where B_m is mass transfer driving force; D_D is mass diffusivity, m^2/s.

The mass transfer Sherwood number without the effect of transverse mass flux is calculated by Equation (13) from an analogy between heat and mass transfer:

$$Sh_0 = \frac{\beta_{D0} \cdot d_e}{D_D} = 0.065 \cdot \mathrm{Re}^{\frac{6}{7}} \cdot \left(\frac{\psi \cdot \zeta}{\Phi_X}\right)^{\frac{3}{7}} \cdot Sc^{0.43}$$

, (15)

where Sc is the Schmidt number, calculated as $Sc = \mu mx/(\rho D_D)$; β_{D0} is mass transfer coefficient in the absence of transverse mass flux.

The friction factor in the evaporator PHE for two-phase stream was calculated according to Equation (8), and the heat transfer coefficient was taken as proposed by Amalfi et al. (2016):

$$h_{fc} = 15.698\left(2.125\left(\frac{\beta}{70}\right)^{9.993} + 0.955\right)We^{-0.475}Bd^{0.255}\left(\frac{\rho_l}{\rho_v}\right)^{-0.571}$$

, (16)

where We is the Weber number, taken 0.5; Bd is the Bond number, $Bd = g(\rho_l - \rho_v d_e^2)/\sigma$, σ is the surface tension, N/m; v index stays for vapour part of the two-phase flow.

Using Equations (8)–(16) it is possible to estimate the optimum of the function presented in Eqiation (5), within the limits presented in Equations (6), and (7). It results in the design of PHE with the minimal heat transfer area, which will satisfy the process defined conditions.

Estimation of Treated Sewage Wastewater Technical Potential and Availability of Heat Energy Produced by Heat Pumps

The amount of heat, which can be recovered from the wastewater, is defined from the amount of wastewater that can be treated and its temperature difference at the inlet and outlet of the evaporator in the heat pump:

$$Q_{ww} = G \cdot c_p \cdot \Delta t_{ev} \quad , \qquad (17)$$

where Q_{ww} is the heat power of available wastewater, W; G is the mass flowrate of the wastewater, kg/s; c_p is specific heat capacity of the wastewater, kJ/(kg·K); Δt_{ev} is the temperature difference at the inlet and outlet of wastewater in the evaporator of heat pump, K.

To estimate the performance of the heat pump, utilising heat from the sewage collector, the following assumptions were made:

- the temperature of the wastewater is constant over the year;
- the waste heat recovery system is located in the urban sewage water collector in proximity to the buildings;
- the sewage collector and inlet pipes are insulated, what prevents the heat losses;
- the flowrate of the sewage water was suggested to be constant over time.

The assumptions made allow to estimate the heat recovery potential for the minimal temperature received from the sewage collector equal to 20°C. To calculate the heat potential, gained through heat pump, the key variable is the temperature difference, which can vary in the range from 3 K to 5 K (Živković and Ivezić, 2022).

To simulate the work of the heat pumps, and to estimate the parameters and working conditions for the heat pump, its simulation was performed in Aspen HYSYS 12 software, presented in Figure 7.

For the heat pump system for the calculations the brine was taken as the working fluid. The Peng-Robinson estimation of process streams, available in Aspen were used for simulation. As the heat carrier to transfer heat from the sewage collector to the evaporator, the water was selected. Its temperature was taken equal to 18°C at inlet and 12.5°C at outlet of the evaporator with flow rate 15.46 kg/s.

Figure 7. The heat pumps simulation in Aspen HYSYS

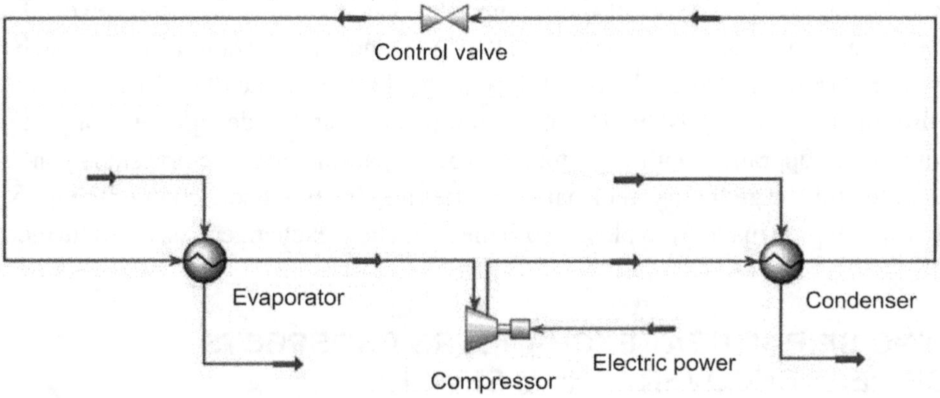

The supply heating loop with circulating water for the radiator heating was assumed to enter the evaporator at 20°C and heated up to 35°C, what suits low-temperature residential heating systems, with flowrate 4.79 kg/s. As the working fluid of the heat pump, the refrigerant R1234ze(E) was taken. The flow rate of refrigerant was equal to 2.93 kg/s, its temperature in the condenser was 65°C, and in the evaporator – 5°C.

The heat demand for the heating of the building was taken equal to 400 kW. Compressor power equal to 140 kW. The calculated parameters of the heat pump: 400 kW for the condenser; 260 kW for the evaporator. The estimated COP of the heat pump was 2.85. The total amount of consumed electricity is 210 MWh.

For such configuration, the optimal design of the evaporator and condenser heat exchangers was done according to mathematical model, presented by Equations (5)–(16):

- **Condenser:** the CB14 PHE with heat transfer area equal 0.7840 m^2 consisted of 56 plates and with 60° inclination angle is selected.
- **Evaporator:** the AlfaVap heat exchanger with 31.68 m^2 of heat transfer area assembled from 24 plates with 30° inclination angle is selected.

The implementation of PHEs with total heat transfer area equal to 32.464 m2 will ensure the COP of heat pump equal to 2.85 with R1234ze(E) refrigerant for wastewater heat application in low-temperature residential heating system. Such heat pump can cover the 400 kW heat demand for heating.

The performed calculations were made for the fixed parameters of the heat pump and waste heat available from the sewage collector. The climate conditions, seasonal and daily variations of wastewater discharge and change of its temperature will affect the considered heat recovery system (Dürrenmatt and Wanner, 2014; Golzar and Silveira, 2021). To model the temperature fluctuations inside the sewer, the sewer pipe properties, soil properties and influent and ambient conditions should be considered (Kretschmer et al., 2016). It will enable to estimate the influence of the environmental conditions on wastewater temperature and to design the heat pump with reliable operation within estimated process parameters. The present research accounted for the geometry, thermal properties and friction coefficient of the innovative sewer pipe made from plastic with the spiral heat exchanger described further.

USING SPIRAL HEAT EXCHANGERS PROSPECTS FOR HEAT RECOVERY FROM SEWERS

Let's consider the prospects for using a thermoplastic collector block, which simultaneously solves two problems – heat extraction from sewage wastewater and sanitization of large-diameter collectors. Such a collection block can be integrated into existing sewer collectors during repair operations. In this case, only a partial opening of the manifold being repaired is required, sufficient to insert a plastic block. The collector block comprises one or more spiral pipes that are tightly connected, created by welding a sealed, hollow rectangular profile made of low-pressure polyethylene. These pipes house a central cylindrical channel along with a helical peripheral channel embedded within the wall.

The collector block features an extra channel running parallel to the main cylindrical one. This additional channel is created by placing another pipeline within the main cylindrical channel's cavity and securing it to the inner wall of the collector block (see Figure 8). Both the screw peripheral channel and the additional channel are interconnected in series, forming a closed heat exchange circuit distinct from the main channel of the collector block. The heat transfer medium from this circuit is supplied to the heat pump.

Figure 8. Outline of the inspected segment of the sewage conduit

In assessing the efficacy of extracting heat from sewage, it is crucial to ascertain the anticipated increase in heat yield during the sanitation process of the collectors. Moreover, identifying strategies to enhance the efficiency of this method and delineating avenues for further theoretical and experimental exploration to improve its effectiveness are essential tasks.

One proposed approach to address the initial challenge is employing direct numerical modeling to study the flow dynamics and heat transfer within both the primary cylindrical channel and the helical peripheral channel of the sewer collector block. However, analyzing the main cylindrical channel would necessitate tackling the flow problem with a free surface, as depicted in Cecconet et al. (2019). Yet, for extensive collector systems, reaching lengths of up to 250 meters (Biletskyi et al., 2023), this method would demand an extensive grid resolution, resulting in substantial computational resources.

Conducting a simulation to assess the attributes of the sewage collector unit is deemed unnecessary. Gravity sewer collectors are required to comply with state design regulations that set forth minimum velocities and filling levels (Biletskyi et al., 2023). These standards define the minimum wastewater flow rate to achieve maximum filling of collector pipes, as outlined in Table 3.

Table 3. The minimum flow rates of wastewater in gravity sewer collectors

Diameter *D*, mm	**Wastewater velocity v_{min}, m/s, for filling *H*/*D***			
	0.60	**0.70**	**0.75**	**0.80**
150–250	0.70	-	-	-
300–400	-	0.80	-	
600–800	-	-	0.90	

continued on following page

Table 3. Continued

Diameter *D*, mm	Wastewater velocity v_{min}, m/s, for filling *H*/*D*			
	0.60	0.70	0.75	0.80
1000–1200	-	-	1.00	1.20
1500	-	-	-	1.30
> 1500	-	-	-	1.50

(Biletskyi et al., 2023)

It's essential to consider potential variations in collector capacity separately. According to state design guidelines, these capacities can be determined using data provided in Table 4. The average wastewater temperature depends on various factors, including geographical location. For instance, based on data from Cecconet et al. (2019), the average temperature in Bologna (Italy) and Zurich (Switzerland) was recorded at 15.2°C. As per regulations, the air in the collector section not occupied by drains must be ventilated. Consequently, its temperature may be lower than that of the wastewater, but it's reasonable to assume it won't drop below the temperature of the heat transfer medium.

Table 4. Coefficients of sewage inflow unevenness in urban areas

General coefficients of unevenness of sewage inflow	Average wastewater consumption, l/s				
	100	300	500	1000	>5000
maximal $K_{gen,max}$	1.60	1.55	1.50	1.47	1.44
minimal $K_{gen,min}$	0.59	0.62	0.66	0.69	0.71

(Biletskyi et al., 2023)

The speed and temperature of the heat transfer medium within the spiral channel are regulated, as per the building regulations of Ukraine. These regulations specify that the velocity of water flow in pipelines made of polymer materials should not exceed 2.5 m/s (Biletskyi et al., 2023). It is assumed that the temperature of the heat transfer medium at the inlet of the spiral channel falls within the range of 5 to 10°C. Consequently, the evaluation of heat recovery efficiency in the analyzed sewage collector block can be conducted without addressing the free-surface flow concern, as the flow level and velocity in the central channel are controlled by regulations. Similarly, the state's regulations also constrain the speed and temperature of the heat transfer medium within the spiral channel. Therefore, a simplified model will be adequate for assessing the characteristics of the sewage collector unit under investigation.

Considering the provided feedback, when assessing the heat recovery characteristics within the studied section of the sewage collector, it's viable to focus on analyzing individual loops of the spirally wound pipe rather than the entire structure (Figure 9, *a*). Consequently, determining the temperature of the heat transfer medium within each pipe loop will mainly vary at the inlet.

Figure 9. Construction the problem calculation model: (a) transitioning to a model featuring a single spiral turn, and (b) shifting to a model incorporating heat exchange through a flat wall.

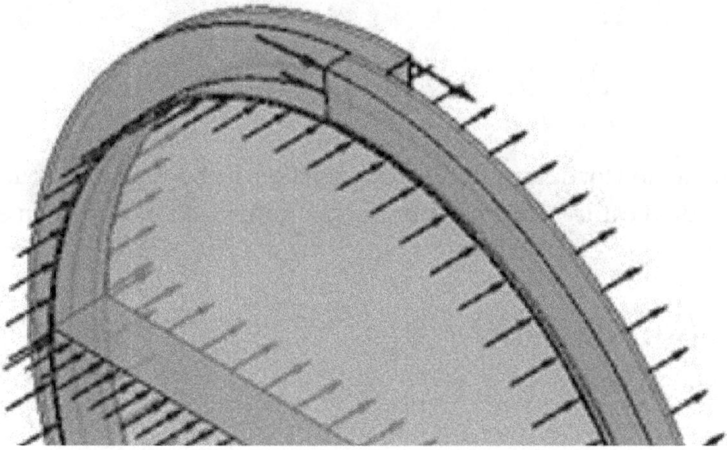

Given the minimal temperature variance between adjacent loops, it's reasonable to assume that the side walls of each spiral turn, where they encounter neighboring turns, behave adiabatically. Considering that the temperature of the air in contact with both the inner and outer ring walls of the coil closely aligns with the heat transfer medium's temperature, and given that the coefficients of heat transfer by free convection between the coil walls and air are notably lower compared to those between the coil walls and water, it follows that, for estimating the upper limit of recoverable heat from wastewater, all coil walls, except those interfacing with water, can be deemed adiabatic. This simplification enables further streamlining of the calculation model, facilitating a transition to an unfolded coil model to assess heat recovery efficiency (Figure 9, *b*).

In this scenario, the aim is to calculate the heat transfer occurring from wastewater at 15°C to a heat carrier at a lower temperature, presumed to be 10°C for assessment purposes. The heat carrier travels through a rectangular pipe with a velocity not exceeding 2.5 m/s.

The sections located at the ends of the tube are not involved in the heat exchange process due to their adiabatic walls. Therefore, to calculate the necessary tube length for achieving a specified temperature increase in the heat carrier, we can simplify the analysis by considering a rectangular tube. In this setup, heat flow occurs along one of the tube walls. To determine the effective length of the tube, accounting for the non-participating sections, we can introduce a correction factor. This correction factor depends on the filling of the collector. One way to define this correction factor is by considering the length of the arc corresponding to the cross-sectional area of the cylindrical part of the collector that interacts with the wastewater:

$$k = \frac{\pi}{\arccos\left(1 - \frac{2H}{D}\right)}, \quad (18)$$

The heat balance equation for describing the phenomenon of heat transfer through a tube of length L as the heat transfer medium flows through can be rewritten as follows:

$$Gc_p\left(T_{w_{out}} - T_{w_{in}}\right) = h\left(T_{sew} - \overline{T}_w\right)S, \quad (19)$$

where $G = \rho_w(b - 2\delta)(a - 2\delta)v_w$ is the mass flow rate of the heat transfer medium; c_p is the heat capacity of the heat transfer medium; $T_{w_{out}}, T_{w_{in}}$ are heat transfer medium temperature at the inlet and outlet of the tube; h is the heat transfer coefficient; T_{sew} is the temperature of waste water in the collector; $\overline{T}_w$ is the average temperature of the heat transfer medium in the tube; $S = L(b - 2\delta)$ is the area on which the heat flow acts; a, b, δ are height, width and wall thickness of the plastic tube.

The expression for the heat transfer coefficient when heat is transferred through a flat wall is determined by Equation (Zohuri, 2017):

$$h = \frac{1}{\frac{1}{h_w} + \frac{1}{h_{sew}} + \frac{\delta}{\lambda_{tube}}}, \quad (20)$$

here the heat transfer coefficients from sewage water and the heat transfer medium to the wall are denoted as h_{sew} and h_w respectively, while λ_{tube} represents the thermal conductivity coefficient of the tube material.

Heat transfer coefficients vary in a nuanced manner based on the attributes of the flow, and can be described as:

$$h = \frac{Nu\,\lambda_w}{D_{eq}}$$

, (21)

where Nu represents the Nusselt number; λ_w denotes the thermal conductivity of the liquid (typically assumed to be that of water in calculations); D_{eq} is the equivalent diameter for a channel, calculated as the square root of $\sqrt{4F/\pi}$, where F represents the cross-sectional area of the channel.

To calculate the Nusselt number in a turbulent flow, a simplified equation is provided in Zohuri (2017):

$$Nu = 0{,}023\,Re^{0,8}\,Pr^{0,4} \text{ when fluid is heated,}$$
$$Nu = 0{,}023\,Re^{0,8}\,Pr^{0,3} \text{ when fluid is cooled,}$$

here the Reynolds number of the flow, denoted as Re, and the Prandtl number for a liquid at average temperature, denoted as Pr, are parameters used to characterize fluid dynamics.

In the scenario being analyzed, the paramount priority lies in obtaining an accurate estimation for the heat transfer coefficient outlined in Equation (20):

$$h \leq \min\left(h_w, h_{sew}, \frac{\lambda_{tube}}{\delta}\right)$$

, (22)

for the given input data of the problem and with a thermal conductivity of polyethylene of

0.2 $Wm^{-1}K^{-1}$, we can estimate $h \leq \frac{\lambda_{tube}}{\delta}$. Considering that the objective is to obtain an upper estimate of heat recovery efficiency, rewrite as follows:

$$h = \frac{\lambda_{tube}}{\delta}$$

. (23)

Let's calculate the average temperature of the heat transfer medium in Equation (19) using the average logarithmic temperature difference:

$$T_w = T_{sew} - \overline{\Delta T} = T_{sew} - \frac{\left(T_{w_{in}} - T_{sew}\right) - \left(T_{w_{out}} - T_{sew}\right)}{\ln\frac{\left(T_{w_{in}} - T_{sew}\right)}{\left(T_{w_{out}} - T_{sew}\right)}}$$

. (24)

By incorporating factors (18), (23), and (24) from Equation (19), we can derive an approximation for the length of the tube required to achieve the specified temperature of the heat transfer medium:

$$L = \frac{\pi}{\arccos\left(1 - \frac{2H}{D}\right)} \times \frac{\rho_w (b - 2\delta)(a - 2\delta) v_w c_p \ln\frac{\left(T_{w_{in}} - T_{sew}\right)}{\left(T_{w_{out}} - T_{sew}\right)} \delta}{b \lambda_{tube}}$$

. (25)

The length of one spiral turn of the collector, given the inner diameter D and pitch b, is calculated as $l = \sqrt{b^2 + (\pi D)^2}$. Consequently, the total number of turns required to achieve the length determined by Equation (25) will be $n = L/l$. Considering that the size of one turn along the axis of the collector is $2b$, the length of the collector block, which ensures a given increase in the temperature of the heat transfer medium, can be determined as

$$L_{col} = \frac{\pi}{\arccos\left(1 - \frac{2H}{D}\right)} \times \frac{2\rho_w (b - 2\delta)(a - 2\delta) v_w c_p \ln\frac{\left(T_{w_{in}} - T_{sew}\right)}{\left(T_{w_{out}} - T_{sew}\right)} \delta}{\sqrt{b^2 + (\pi D)^2}\, \lambda_{tube}}$$

. (26)

Equation (26) provides a method for determining the necessary length of the collector to achieve a desired temperature rise in the heat carrier. This calculation considers factors such as the dimensions of the wound plastic profile, the filling of the collector, the thermal conductivity of materials involved, and the inner diameter of the collector. This equation will be applied in upcoming evaluations to gauge the efficiency of heat recovery using the sewage collector block in the design currently under investigation.

Using Equation (26), we conducted calculations considering the initial temperature of the heat carrier as 10°C and the initial temperature as 13°C. The dimensions of the rectangular profile used for winding the collector pipe are as follows: $a = 50$ mm, $b = 40$ mm, and $\delta = 3$ mm. We considered three typical sizes of the collector pipe: $D = 400$ mm, 700 mm, and 1000 mm. We took the speed of the coolant to be 1 m/s. The results of the calculations are presented in Table 5.

Table 5. Ensuring a 3°C increase in heat carrier temperature requires extending the length of the collector

D, m	***H/D***	L_{col}**, m**	***Q*, kW**
0.4	0.5	273.55	18.767
	0.7	216.76	
0.7	0.5	156.37	
	0.7	123.90	
1.0	0.5	109.47	
	0.7	86.74	

(Biletskyi et al., 2023)

Based on the data gathered, none of the options investigated were able to achieve the desired increase in heat carrier temperature over the length of the collector, which corresponds to distances between sewer hatches (50 or 100 m). While reducing the speed of the heat carrier could potentially achieve the desired temperature increase, it would also lead to a proportional decrease in the amount of heat extracted. Therefore, the most effective strategy to improve heat recovery efficiency is to utilize materials with higher thermal conductivity. Introducing heat-conducting fillers into polymers can enhance their thermal conductivity (Wei and Yang, 2021; Wolfsgruber et al., 2023). Recent studies have shown that this approach can raise the thermal conductivity of polypropylene (Wolfsgruber et al., 2023) to 3.5 to 4 W/(mK) and polyethylene (Wei and Yang, 2021) to 8.9 to 19.6 W/(mK). By employing polymers with higher thermal conductivity, the calculated value of L_{col}, as determined by Equation (9), will decrease proportionally (Table 6).

Table 6. Ensuring a 3°C temperature rise along the collector length, varying with the thermal conductivity of the pipe material

λ, W/(mK)	***D*, m**	***H/D***	L_{col}**, m**
2.0	0.4	0.5	27.35
		0.7	21.68
	0.7	0.5	15.64
		0.7	12.39
	1.0	0.5	10.95
		0.7	8.67

continued on following page

Table 6. Continued

λ, W/(mK)	*D*, m	*H/D*	L_{col}, m
3.0	0.4	0.5	18.24
		0.7	14.45
	0.7	0.5	10.42
		0.7	8.26
	1.0	0.5	7.30
		0.7	5.78

(Biletskyi et al., 2023)

In Table 6, the thermal conductivity coefficient of the collector material was fixed at 3.0 W/(mK), despite recent advancements suggesting the potential for increasing it by at least one order of magnitude. This constraint was implemented because, as previously mentioned, the heat transfer coefficient is bounded by the minimum value of h_w, h_{sew}, $\frac{\lambda_{tube}}{\delta}$ (22). Consequently, a significant rise in the thermal conductivity coefficient of the plastic used for manufacturing the collector could result in escalated costs and the heat transfer coefficient being constrained by h_w instead of the collector pipe material's thermal conductivity. However, augmenting the coolant pumping speed can improve this coefficient. Nonetheless, plastic pipelines are governed by existing standards limiting flow speed to 2.5 m/s. Hence, to further enhance the heat recovery efficiency using collectors of this type, alternative strategies must be explored.

Lowering the temperature of the coolant at the entrance to the spiral channel is feasible. In water-to-water heat pumps, it has been observed that the temperature difference of the coolant at the outlet and inlet of the primary circuit's heat exchanger can reach up to 11°C (Wang et al., 2021). By restricting this difference to 9°C and considering the initial coolant temperature as 5°C, as per Equation (9), the following data can be derived (refer to Table 7).

Table 7. Ensuring a 9°C increase in heat carrier temperature requires extending the length of the collector

D, m	*H/D*	L_{col}, m	*Q*, W
0.4	0.5	45.83	56.301
	0.7	36.31	
0.7	0.5	26.20	
	0.7	20.76	

continued on following page

Table 7. Continued

***D*, m**	***H/D***	**L_{col}, m**	***Q*, W**
1.0	0.5	18.34	
	0.7	14.53	

(Biletskyi et al., 2023)

The data presented in Table 7 demonstrates that simply raising the temperature difference between the outlet and inlet of the heat exchange channel doesn't effectively utilize the entire length of the collector, which is defined by the distance between the sewer hatches. However, this issue can be resolved by adopting a multi-pass winding configuration for the collector pipe, as depicted in Figure 10. This method enables an expanded winding step (*nb* instead of *b* in Equation (26)), leading to a proportional increase in coolant consumption.

Figure 10. Pipe collector with four-way winding configuration

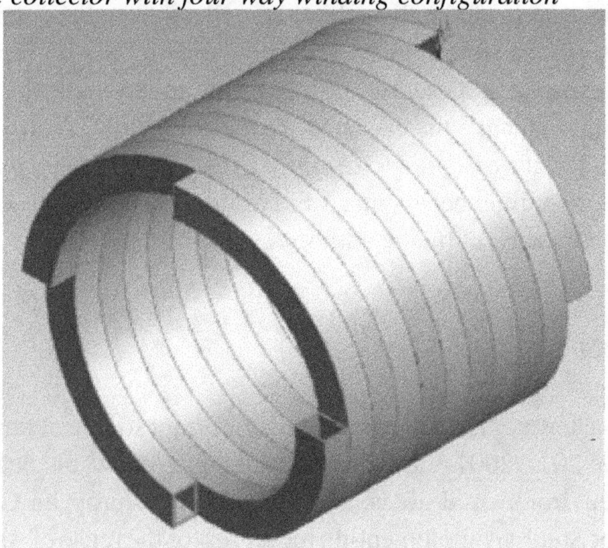

This solution is suitable for collectors with a large diameter and effectively enhances the heat dissipation by the coolant. The calculations result for a collector pipe wound in a four-way configuration are presented in Table 8, showcasing the significant increase in heat removal capacity.

Table 8. Ensure a 9°C rise in heat carrier temperature by adjusting the length of the four-way winding collector

D, m	**H/D**	L_{col}, **m**	**Q, W**
0.7	0.5	65.33	225.21
	0.7	51.77	
1.0	0.5	45.79	
	0.7	36.28	

(Biletskyi et al., 2023)

Incorporating materials with enhanced thermal conductivity and employing multi-step winding techniques can lead to achieving high heat recovery efficiency from sewage effluents using the collector block design under study. With modern heat pumps reaching a Coefficient of Performance of up to 3.74 (Wang et al., 2021), it is plausible to extract between 50 to 200 kW of heat from a 50 m collector of the examined type, contingent upon its diameter. However, it's imperative to conduct a thorough investigation to optimally select the collector material, considering both thermal conductivity and mechanical characteristics to ensure reliable operation over a 50-year lifespan. Additionally, optimizing the collector block design to reduce hydraulic losses and manage the effects of heat carrier flow reversal via numerical simulation is crucial. It's essential to recognize that the presented estimates rely on assumptions that require experimental validation. These aspects will be the focus of future research endeavors by the authors.

CONCLUSION

Nowadays, wastewater is recognized as an important source of renewable energy. The EU Directive 2018/2001 defined wastewater as a renewable heat source in line with European environmental goals. Additionally, the European Green Deal plan supports Member States in implementing measures for the reuse of waste heat. Based on an analysis of investment and maintenance costs, it is possible to calculate the price per kWh recovered by the wastewater heat recovery system over its lifetime. According to available information, when combined with heat pumps, the best estimates give a cost of only €0.06/kWh, which is almost a quarter of the EU average electricity price of €0.232/kWh in the first quarter of 2021 (Sevela et al., 2022).

As of today, there is no data on the analysis of the heat recovery potential throughout the entire cycle of the sewage system, from the exit from the building to discharge into water bodies after cleaning. Such an analysis is essential to present a complete picture of the effectiveness of wastewater energy recovery, as well as the shortcomings and problems associated with its use. Therefore, existing evaluations of the efficiency of using wastewater heat recovery systems may be overly optimistic.

Heat exchangers are the key element that determines the efficiency of wastewater heat recovery systems, and the thermal characteristics of these systems are highly dependent on the choice of heat exchangers. The approach proposed in this study allows for the design of heat pumps for wastewater utilization with optimal heat transfer equipment design. The investigation of influence of the wastewater changes in time on the performance of the heat pump and proper estimation of PHEs used as evaporator and condenser, performing the reliable operation for all regimes, is the subject of further work.

Hollow plastic profile pipes wound in a spiral configuration offer a practical solution for sewer collector repair and heat recovery from sewage effluents. A simplified calculation model has been developed to evaluate the efficiency of heat recovery using this type of sewage collector block. This model allows for estimating the necessary collector length to achieve a desired increase in the temperature of the heat carrier. The calculations conducted confirm the effectiveness of wastewater heat recuperation through the implementation of these sewage collector blocks. The use of modified plastics with a thermal conductivity coefficient ranging from 2.0 to 3.0 W/(mK), along with multi-pass winding pipes, holds potential for enhanced performance. By combining these solutions, heat extraction ranging from 50 to 200 kW can be achieved, depending on the collector's diameter.

To evaluate the efficiency of heat recovery using proposed spiral heat exchangers, it is important to consider that reducing the temperature of sewage effluents is restricted. This is because the temperature of wastewater affects biological treatment processes. Lower temperatures can slow down the activity of microorganisms used in wastewater treatment, potentially reducing effectiveness.

In addition, the overall efficiency of heat recovery from large-diameter sewer collectors can be limited by technological losses. The resulting heat must be transferred to consumers, and energy costs during transmission may limit the maximum economically feasible distance for such transmission. Therefore, future studies should evaluate the feasibility of creating combined systems for the transmission and storage of recovered thermal energy. Specifically, these studies should explore the possibility of including the generation of electrical energy through the organic Rankine cycle and the use of thermal accumulators in such systems.

ACKNOWLEDGMENT

The study was supported by the Ministry of Education and Science of Ukraine (Research Project No. 0125U001556), and partially by the European Union Assistance Instrument for the Fulfillment of Ukraine's Commitments in the Horizon 2020 Framework Program for Research and Innovation of the European Union (Research Project No. 0123U102775).

REFERENCES

Amalfi, R. L., Vakili-Farahani, F., & Thome, J. R. (2016). Flow boiling and frictional pressure gradients in plate heat exchangers. Part 2: Comparison of literature methods to database and new prediction methods. *International Journal of Refrigeration*, *61*, 185–203. DOI: 10.1016/j.ijrefrig.2015.07.009

Arsenyeva, O., Klemeš, J. J., Plankovskyy, S., & Kapustenko, P. (2022). The influence of plate corrugation geometry on heat and mass transfer performance of plate heat exchangers for condensation of steam in the presence of air. *Thermal Science and Engineering Progress*, *30*, 101248. DOI: 10.1016/j.tsep.2022.101248

Arsenyeva, O. P., Tovazhnyanskyy, L. L., Kapustenko, P. O., & Khavin, G. L. (2011). The generalized correlation for friction factor in crisscross flow channels of plate heat exchangers. *Chemical Engineering Transactions*, *25*, 399–404. DOI: 10.3303/CET1125067

Averfalk, H., Ingvarsson, P., Persson, U., Gong, M., & Werner, S. (2017). Large heat pumps in Swedish district heating systems. *Renewable & Sustainable Energy Reviews*, *79*, 1275–1284. DOI: 10.1016/j.rser.2017.05.135

Averfalk, H., & Werner, S. (2017). Essential improvements in future district heating systems. *Energy Procedia*, *116*, 217–225. DOI: 10.1016/j.egypro.2017.05.069

Babaev, V., Sukhonos, M., Starostina, A., & Beletsky, I. (2016). Improving the processes of cost management in the construction and energy projects. *Eastern-European Journal of Enterprise Technologies*, *4*(3), 10–17. DOI: 10.15587/1729-4061.2016.75515

Biletskyi, I., Plankovskyy, S., Tsegelnyk, Y., Teliura, N., & Osinska, M. (2023). Examining the possibility of heat recovery during sanitation of sewage collectors. In Arsenyeva, O. (Eds.), *Smart Technologies in Urban Engineering* (pp. 350–362). Springer., DOI: 10.1007/978-3-031-46877-3_31

Cecconet, D., Raček, J., Callegari, A., & Hlavínek, P. (2019). Energy recovery from wastewater: A study on heating and cooling of a multipurpose building with sewage-reclaimed heat energy. *Sustainability (Basel), 12*(1), 116. DOI: 10.3390/su12010116

Dacquay, C., Holländer, H. M., Kavgic, M., Maghoul, P., Liu, H., & Fujii, H. (2020). Evaluation of an integrated sewage pipe with ground heat exchanger for long-term efficiency estimation. *Geothermics, 86*, 101796. DOI: 10.1016/j.geothermics.2019.101796

Delil, A. A. M., Baturkin, V., Gorbenko, G., Gakal, P., & Ruzaykin, V. (2002). Modeling of a miniature loop heat pipe with a flat evaporator. *SAE Technical Paper* 2002-01-2506. https://doi.org/DOI: 10.4271/2002-01-2506

Dürrenmatt, D. J., & Wanner, O. (2014). A mathematical model to predict the effect of heat recovery on the wastewater temperature in sewers. *Water Research, 48*, 548–558. DOI: 10.1016/j.watres.2013.10.017 PMID: 24216228

Farhat, O., Faraj, J., Hachem, F., Castelain, C., & Khaled, M. (2022). A recent review on waste heat recovery methodologies and applications: Comprehensive review, critical analysis and potential recommendations. *Cleaner Engineering and Technology, 6*, 100387. DOI: 10.1016/j.clet.2021.100387

Golzar, F., & Silveira, S. (2021). Impact of wastewater heat recovery in buildings on the performance of centralized energy recovery – A case study of Stockholm. *Applied Energy, 297*, 117141. DOI: 10.1016/j.apenergy.2021.117141

Jesper, M., Schlosser, F., Pag, F., Walmsley, T. G., Schmitt, B., & Vajen, K. (2021). Large-scale heat pumps: Uptake and performance modelling of market-available devices. *Renewable & Sustainable Energy Reviews, 137*, 110646. DOI: 10.1016/j.rser.2020.110646

Kapustenko, P., Arsenyeva, O., Fedorenko, O., & Kusakov, S. (2022). Integration of low-grade heat from exhaust gases into energy system of the enterprise. *Clean Technologies and Environmental Policy, 24*(1), 67–76. DOI: 10.1007/s10098-021-02082-3

Kivirenko, O., Kostenko, S., Sorokin, V., Tsegelnyk, Y., Plankovskyy, S., & Kombarov, V. (2023). Fiberglass pipeline continuous filament winding automation. *International Journal of Mechatronics and Applied Mechanics, 2023*(14), 241–247. DOI: 10.17683/ijomam/issue14.28

Klemes, J. J., Arsenyeva, O., Kapustenko, P., & Tovazhnyanskyy, L. (2015). *Compact Heat Exchangers for Energy Transfer Intensification: Low Grade Heat and Fouling Mitigation*. CRC Press. DOI: 10.1201/b18862

Kombarov, V., Kryzhyvets, Y., Biletskyi, I., Tsegelnyk, Y., Aksonov, Y., & Piddubna, L. (2021). *Numerical control of fiberglass pipe bends manufacturing. In 2021 IEEE 2nd KhPI Week on Advanced Technology (KhPIWeek)*. IEEE., DOI: 10.1109/KhPIWeek53812.2021.9570068

Kretschmer, F., Simperler, L., & Ertl, T. (2016). Analysing wastewater temperature development in a sewer system as a basis for the evaluation of wastewater heat recovery potentials. *Energy and Building*, *128*, 639–648. DOI: 10.1016/j.enbuild.2016.07.024

Krogness, C., Britt, K., & Owen, M. (2020). Trenchless structural rehabilitation of non-circular pipe in congested utility corridor of commercial business district. In Pulido, J. F., & Poppe, M. (Eds.), *Pipelines 2020* (pp. 466–474). American Society of Civil Engineers., DOI: 10.1061/9780784483206.053

Kugler, T., Schittenhelm, C., Volkmer, S., Ryba, M., Moormann, C., Kurth, D., & Koenigsdorff, R. (2022). Sustainable heating and cooling management of urban quarters. *Sustainability (Basel)*, *14*(7), 4353. DOI: 10.3390/su14074353

Lu, C., Wang, S., Wang, K., Gao, Y., & Zhang, R. (2020). Uncovering the benefits of integrating industrial symbiosis and urban symbiosis targeting a resource-dependent city: A case study of Yongcheng, China. *Journal of Cleaner Production*, *255*, 120210. DOI: 10.1016/j.jclepro.2020.120210

Morishita, N., Heidenreich, M., Hemmers, R., Vankann, M., Sahakari, T., Vainio, T., & Österreicher, D. (2017). EU-GUGLE: A Sustainable Renovation for Smarter Cities from a Pilot Project. In Bisello, A. (Eds.), *Smart and Sustainable Planning for Cities and Regions* (pp. 353–382). Springer., DOI: 10.1007/978-3-319-44899-2_21

Papagiannis, F., Gazzola, P., Burak, O., & Pokutsa, I. (2018). Overhauls in water supply systems in Ukraine: A hydro-economic model of socially responsible planning and cost management. *Journal of Cleaner Production*, *183*, 358–369. DOI: 10.1016/j.jclepro.2018.02.156

Piotrowska, B., Słyś, D., Kordana-Obuch, S., & Pochwat, K. (2020). Critical analysis of the current state of knowledge in the field of waste heat recovery in sewage systems. *Resources*, *9*(6), 72. DOI: 10.3390/resources9060072

Pliuhin, V., Tsegelnyk, Y., Plankovskyy, S., Aksonov, O., & Kombarov, V. (2023). Implementation of induction motor speed and torque control system with reduced order model in ANSYS Twin Builder. In Cioboată, D.D. (eds.) *International Conference on Reliable Systems Engineering (ICoRSE)*– 2023 (pp. 514–531). Springer. https://doi.org/DOI: 10.1007/978-3-031-40628-7_42

Pokhrel, S., Amiri, L., Poncet, S., Sasmito, A. P., & Ghoreishi-Madiseh, S. A. (2022). Renewable heating solutions for buildings; a techno-economic comparative study of sewage heat recovery and Solar Borehole Thermal Energy Storage System. *Energy and Building*, *259*, 111892. DOI: 10.1016/j.enbuild.2022.111892

Pomianowski, M. Z., Johra, H., Marszal-Pomianowska, A., & Zhang, C. (2020). Sustainable and energy-efficient domestic hot water systems: A review. *Renewable & Sustainable Energy Reviews*, *128*, 109900. DOI: 10.1016/j.rser.2020.109900

Qin, N., & Tian, H. (2022). Operational analysis of a newly untreated sewage source heat pump with a plate heat exchanger. *Heat and Mass Transfer*, *58*(4), 683–693. DOI: 10.1007/s00231-021-03141-7

Ruzaikin, V., & Lukashov, I. (2023). Pump assistant ground-source two-phase thermosiphon for the heating and cooling of households. In Nechyporuk, M. (Eds.), *Integrated Computer Technologies in Mechanical Engineering – 2022* (pp. 665–676). Springer., DOI: 10.1007/978-3-031-36201-9_55

Sayegh, M. A., Jadwiszczak, P., Axcell, B. P., Niemierka, E., Bryś, K., & Jouhara, H. (2018). Heat pump placement, connection and operational modes in European district heating. *Energy and Building*, *166*, 122–144. DOI: 10.1016/j.enbuild.2018.02.006

Selvakumar, A., & Tafuri, A. N. (2012). Rehabilitation of aging water infrastructure systems: Key challenges and issues. *Journal of Infrastructure Systems*, *18*(3), 202–209. DOI: 10.1061/(ASCE)IS.1943-555X.0000091

Sevela, P., Frenger, J., Schnieders, J., & Pfluger, R. (2022). Potential of WASTE WATER HEAT RECOVERY in reducing the EU's energy need. In *CLIMA 2022 Conference Proceedings* (pp. 2099–2105). Delft University of Technology. https://doi.org/DOI: 10.34641/clima.2022.439

Shen, C., Wang, M., Zhang, C., Dong, Y., Ni, L., & Jiang, Y. (2020). Experimental investigation on a novel sewage-resource-based system with functions of heat recovery and water purification. *Applied Thermal Engineering*, *165*, 114580. DOI: 10.1016/j.applthermaleng.2019.114580

Spriet, J., McNabola, A., Neugebauer, G., Stoeglehner, G., Ertl, T., & Kretschmer, F. (2020). Spatial and temporal considerations in the performance of wastewater heat recovery systems. *Journal of Cleaner Production*, *247*, 119583. DOI: 10.1016/j.jclepro.2019.119583

Tovazhnyanskyy, L., Klemeš, J. J., Kapustenko, P., Arsenyeva, O., Perevertaylenko, O., & Arsenyev, P. (2020). Optimal design of welded plate heat exchanger for ammonia synthesis column: An experimental study with mathematical optimisation. *Energies*, *13*(11), 2847. DOI: 10.3390/en13112847

Usatenko, D. V., & Biletskyi, I. V. (2023). *Block of sewage collector* (UA Patent 154296). Ukrainian National Office for Intellectual Property and Innovations. https://iprop-ua.com/inv/ew1g50e8/ [in Ukrainian]

Vystavna, Y., Schmidt, S. I., Diadin, D., Rossi, P. M., Vergeles, Y., Erostate, M., Yermakovych, I., Yakovlev, V., Knöller, K., & Vadillo, I. (2019). Multi-tracing of recharge seasonality and contamination in groundwater: A tool for urban water resource management. *Water Research*, *161*, 413–422. DOI: 10.1016/j.watres.2019.06.028 PMID: 31226539

Wang, B., Arsenyeva, O., Zeng, M., Klemeš, J. J., & Varbanov, P. S. (2022). An advanced Grid Diagram for heat exchanger network retrofit with detailed plate heat exchanger design. *Energy*, *248*, 123485. DOI: 10.1016/j.energy.2022.123485

Wang, Z., Luther, M. B., Amirkhani, M., Liu, C., & Horan, P. (2021). State of the art on heat pumps for residential buildings. *Buildings (Basel, Switzerland)*, *11*(8), 350. DOI: 10.3390/buildings11080350

Wehbi, Z., Taher, R., Faraj, J., Ramadan, M., Castelain, C., & Khaled, M. (2022). A short review of recent studies on wastewater heat recovery systems: Types and applications. *Energy Reports*, *8*, 896–907. DOI: 10.1016/j.egyr.2022.07.104

Wei, B., & Yang, S. (2021). Polymer composites with expanded graphite network with superior thermal conductivity and electromagnetic interference shielding performance. *Chemical Engineering Journal*, *404*, 126437. DOI: 10.1016/j.cej.2020.126437

Wilson, M. P., & Worrall, F. (2021). The heat recovery potential of 'wastewater': A national analysis of sewage effluent discharge temperatures. *Environmental Science. Water Research & Technology*, *7*(10), 1760–1777. DOI: 10.1039/D1EW00411E

Wolfsgruber, N., Tanda, A., Archodoulaki, V. M., & Burgstaller, C. (2023). Influence of filler type and content on thermal conductivity and mechanical properties of thermoplastic compounds. *Polymer Engineering and Science*, *63*(4), 1094–1105. DOI: 10.1002/pen.26266

Yesina, V., Slavuta, O., Matvieieva, N., Bazetska, H., & Chala, T. (2023). Water resources of EU countries and Ukraine: Sustainable development goals achieving by statistical analysis. In Arsenyeva, O. (Eds.), *Smart Technologies in Urban Engineering* (pp. 189–199). Springer., DOI: 10.1007/978-3-031-46877-3_17

Zhu, H., Wang, T., Wang, Y., & Li, V. C. (2021). Trenchless rehabilitation for concrete pipelines of water infrastructure: A review from the structural perspective. *Cement and Concrete Composites*, *123*, 104193. DOI: 10.1016/j.cemconcomp.2021.104193

Živković, M., & Ivezić, D. (2022). Utilizing sewage wastewater heat in district heating systems in Serbia: Effects on sustainability. *Clean Technologies and Environmental Policy*, *24*(2), 579–593. DOI: 10.1007/s10098-021-02063-6

Zohuri, B. (2017). Forced convection heat transfer. In *Thermal-Hydraulic Analysis of Nuclear Reactors* (pp. 323–345). Springer., DOI: 10.1007/978-3-319-53829-7_9

Chapter 4
Innovations and Sustainability in Geotechnical Engineering for Resilient Infrastructure:
Sustainable Development and the Evolution

Ali Akbar Firoozi
University of Botswana, Botswana

Ali Asghar Firoozi
University of Botswana, Botswana

ABSTRACT

This chapter explores sustainable advancements in geotechnical engineering, focusing on integrating environmental stewardship, economic viability, and social equity into infrastructure projects. It covers theoretical frameworks, innovative materials, advanced soil stabilization techniques, and renewable energy integration. Through global case studies, the chapter highlights practical applications and challenges, offering insights into best practices. It also discusses emerging trends and the importance of policy support and education. The conclusion underscores the need for ongoing commitment from all stakeholders to achieve global sustainability goals and enhance infrastructure resilience.

DOI: 10.4018/979-8-3693-2448-6.ch004

INTRODUCTION

The field of geotechnical engineering, a pivotal branch of civil engineering, plays a fundamental role in the design and construction of infrastructure projects. It encompasses the science that explains the mechanical behavior of earth materials and how they can be utilized or modified to support various types of structures, from skyscrapers and bridges to dams and highways. The discipline's evolution has been marked by continuous advancements in understanding soil mechanics, rock mechanics, and related fields, leading to innovative solutions for complex engineering challenges. However, the increasing urgency of addressing environmental degradation and promoting sustainability in construction practices has steered geotechnical engineering towards new frontiers. This shift is not merely a response to regulatory pressures or societal demand but a fundamental reevaluation of the principles that underlie the discipline (Chang et al., 2019; Dalal et al., 2021).

The concept of sustainability in geotechnical engineering integrates traditional engineering concerns with environmental stewardship, economic feasibility, and social responsibility. It emphasizes the use of resources in a manner that does not diminish their availability for future generations, advocating for minimal environmental impact and the promotion of social well-being through infrastructure development (Sivakumar Babu, 2024). This approach has led to a broad rethinking of materials, methods, and technologies used in geotechnical projects, aligning them with the principles of sustainable development.

Sustainable geotechnical engineering is not merely an academic ideal but a practical necessity in the face of global challenges such as climate change, resource depletion, and urbanization. The adaptation of geotechnical practices to embrace sustainability principles is evidenced by the increasing adoption of recycled materials, the development of bioengineering methods for soil stabilization, and the exploration of geotechnical applications for renewable energy harvesting (Tahwia et al., 2022; Tasalloti et al., 2021). These innovations reflect a growing recognition of the need to mitigate the environmental footprint of construction activities while enhancing the resilience and longevity of infrastructure systems.

The scope of this chapter encompasses a comprehensive examination of sustainable practices in geotechnical engineering, from theoretical underpinnings to practical applications. It aims to elucidate the advancements that have been made in the field, highlighting the integration of environmentally friendly materials, innovative soil stabilization techniques, and renewable energy solutions in geotechnical projects. Through an exploration of case studies and practical examples, the chapter will demonstrate how sustainable practices have been successfully implemented in projects worldwide, overcoming challenges and showcasing the viability of these approaches (Cunha & Bourne-Webb, 2022; Giupponi et al., 2019).

Moreover, the chapter will discuss the future directions of sustainable geotechnical engineering, considering the potential for new materials, technologies, and methodologies to further enhance the sustainability of geotechnical projects. It will address the ongoing challenges in the field, such as the need for comprehensive lifecycle assessments, the integration of sustainability into educational curricula, and the development of policies that promote sustainable practices (Al-Naje et al., 2020; Phoon & Zhang, 2023).

In conclusion, the introduction sets the stage for a detailed exploration of sustainable geotechnical engineering, highlighting its significance, current advancements, and prospects. This chapter aims to contribute to the body of knowledge by providing a thorough analysis of sustainable practices in geotechnical engineering, serving as a valuable resource for researchers, practitioners, and policymakers engaged in the pursuit of sustainable infrastructure development.

THEORETICAL FRAMEWORK OF SUSTAINABLE GEOTECHNICAL ENGINEERING

The transition towards sustainable geotechnical engineering is grounded in a multifaceted theoretical framework that integrates environmental ethics, economic viability, and social equity—collectively known as the triple bottom line (TBL) of sustainability. This section delves into the principles that constitute the backbone of sustainability in geotechnical engineering, exploring their implications and applications in this specialized field.

Defining Sustainability in the Context of Geotechnical Engineering

Sustainability in geotechnical engineering transcends the simple incorporation of eco-friendly materials or practices; it embodies a holistic approach to designing and executing projects that are environmentally responsible, economically feasible, and socially beneficial over their entire lifecycle. Central to this definition is the concept of lifecycle assessment (LCA), which evaluates the environmental impacts associated with all stages of a project's life, from material extraction through construction, operation, and eventual decommissioning (Zannin et al., 2022). This comprehensive perspective ensures that decisions made at the design phase consider long-term environmental, economic, and social ramifications (Mohamed et al., 2023).

Equation 1 provides a conceptual framework for quantifying the environmental impact of a project or material within the context of a Life Cycle Assessment (LCA). This equation consolidates the impact of carbon footprint, resource consumption, and waste production into a single environmental impact score. The weighting factors *a*, *b*, and *c* are crucial as they adjust the influence of each variable based on its relative environmental significance. For instance, a higher weighting factor might be assigned to carbon footprint if the project is in a sector or region where greenhouse gas emissions are a significant concern. Similarly, resource consumption and waste production are quantified and incorporated into the overall environmental impact calculation, allowing for a comprehensive assessment of a project's or material's environmental footprint. This equation underscores the multi-faceted nature of environmental sustainability and the importance of considering various impact categories in the decision-making process for geotechnical engineering projects.

$$Environmental\ Impact\ (EI) = a(CF) + b(RC) + c(WP) \quad (1)$$

where:

EI = Total environmental impact score
CF = Carbon footprint (measured in tons of CO_2 equivalent)
RC = Resource consumption (measured in tons or cubic meters, depending on the resource)
WP = Waste production (measured in tons)
a, b, c = Weighting factors assigned to each component based on their relative environmental impact (dimensionless)

Figure 1 illustrates the Lifecycle Assessment (LCA) process tailored for geotechnical projects, depicting the sequential stages from material extraction through to decommissioning. This flowchart underscores the holistic approach required in evaluating the environmental impacts at each phase of a project's lifespan, thereby ensuring sustainable engineering practices.

Figure 1. Lifecycle Assessment (LCA) process flowchart for geotechnical projects

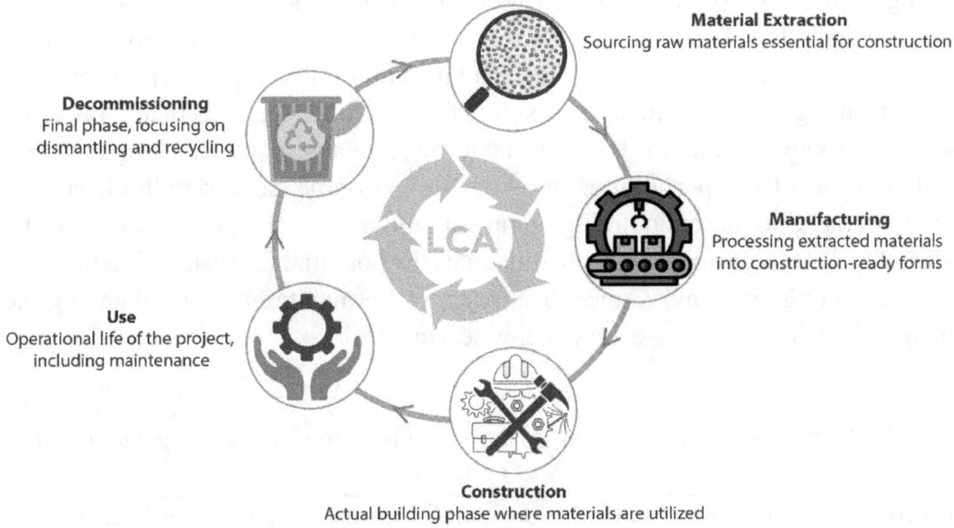

Principles of Eco-Friendly Design and Construction

The application of sustainability principles in geotechnical engineering involves several key strategies: minimizing the use of non-renewable resources, reducing emissions and waste, enhancing energy efficiency, and preserving natural habitats and biodiversity. These principles guide the selection of materials and methods in geotechnical projects, encouraging the use of recycled materials, such as crushed concrete and glass, and natural materials, like bamboo and timber, that have lower carbon footprints compared to traditional geotechnical materials (Moni et al., 2020; Okorie et al., 2021).

Innovative design approaches, such as the utilization of energy piles that serve both structural and energy-harvesting functions, exemplify how geotechnical engineering can contribute to energy efficiency and renewable energy goals (Taherian et al., 2024). Additionally, bioengineering techniques for soil stabilization, which integrate biological and engineering solutions, enhance soil strength while promoting ecological restoration and carbon sequestration (Ok et al., 2020).

Table 1 provides a comparative analysis between traditional materials and their sustainable alternatives used in geotechnical engineering projects. For each material type, the table evaluates traditional materials against sustainable ones based on three key criteria: carbon footprint, longevity, and performance. Recycled concrete aggregate (RCA) and glass cullet serve as eco-friendly alternatives to crushed stone and quarry dust, respectively, offering a lower carbon footprint without compromising on longevity and performance. In the stabilizer category, industrial by-products such as fly ash and slag significantly reduce the carbon footprint when used instead of Portland cement and offer comparable longevity with some variation in performance. Bamboo emerges as a sustainable substitute for structural steel, drastically lowering the carbon footprint and offering high flexibility, albeit with reduced longevity. For insulation, recycled paper cellulose provides an environmentally friendly alternative to polyurethane foam, maintaining comparable longevity and performance with a reduced carbon footprint. This table illustrates the potential for sustainable material selection in reducing environmental impacts while maintaining or enhancing the functional attributes of geotechnical engineering projects.

Table 1. Comparison of traditional vs. sustainable materials in geotechnical engineering

Material Type	Traditional Material	Sustainable Alternative	Carbon Footprint	Longevity	Performance
Aggregate	Crushed stone	Recycled concrete aggregate (RCA)	Lower	Comparable	Comparable
Filler	Quarry dust	Glass cullet	Lower	Higher	Comparable
Stabilizer	Portland cement	Fly ash/slag	Significantly lower	Comparable	Slightly varies
Structural	Steel	Bamboo	Significantly lower	Lower	High flexibility
Insulation	Polyurethane foam	Recycled paper cellulose	Lower	Comparable	Comparable

The Triple Bottom Line: Environmental, Economic, and Social Considerations

The TBL framework ensures that sustainable geotechnical engineering projects not only protect the environment but also are economically viable and socially equitable. Economic viability involves analyzing the cost-effectiveness of sustainable practices, considering both initial investment and long-term savings derived from reduced maintenance and operational costs (Zarte et al., 2019). Social equity focuses on the impacts of geotechnical projects on local communities and stakeholders, ensuring access to the benefits of infrastructure developments and minimizing adverse effects, such as displacement or environmental degradation (Manandhar et al., 2019).

Integrating TBL into geotechnical engineering requires a shift in project evaluation metrics, moving beyond traditional cost-benefit analyses to include measures of environmental and social impact. This shift can be facilitated by the adoption of sustainability assessment frameworks, such as the Envision Sustainability Rating System, which provides a comprehensive set of criteria for assessing the sustainability of infrastructure projects across multiple dimensions (Raymond et al., 2021).

Table 2 showcases examples of geotechnical projects across the globe that exemplify the successful integration of Triple Bottom Line (TBL) considerations. The projects are evaluated across environmental, economic, and social dimensions, demonstrating the multifaceted benefits of incorporating sustainability into geotechnical engineering. The Green Roof Initiative in Chicago showcases environmental benefits through biodiversity enhancement and mitigation of the urban heat island effect, leading to economic savings on cooling costs and enhanced property values, alongside significant social benefits by improving community well-being. The Coastal Erosion Control project in Lyme Regis illustrates the effectiveness of sustainable practices in preserving natural habitats and reducing maintenance costs, while also enhancing local safety and providing recreational spaces. In Zurich, the Energy Piles Project highlights the reduction of greenhouse gas emissions and energy conservation as key environmental benefits, with economic savings on heating and cooling and a social push towards renewable energy adoption. Finally, Copenhagen's implementation of Sustainable Drainage Systems showcases improved water management and flood risk reduction, leading to infrastructure cost savings and a better urban environment, thus enhancing public spaces and the community's quality of life. These examples underline the potential of TBL to guide geotechnical projects toward outcomes that are environmentally responsible, economically viable, and socially beneficial.

Table 2. Examples of TBL application in geotechnical projects

Project Name	Location	Environmental Benefits	Economic Benefits	Social Benefits
Green Roof Initiative	Chicago, USA	Reduced urban heat island effect, enhanced biodiversity	Lowered cooling costs, increased property value	Improved community well-being, created green spaces
Coastal Erosion Control	Lyme Regis, UK	Stabilized coastline, preserved natural habitats	Cost savings from reduced maintenance	Enhanced safety and recreational opportunities for locals
Energy Piles Project	Zurich, Switzerland	Reduced greenhouse gas emissions, energy conservation	Lowered heating and cooling costs	Increased awareness and adoption of renewable energy
Sustainable Drainage Systems	Copenhagen, Denmark	Improved water quality, flood risk reduction	Infrastructure cost savings reduced flood damage costs	Increased public spaces, improved urban environment

The theoretical framework of sustainable geotechnical engineering sets the stage for practical applications and innovations that embody these principles. As the field progresses, continuous refinement of these theories, driven by research, technological advancements, and project outcomes, will be crucial in advancing the sustainability agenda within geotechnical engineering.

Figure 2 illustrates the Triple Bottom Line (TBL) Framework, highlighting the three core components—Environmental, Economic, and Social—and their synergistic interactions. This diagram emphasizes the need for a holistic approach in sustainable geotechnical engineering, showing how integrated practices can foster environmental integrity, economic prosperity, and social equity simultaneously, leading to sustainable development.

Figure 2. The triple bottom line framework in geotechnical engineering

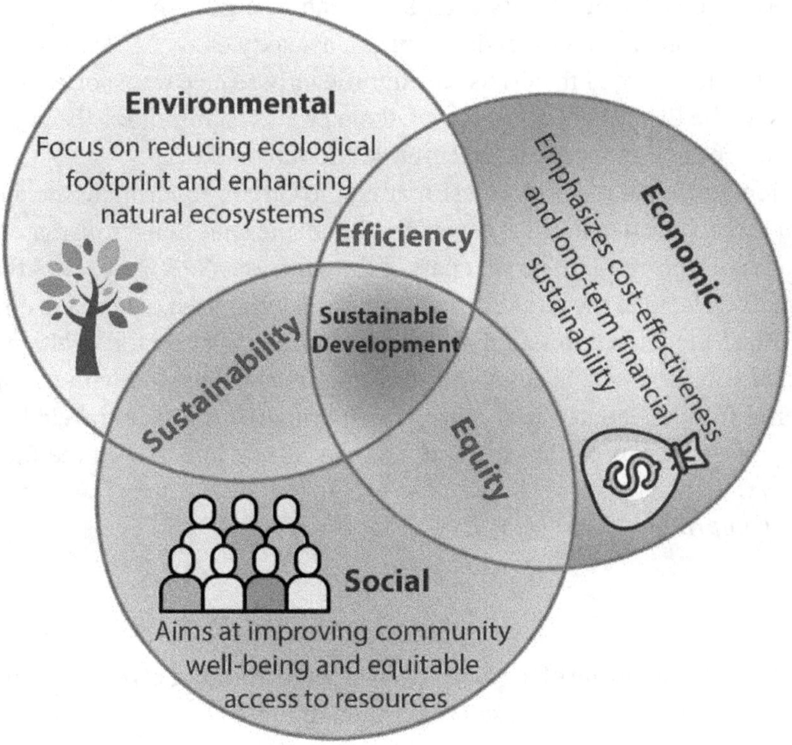

SUSTAINABILITY IN MATERIAL SELECTION AND USE

The selection and utilization of materials in geotechnical engineering significantly influences both the environmental impact and the sustainability of infrastructure projects. This section explores the considerations for choosing sustainable materials, the innovations in this arena, and the environmental impact assessments that guide such decisions.

Evaluation and Lifecycle Assessment of Geotechnical Materials

Lifecycle assessment (LCA) is a critical tool in evaluating the environmental impacts of materials from extraction through processing, use, and disposal. In geotechnical engineering, the choice of materials can have profound implications on the carbon footprint, energy consumption, and overall sustainability of a project.

LCA provides a framework for comparing materials and designs on an environmental basis, considering factors such as greenhouse gas emissions, energy use, water consumption, and biodiversity impact (Jayakody et al., 2019). For example, the use of locally sourced materials can significantly reduce transportation-related emissions, while the utilization of recycled materials can minimize the demand for virgin resources and reduce waste (Patton & McMahon, 2021).

Equation 2 offers a method for estimating the carbon footprint associated with the use of geotechnical materials, considering both the emissions from material production and transportation. The first part of the equation, $Q \times E$, calculates the emissions directly related to the quantity and type of material used, where the emissions factor (E) reflects the environmental impact of producing one ton of the material. The second part, $D \times F \times T$, accounts for the emissions associated with transporting the material from the source to the site, considering the distance traveled, the fuel efficiency of the transportation method, and the carbon intensity of the fuel used.

$$Carboon\ Footprint\ (CF) = (Q \times E) + (D \times F \times T) \tag{2}$$

where:

CF = Carbon footprint of the material (measured in kilograms of CO_2 equivalent)
Q = Quantity of material used (measured in tons)
E = Emissions factor of the material (measured in kilograms of CO_2 equivalent per ton of material)
D = Transportation distance (measured in kilometers)
F = Fuel efficiency of the transport mode (measured in liters of fuel per kilometer)
T = Emissions factor of the fuel (measured in kilograms of CO_2 equivalent per liter of fuel)

Innovations in Using Recycled and Locally Sourced Materials

The geotechnical engineering field has seen significant advancements in the use of recycled and locally sourced materials. Recycled concrete aggregate, glass cullet, and rubber tires have been successfully used in various applications, including as backfill materials and in the construction of road bases and embankments. These materials not only reduce the environmental impact but also offer cost savings and improved performance in certain conditions (Roque et al., 2021; Sawarkar et al., 2020). Additionally, the use of natural materials such as bamboo and timber, particularly in regions where these resources are abundant, provides a sustainable alternative to conventional steel and concrete, with the added benefits of carbon sequestration and renewable sourcing (Lima et al., 2021).

Table 3 provides a comprehensive comparison between recycled, locally sourced, and traditional materials used in geotechnical engineering projects based on cost, performance, carbon footprint, and availability. Recycled materials typically offer a cost advantage and a significantly lower carbon footprint by virtue of the recycling process and reduced transportation needs, although their performance can vary depending on the material type and application. Locally sourced materials present variable costs, which are often competitive due to minimal transportation; their performance is dependent on the quality of local resources but can be on par with or better than traditional materials. The carbon footprint of locally sourced materials is generally lower, attributed to shorter transportation routes. Traditional materials usually incur higher costs related to extensive processing and transportation, offer consistent and reliable performance, but have a higher carbon footprint due to energy-intensive production and logistics. Availability varies, with recycled materials depending on local infrastructure and policies, locally sourced materials on regional resources, and traditional materials widely available but potentially requiring significant transportation. This table underscores the importance of considering various factors when selecting materials for geotechnical engineering projects, highlighting the potential environmental and economic benefits of choosing recycled or locally sourced options.

Table 3. Comparison of recycled, locally sourced, and traditional materials

Criteria	Recycled Materials	Locally Sourced Materials	Traditional Materials
Cost	Often lower due to reduced processing and transportation	Variable can be lower due to reduced transportation	Generally higher due to processing and transportation costs
Performance	Can vary; generally comparable to traditional materials	Highly dependent on local quality; can match or exceed traditional materials	Consistently high, well-understood performance metrics
Carbon Footprint	Significantly lower due to the recycling process and reduced transportation	Lower due to reduced transportation distances	Higher due to energy-intensive production processes and longer transportation distances
Availability	Dependent on local recycling capabilities and policies	High, subject to regional resource availability	High, but may require long-distance transportation

Figure 3 illustrates various innovative materials employed in geotechnical engineering, highlighting their practical applications and sustainability benefits. Recycled concrete aggregate is used for backfilling to minimize costs and environmental impact. Glass cullet is incorporated in decorative concrete to enhance aesthetic value and recycle waste. Bamboo serves as structural reinforcement, providing strength and flexibility while capturing significant carbon. Additionally, repurposed rubber tires are used in road-based applications, demonstrating durability and innovative recycling. Each example underscores the potential for creative material solutions in enhancing sustainability in geotechnical projects.

Figure 3. Innovative material use in geotechnical engineering projects

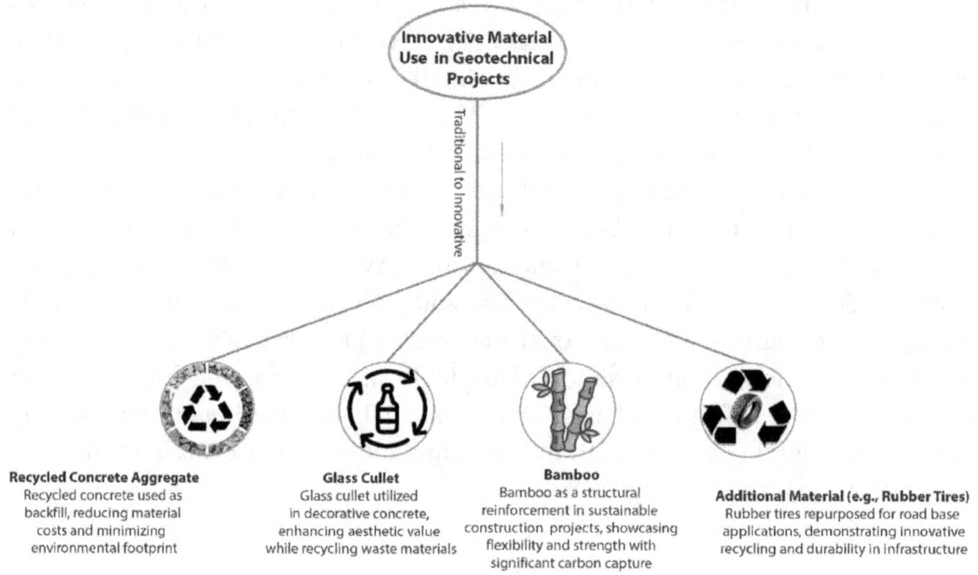

Assessing the Environmental Impact

Assessing the environmental impact of materials involves not only evaluating their lifecycle emissions but also considering their effects on local ecosystems and biodiversity. Sustainable geotechnical engineering seeks to minimize these impacts through careful material selection and project planning. For instance, the use of invasive species for bioengineering purposes may offer short-term soil stabilization benefits but can lead to long-term ecological harm (Shah & Dahal, 2023). Similarly, the extraction of natural materials must be managed carefully to prevent habitat destruction and water pollution. Environmental impact assessments (EIA) are thus integral to the project planning process, ensuring that potential adverse effects are identified and mitigated through design modifications or the selection of alternative materials (Ghadir et al., 2021).

The move towards sustainable material selection and use in geotechnical engineering represents a shift in how projects are conceived and executed. By prioritizing environmental considerations and utilizing innovative materials and methods, the field can significantly reduce its ecological footprint while still meeting the structural and economic demands of infrastructure projects. This approach not only contributes to the broader goals of sustainability but also enhances the resilience and longevity of the construction, ensuring they serve present and future generations with minimal environmental impact.

Table 4 outlines the common environmental impacts associated with the use of geotechnical materials and activities, alongside a set of targeted mitigation strategies designed to minimize ecological harm. Habitat destruction, often a consequence of expansive construction activities, can be mitigated by adopting construction techniques that minimize physical intrusion and by undertaking habitat rehabilitation efforts post-project completion. Soil erosion and sedimentation, risks associated with land disturbance, can be effectively managed through the implementation of erosion control measures and prompt revegetation of disturbed areas. Water pollution, arising from construction runoff, necessitates the use of containment and treatment solutions to prevent harmful substances from entering water bodies. Air pollution, primarily dust, and emissions from construction machinery, can be curtailed through dust suppression techniques and the use of cleaner technology. Noise pollution, a common issue in construction zones, can be mitigated through the use of noise barriers, operation time restrictions, and quieter equipment. The carbon footprint associated with construction activities and materials can be reduced by selecting low-carbon materials and optimizing energy use across operations. Lastly, resource depletion concerns can be addressed by prioritizing the use of materials that are recycled, renewable, or sourced locally, thereby reducing the demand for virgin resources. This table serves as a guide for integrating environmental considerations into geotechnical engineering projects, highlighting practical strategies for reducing the ecological impact of construction activities.

Table 4. Potential environmental impacts and mitigation strategies

Environmental Impact	Mitigation Strategies
Habitat Destruction	Utilize minimally invasive construction techniques; rehabilitate areas post-construction.
Soil Erosion and Sedimentation	Implement erosion control measures such as silt fences, sediment traps, and revegetation.
Water Pollution	Use containment systems for runoff; treat water before release; apply sustainable drainage systems.
Air Pollution	Limit dust through water spraying; use low-emission vehicles and machinery.
Noise Pollution	Implement noise barriers; restrict hours of operation; use modern, quieter equipment.
Carbon Footprint	Opt for materials with lower embodied carbon; enhance energy efficiency of operations.
Resource Depletion	Prioritize the use of recycled, renewable, or locally sourced materials.

Figure 4 outlines the sequential steps involved in conducting an Environmental Impact Assessment (EIA) for geotechnical projects. Starting with project initiation and material selection, the process progresses through evaluating lifecycle emissions and impacts on local ecosystems. It concludes with proposing mitigation strategies, engaging in public and expert consultation, and revising the project plan based on received feedback. This flowchart emphasizes the comprehensive approach required to ensure environmental considerations are integrated throughout the project planning and execution phases.

Figure 4. Environmental Impact assessment process for geotechnical projects

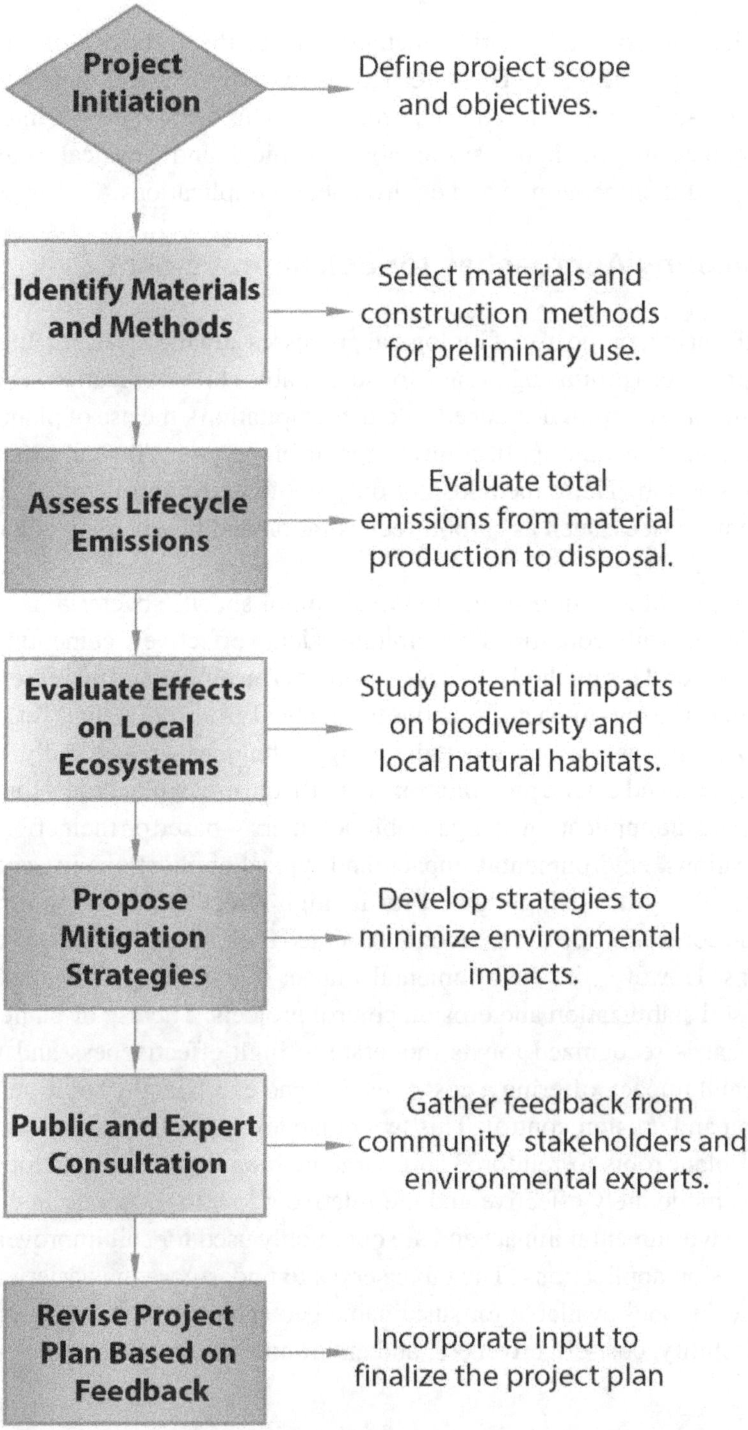

INNOVATIVE SOIL STABILIZATION TECHNIQUES

Innovations in soil stabilization methods are at the forefront of sustainable geotechnical engineering, offering alternatives to traditional techniques that often involve energy-intensive materials and processes. This section delves into cutting-edge bioengineering methods, sustainable chemical and physical stabilization techniques, and their economic and environmental implications.

Bioengineering Approaches for Soil Improvement

Bioengineering, or the use of biological processes and materials for engineering purposes, presents a promising avenue for sustainable soil stabilization. Techniques such as bio-grout (microbial-induced calcite precipitation), the use of plant roots for soil reinforcement, and the application of organic biopolymers for soil consolidation are gaining traction. These methods not only stabilize the soil but also contribute to ecosystem services, such as carbon sequestration and biodiversity enhancement (Tiwari et al., 2021).

For example, bio-grout involves the injection of specific bacteria into the soil, which, under the right conditions, precipitate calcite, effectively cementing the soil particles together. This method reduces the soil's permeability and increases its shear strength, offering a low-impact alternative to chemical grouting (Osinubi et al., 2020).

Table 5 evaluates various bioengineering techniques—specifically bio-grout (microbially induced calcite precipitation, MICP), the use of plant roots for soil reinforcement, and the application of organic biopolymers—based on their effectiveness, cost implications, environmental impact, and typical applications in geotechnical engineering. Bio-grout is highlighted for its high effectiveness in improving soil stability and controlling erosion, though it is noted for having moderate to high-cost implications. However, its environmental impact is low, making it a sustainable choice for soil stabilization and erosion control projects. The use of plant roots for reinforcement is recognized for its moderate to high effectiveness and very low environmental impact, offering a cost-effective and eco-friendly solution for slope stabilization and erosion control. This technique leverages the natural growth and strength of plant roots to reinforce soil. Organic biopolymers, on the other hand, are rated as moderately effective and are relatively low to moderate in cost. They have a low environmental impact and are commonly used for soil improvement and dust suppression applications. The table serves to underscore the variety of bioengineering techniques available for sustainable geotechnical solutions, highlighting their applicability, cost-effectiveness, and environmental benefits.

Table 5. Comparison of bioengineering techniques

Technique	Effectiveness	Cost	Environmental Impact	Typical Applications
Bio-grout (MICP)	High	Moderate-High	Low	Soil stabilization, erosion control
Plant Roots for Reinforcement	Moderate-High	Low	Very Low	Slope stabilization, erosion control
Organic Biopolymers	Moderate	Low-Moderate	Low	Soil improvement, dust suppression

Sustainable Chemical and Physical Stabilization Methods

Sustainable chemical stabilization techniques often involve the use of industrial by-products, such as fly ash and blast furnace slag, which can partially or fully replace cement in soil stabilization projects. These materials not only recycle waste but also significantly reduce the carbon footprint associated with cement production (Dorairaj & Osman, 2021).

Physical stabilization methods have also evolved to incorporate sustainable practices. Ground improvement techniques, such as rammed earth and controlled low-strength material (CLSM), utilize local materials, and require less energy compared to traditional soil stabilization methods. These techniques not only improve soil properties but also promote the use of local resources, reducing the environmental impact associated with material transportation (Sahoo & Singh, 2022).

Table 6 presents a comparison of sustainable stabilization materials like fly ash, blast furnace slag, rice husk ash, glass cullet, and crumb rubber, focusing on their properties, benefits, and common applications in geotechnical engineering projects. Fly ash, recognized for its pozzolanic nature, improves durability and reduces the permeability of stabilized soils, offering an environmentally friendly alternative by reducing landfill waste and conserving natural resources. Blast furnace slag, with its hydraulic properties, enhances the strength and durability of constructions, utilizing an industrial by-product effectively and contributing to CO_2 emissions reduction. Rice husk ash, a product of agricultural waste, is notable for its high silica content and pozzolanic properties, improving the durability of concrete structures while providing an eco-friendly use for waste material. Glass cullet, used for its non-reactive and abrasive characteristics, offers a way to reduce landfill waste and conserve natural sand, serving as an aggregate replacement in drainage layers among other applications. Crumb rubber, derived from recycled waste tires, is valued for its flexibility and vibration absorption, making it suitable for lightweight fill and embankment construction, while also reducing noise pollution and recycling waste. This table underscores the versatility and environmental benefits of incorporating sustainable materials in soil stabilization efforts, highlighting their contribution to more resilient and sustainable infrastructure.

Table 6. Properties and applications of sustainable stabilization materials

Material	Properties	Benefits	Common Applications
Fly Ash	Pozzolanic, increases durability, reduces permeability	Reduces landfill waste, low cost, conserves natural resources	Soil stabilization, road base construction
Blast Furnace Slag	Hydraulic properties improve strength and durability	Utilizes industrial by-products, reduces CO_2 emissions	Subgrade stabilization, embankments
Rice Husk Ash	High silica content, pozzolanic	Agricultural waste product improves concrete durability	Soil stabilization, lightweight fill material
Glass Cullet	Non-reactive, abrasive	Reduces landfill waste, conserves natural sand	Drainage layers, aggregate replacement
Crumb Rubber	Flexible, absorbs vibrations	Recycles waste tires, reduces noise pollution	Lightweight fill, embankment construction

Figure 5 demonstrates the circular lifecycle of sustainable stabilization materials used in geotechnical engineering. It begins with their production from industrial processes like steel manufacturing, followed by their generation as by-products such as fly ash and slag. These materials are then processed to enhance their properties, making them suitable for applications like soil stabilization, which improves strength and reduces environmental impact. This cycle exemplifies the principles of the circular economy by utilizing waste products in environmentally beneficial ways.

Figure 5. Lifecycle of sustainable stabilization materials

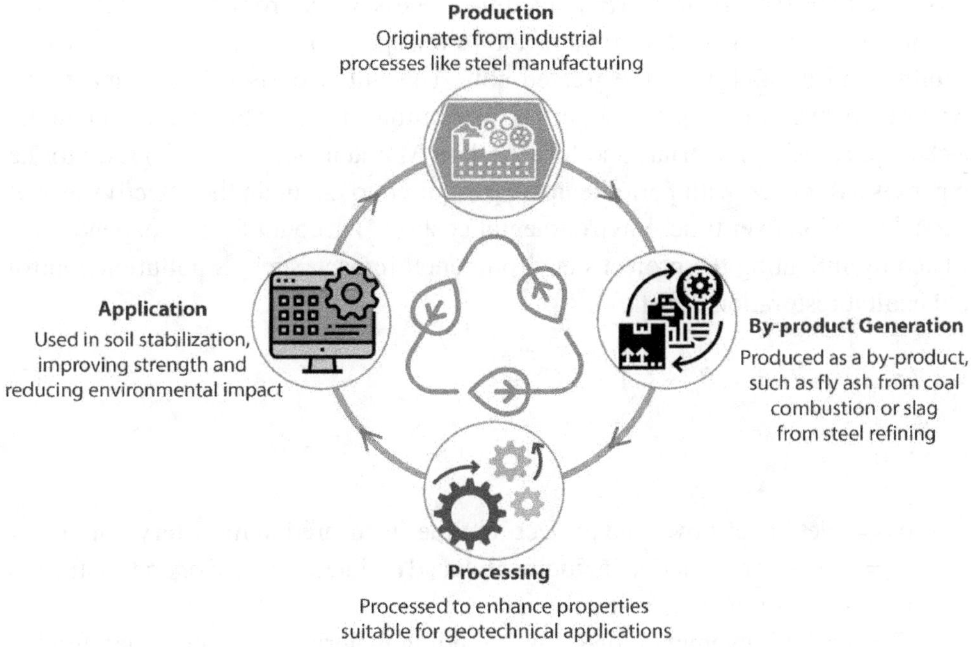

Techno-Economic Analysis of Soil Stabilization Techniques

Adopting sustainable soil stabilization techniques requires a careful examination of their technical feasibility and economic viability. Techno-economic analysis (TEA) offers a comprehensive framework for evaluating these aspects, considering both the direct costs associated with material and labor and the long-term environmental and social costs (or benefits) of different stabilization methods. For instance, while the initial cost of bioengineering techniques might be higher than traditional methods, their long-term benefits in terms of reduced environmental impact and enhanced ecosystem services can offset these initial investments (Jiang et al., 2020).

Moreover, the implementation of sustainable soil stabilization techniques can lead to indirect economic benefits, such as improved public health outcomes from reduced air and water pollution and enhanced resilience of infrastructure to climate change impacts, thereby reducing maintenance and repair costs over the infrastructure's lifecycle (Mocharla et al., 2022).

Equation 3 outlines a basic techno-economic analysis model for evaluating the cost-effectiveness of soil stabilization techniques. This model calculates the net benefit of a stabilization project by subtracting the sum of initial implementation costs and ongoing maintenance and environmental costs from the long-term benefits. The long-term benefits (B_{lt}) are anticipated savings from the project, which may include reduced need for repairs and lower operational costs due to improved stability and durability of the treated soil. The initial costs (C_i) encompass all expenses incurred during the setup and execution of the stabilization technique, such as purchasing materials and hiring labor. Maintenance costs (C_m) refer to the expenses associated with periodic upkeep required to maintain the effectiveness of the stabilization over time. Environmental costs (C_e) account for any expenditures related to mitigating the project's environmental impact, such as pollution control and habitat restoration efforts.

$$Net\ Benefit\ (NB) = (B_{lt} - C_i) - (C_m + C_e) \quad (3)$$

where:

NB = Net benefit over the project lifetime (measured in monetary units)
B_{lt} = Long-term benefits, including reduced maintenance and operational costs (measured in monetary units)
C_i = Initial implementation costs, including materials and labor (measured in monetary units)
C_m = Maintenance costs over the project lifetime (measured in monetary units)
C_e = Environmental remediation or mitigation costs, if applicable (measured in monetary units)

Comparative Case Studies

Several case studies highlight the successful application of innovative soil stabilization techniques in diverse geotechnical projects. For example, a project in the Netherlands successfully utilized bio-grout for dike reinforcement, demonstrating the technique's effectiveness in improving soil strength and reducing erosion potential with minimal environmental impact (DeJong et al., 2013). Another study in Texas explored the use of fly ash and blast furnace slag for the stabilization of expansive soils, resulting in significant improvements in soil properties and a reduction in the project's carbon footprint (Salehi et al., 2021).

These case studies exemplify how innovative soil stabilization techniques can be effectively implemented in real-world projects, offering insights into their practical challenges, benefits, and the conditions under which they can be most effectively applied.

Table 7 provides a concise summary of comparative case studies, highlighting innovative applications of sustainable geotechnical engineering across various global projects. Each case study is evaluated based on its location, the specific technique used, the outcomes achieved, and the key lessons learned. From the integration of energy piles at Marina Bay Sands in Singapore and Luleå University in Sweden, which demonstrated significant reductions in energy consumption and carbon footprint, to the bioengineering approach at Lyme Regis in the UK, where coastal erosion was effectively managed while enhancing local biodiversity, these examples illustrate the broad applicability and benefits of sustainable practices. The Solar Highway in the Netherlands showcases how renewable energy can be integrated into infrastructure projects, further emphasizing the role of sustainable engineering in addressing climate change. Lastly, the St. Gallen University project underscores the importance of incorporating sustainable solutions from the design phase to realize efficiency and environmental benefits. Collectively, these case studies offer valuable insights into the practical applications and advantages of innovative soil stabilization techniques, reinforcing the potential of sustainable geotechnical engineering to contribute to more resilient and environmentally friendly infrastructure development.

Table 7. Summary of comparative case studies

Case Study	Location	Technique Used	Outcomes	Lessons Learned
Marina Bay Sands	Singapore	Energy Piles	Reduced energy consumption, enhanced structural support	Integration of renewable energy can significantly contribute to sustainability and efficiency in urban infrastructure.
Lyme Regis Coastal Protection	United Kingdom	Bioengineering (Vegetation)	Stabilized coast, preserved natural habitat, enhanced biodiversity	Bioengineering solutions offer effective erosion control while promoting ecological restoration and biodiversity.
Luleå University Energy Piles	Sweden	Energy Piles in Cold Climates	Improved heating efficiency, reduced carbon footprint	Energy piles are adaptable to various climates, providing sustainable heating and cooling solutions.
Solar Highway	Netherlands	Solar Panels on Noise Barriers	Generated renewable energy, reduced carbon emissions	Renewable energy can be seamlessly integrated into existing infrastructure, contributing to energy sustainability.
St. Gallen University Foundation	Switzerland	Foundation Energy Piles	Lowered operational costs, reduced greenhouse gas emissions	Early planning and integration of geotechnical engineering with renewable energy sources enhance project sustainability.

The development and application of sustainable soil stabilization methods represent a significant advancement in geotechnical engineering, aligning the field with broader sustainability goals. By adopting these innovative techniques, geotechnical engineers can contribute to the construction of more sustainable, resilient, and environmentally friendly infrastructure.

INTEGRATION OF RENEWABLE ENERGY SYSTEMS IN GEOTECHNICAL ENGINEERING

The integration of renewable energy systems within geotechnical engineering projects represents a transformative approach to enhancing infrastructure sustainability. This section explores the innovative use of energy piles, the incorporation of solar and wind power into geotechnical structures, and global case studies that exemplify the successful application of these technologies.

Fundamentals of Energy Piles and Their Functionality

Energy piles are a type of ground source heat pump system that utilizes the thermal mass of foundation piles to provide heating, cooling, and hot water to buildings. By circulating fluid through pipes embedded in the piles, heat can be exchanged between the building and the ground, leveraging the relatively constant underground temperatures to reduce energy consumption for heating and cooling. This technology not only stabilizes the structure but also contributes to the building's energy efficiency, offering a dual function that enhances the sustainability of infrastructure projects (Bujor et al., 2021; Cossel, 2019).

Table 8 outlines the multifaceted benefits of employing energy piles in construction projects, focusing on their advantages over traditional foundation methods and HVAC systems. Energy efficiency is a hallmark of energy piles, utilizing the earth's stable temperatures for heating and cooling, thereby diminishing reliance on conventional energy sources, and enhancing the building's energy performance. This approach not only offers a sustainable alternative but also significantly reduces the carbon footprint of buildings by minimizing greenhouse gas emissions. Furthermore, energy piles maintain the primary function of providing structural stability, ensuring the building's foundational integrity while simultaneously serving an environmental purpose. The long-term financial benefits are notable, with substantial cost savings realized through lower energy consumption and decreased maintenance needs. Additionally, the installation of energy piles typically results in less environmental disruption compared to other geotechnical interventions, supporting broader environmental conservation efforts. Lastly, the ability of energy piles to maintain a consistent indoor temperature enhances occupant comfort, making it an attractive option for sustainable building design. This table encapsulates the comprehensive benefits of integrating energy piles into modern construction, highlighting their role in advancing sustainable infrastructure development.

Table 8. Benefits of energy piles

Benefit	Description
Energy Efficiency	Energy piles utilize the constant underground temperature to efficiently heat and cool buildings, reducing the need for traditional HVAC systems and lowering energy consumption.
Reduced Carbon Footprint	By leveraging geothermal energy, energy piles significantly cut down on greenhouse gas emissions associated with heating and cooling, contributing to a lower carbon footprint for the infrastructure.
Structural Stability	Besides their thermal functions, energy piles serve as foundational supports, offering the dual benefit of structural stability to the building, akin to traditional piles.
Cost Savings in the Long Term	Although initial costs may be higher, energy piles offer considerable savings over time through reduced operational costs associated with heating, cooling, and maintenance.
Environmental Conservation	The reduced need for external energy sources and minimal environmental disturbance during installation make energy piles a more sustainable and conservation-friendly option.
Enhanced Indoor Comfort	The consistent temperature control provided by energy piles contributes to improved indoor environmental quality and comfort for occupants.

Integration of Solar and Wind Power in Geotechnical Structures

Beyond energy piles, geotechnical engineering is increasingly incorporating solar and wind energy solutions into its projects. Solar panels can be installed on noise barriers along highways, on the surfaces of embankments, or integrated into the facades and roofs of tunnels and other structures. Similarly, wind turbines can be mounted on bridges and other large structures, using the built environment to generate renewable energy. These integrations not only produce clean energy but also maximize land use efficiency, particularly in urban areas where space is at a premium (Lv et al., 2020).

Table 9 provides a comparative analysis of solar and wind energy as renewable energy sources for geotechnical projects, examining their potential energy output, initial and ongoing costs, suitability for various project types, environmental impacts, and land use efficiency. Solar energy offers high potential output in sunny areas and is particularly adaptable to a wide range of applications, including urban environments and existing structures, making it versatile for geotechnical integration.

The costs associated with solar installations are becoming more favorable thanks to technological advancements. Wind energy, while potentially offering higher energy output in optimal locations, involves higher initial costs and maintenance considerations due to its mechanical complexity. It is most efficient in specific locations that may not be as universally available as those suitable for solar projects. Both solar and wind energy have minimal environmental impacts compared to fossil fuels, though site-specific considerations such as wildlife disruption for wind and space utilization must be considered. This table underscores the importance of site-specific evaluations when choosing between solar and wind energy for geotechnical projects, highlighting the need to balance energy output potential, cost, project suitability, and environmental considerations.

Table 9. Comparison of renewable energy sources in geotechnical projects

Criteria	Solar Energy	Wind Energy
Potential Energy Output	High in areas with strong sunlight exposure; varies seasonally	Highest in open, elevated, or coastal areas; consistent year-round
Installation Cost	Moderate to high, decreasing with advances in technology	High, but can be offset by higher energy output in optimal locations
Maintenance Cost	Low to moderate, depending on system size and location	Moderate to high, due to moving parts and potential for wear
Suitability for Geotechnical Projects	Ideal for urban and suburban projects, noise barriers, and small-footprint areas	Best suited for large-scale projects in areas with high wind speeds, such as coastal or open plains regions
Environmental Impact	Minimal; primarily during manufacture and installation	Minimal; primarily in wildlife disruption and noise during operation
Land Use Efficiency	High; can be integrated into existing structures and surfaces	Lower; requires significant space for optimal operation, though land can often still be used for agriculture

Case Studies: Global Examples of Renewable Energy Applications in Geotechnical Engineering

Several innovative projects around the world exemplify the successful integration of renewable energy systems within geotechnical engineering. One notable example is the Solar Highway in the Netherlands, where solar panels have been installed along a highway noise barrier to generate electricity. This project demonstrates the potential for combining infrastructure with renewable energy generation, contributing to the national grid while reducing environmental impact (Coughlan et al., 2020).

In Switzerland, the first energy pile system was implemented as part of the foundation for the St. Gallen University of Applied Sciences. This project showcased the viability of using geotechnical structures for both support and energy generation, leading to reduced greenhouse gas emissions and operational costs for the building (Coban et al., 2022).

These case studies illustrate the diverse opportunities for incorporating renewable energy into geotechnical projects, from energy piles that provide sustainable heating and cooling solutions to the integration of solar and wind power in infrastructure. Through such innovations, geotechnical engineering can significantly contribute to the reduction of fossil fuel dependence and the advancement of global sustainability goals.

The integration of renewable energy systems into geotechnical engineering not only showcases the field's innovation and adaptability but also aligns with broader environmental and sustainability objectives. As technology advances and the imperative for sustainable development grows, the role of renewable energy in geotechnical engineering is set to become increasingly central, offering new possibilities for enhancing the sustainability and resilience of infrastructure worldwide.

CASE STUDIES AND PRACTICAL APPLICATIONS

Exploring case studies and practical applications illuminates the real-world implementation of sustainable geotechnical engineering practices. This section delves into specific projects that exemplify the integration of sustainability principles, showcasing innovations, overcoming challenges, and deriving lessons learned.

Sustainable Urban Infrastructure in Singapore

Singapore's Marina Bay Sands is a pioneering example of integrating sustainable geotechnical engineering within an urban development project. The complex's foundation incorporates energy piles, which significantly contribute to its heating and cooling systems. This innovation not only supports the structural integrity of the massive complex but also enhances its energy efficiency, reducing the carbon footprint of one of the most iconic landmarks in Singapore. The project's success demonstrates the feasibility of large-scale implementation of energy piles in urban infrastructure, contributing to Singapore's vision of becoming a "City in a Garden" (Sliwa et al., 2021).

Coastal Erosion Control Using Bioengineering in the United Kingdom

The Lyme Regis coastal protection project in the United Kingdom employed bioengineering techniques to stabilize the coast and protect it from erosion. The project utilized native vegetation to reinforce soil and rock, creating a natural barrier against wave action and preventing land slippage. This approach not only preserved the natural landscape but also enhanced biodiversity by providing habitats for local wildlife. The project exemplifies how bioengineering can offer effective, sustainable solutions for coastal erosion, balancing environmental conservation with infrastructure needs (Rauch et al., 2022).

Energy Piles in Cold Regions - The Luleå University of Technology in Sweden

The Luleå University of Technology in Sweden incorporated energy piles into the foundation of its new buildings to leverage the constant temperatures beneath the Earth's surface for heating during the winter and cooling during the summer. This approach proved particularly effective in the cold climate of northern Sweden, showcasing the adaptability of energy pile technology to different environmental conditions. The project highlights the potential of geotechnical engineering to contribute to energy efficiency and sustainability in regions with extreme temperatures (Montefrio et al., 2021).

Lessons Learned and Best Practices

The case studies underscore several key lessons and best practices for sustainable geotechnical engineering:

- Multifunctional Design: Infrastructure can and should serve multiple purposes, such as supporting structural loads and contributing to energy efficiency, as demonstrated by energy piles.
- Local Adaptation: Sustainable solutions must be adapted to local conditions, whether it's leveraging local materials or tailoring bioengineering techniques to the local ecology.
- Stakeholder Engagement: Successful projects often involve collaboration among engineers, environmental scientists, local communities, and policymakers, ensuring that projects meet technical, environmental, and social objectives.

- Innovation and Research: Ongoing research and innovation are crucial for developing and refining sustainable geotechnical engineering practices, as evidenced by the continual advancement of techniques like bio-grout and energy piles.

These case studies not only demonstrate the practical application of sustainable geotechnical engineering principles but also inspire future projects to incorporate sustainability at their core, driving the field towards more environmentally responsible and innovative practices.

Figure 6 illustrates the fundamental principles that unde rpin sustainable geotechnical engineering, encapsulating multifunctional design, local adaptation, stakeholder engagement, and innovation and research. This infographic highlights how integrating these principles ensures that infrastructure projects are not only efficient and effective but also harmoniously aligned with environmental and social values, driving the field toward innovative and responsible engineering solutions.

Figure 6. Core principles of sustainable geotechnical engineering

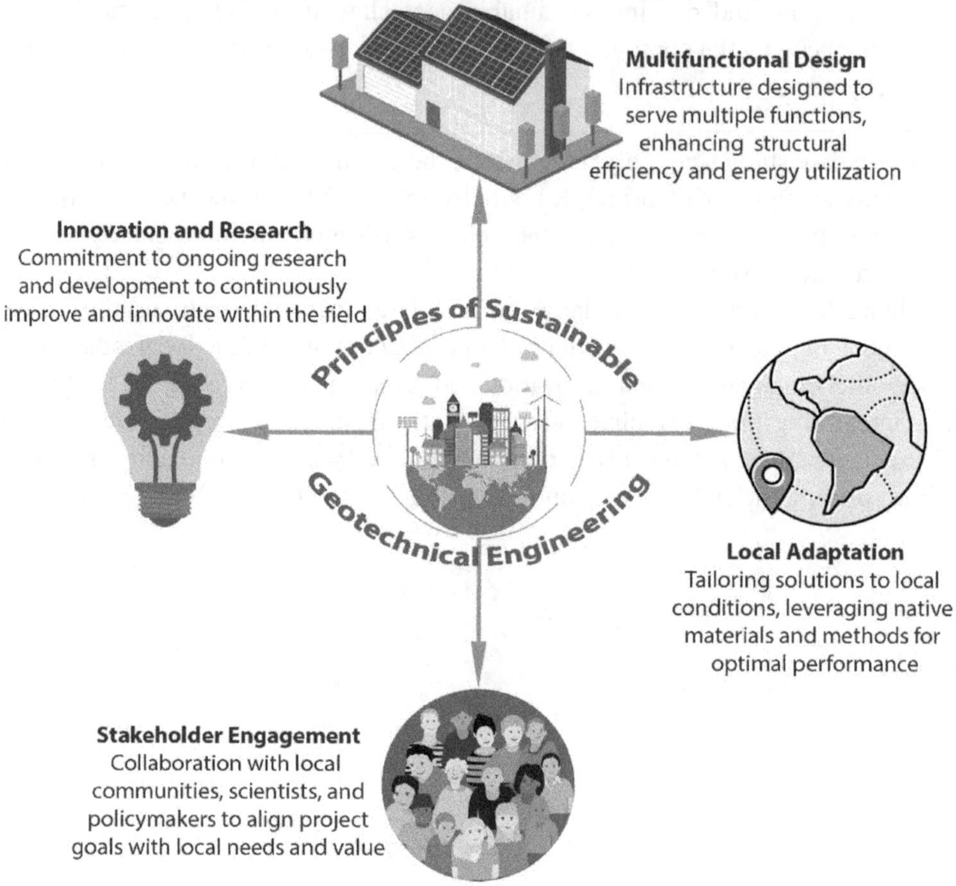

CHALLENGES, OPPORTUNITIES, AND FUTURE DIRECTIONS

The integration of sustainability into geotechnical engineering, while promising, presents a complex array of challenges and opportunities. This section explores these aspects, highlighting the current limitations, emerging trends, and potential future pathways for sustainable practices in geotechnical engineering.

Overview of Current Limitations and Challenges

One of the primary challenges facing sustainable geotechnical engineering is the balance between economic viability and environmental responsibility. Initial costs for implementing sustainable technologies, such as energy piles or bioengineering solutions, can be higher than traditional methods, posing a barrier to widespread adoption (Martinez et al., 2022). Additionally, the lack of standardized methods for assessing the environmental impact of geotechnical projects complicates the evaluation of their sustainability (Chang et al., 2020).

Another significant challenge is the resistance to change within industry. Despite the clear benefits of sustainable practices, there is often hesitation to adopt new technologies or methods due to uncertainties regarding their performance, longevity, and maintenance requirements (Punetha et al., 2019).

Table 10 outlines the common factors contributing to industry resistance against adopting new technologies or methods in sustainable geotechnical engineering, paired with strategies aimed at mitigating these challenges. High initial costs often deter investment in sustainable solutions; however, emphasizing the long-term financial and environmental benefits can help shift perspectives. Uncertainty about the performance of new technologies can be addressed by disseminating successful case studies and empirical data. A lack of knowledge or expertise can hinder the adoption of innovative practices, necessitating enhanced educational initiatives and professional training. Regulatory and policy barriers may also impede the implementation of novel solutions, calling for active dialogue with policymakers to evolve standards that support sustainable development.

Table 10. Summary of industry resistance factors

Resistance Factor	Description	Overcoming Strategies
High Initial Costs	The perception that sustainable solutions are more expensive upfront.	Highlight long-term cost savings and potential for return on investment.
Uncertainty About Performance	Concerns about the reliability and effectiveness of new technologies.	Provide case studies, research, and pilot project results to demonstrate effectiveness.
Lack of Knowledge or Expertise	Insufficient understanding of how to implement sustainable practices.	Increase training and education programs for professionals.
Regulatory and Policy Barriers	Existing codes and standards may not accommodate innovative solutions.	Work with regulatory bodies to update standards and advocate for supportive policies.
Cultural Inertia	Preference for traditional methods due to familiarity and proven track record.	Foster a culture of innovation and highlight successful global examples of sustainable practices.

continued on following page

Table 10. Continued

Resistance Factor	Description	Overcoming Strategies
Investment Risk	Fear of investing in technologies that may not become mainstream or adopted widely.	Encourage government incentives and subsidies to mitigate initial investment risks.
Supply Chain Limitations	Difficulty in sourcing sustainable materials or technologies.	Develop partnerships with suppliers of sustainable materials and technologies.
Stakeholder Misalignment	Differing priorities among project stakeholders, including investors, contractors, and end-users.	Facilitate early engagement and continuous communication among all stakeholders.

Figure 7 visualizes the primary obstacles to the widespread adoption of sustainable geotechnical engineering practices. Highlighting challenges such as higher initial costs, cultural inertia, regulatory and policy barriers, and a lack of standardized assessment methods, this diagram underscores the complexity of integrating sustainability into existing frameworks. By identifying these barriers, the engineering community can focus on developing targeted strategies to overcome them and advance the adoption of sustainable technologies and methods.

Figure 7. Barriers to sustainable geotechnical engineering adoption

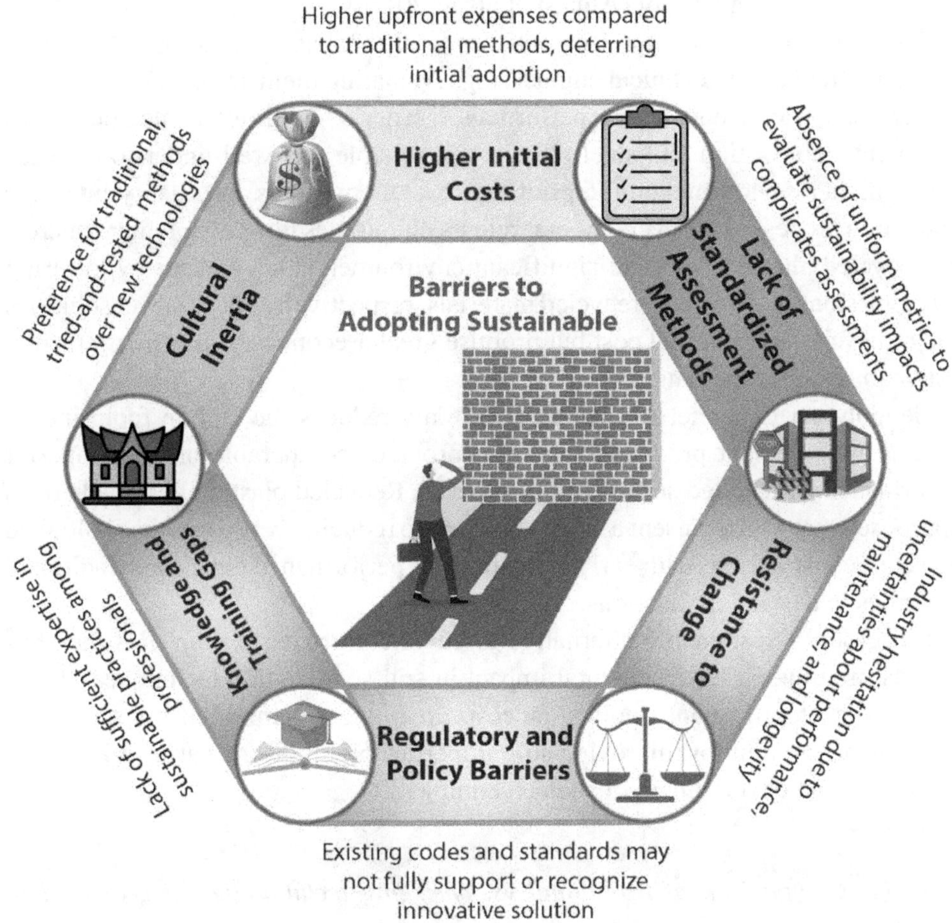

Emerging Materials and Technologies

The future of sustainable geotechnical engineering is bright, with several emerging materials and technologies poised to address current limitations. Innovations such as geopolymer concrete, which has a significantly lower carbon footprint than traditional concrete, and the use of nanotechnology for soil stabilization, offer environmentally friendly alternatives for construction (Raymond et al., 2020; Russo et al., 2021). Additionally, the development of smart geotechnical structures equipped with sensors can facilitate real-time monitoring and maintenance, enhancing their efficiency and sustainability (Alqaisi et al., 2020).

Table 11 provides a comparison between traditional and emerging technologies within the field of geotechnical engineering, evaluating them across environmental impact, cost-effectiveness, and performance criteria. Traditional methods, such as cement stabilization and steel piles, while reliable and well-understood, often carry a higher environmental footprint due to CO_2 emissions and energy-intensive production processes. Conversely, emerging technologies like geopolymer concrete and recycled plastic piles offer significant environmental advantages by lowering CO2 emissions and utilizing recycled materials, respectively. These newer solutions tend to have a higher initial cost but promise greater economic and environmental sustainability over the long term.

Geopolymer concrete, for instance, not only reduces the carbon footprint associated with cement production but also provides comparable or even superior performance with added sustainability benefits. Recycled plastic piles, made from repurposed materials, present a novel approach to reducing waste and the ecological impact of construction, with early indications of performance that rivals traditional steel piles in certain applications.

Bio-grout, a sustainable alternative to chemical grouts, exemplifies the shift towards minimizing environmental impact in soil stabilization techniques. While its initial development and application costs are higher, its minimal environmental impact and potential for improving soil properties highlight the value of investing in research and development of such technologies.

Table 11. Comparison of traditional vs. emerging technologies in geotechnical engineering

Technology	Environmental Impact	Cost-Effectiveness	Performance
Traditional (e.g., Cement Stabilization)	Higher due to CO_2 emissions from cement production	Initially lower but less sustainable long-term	High reliability and predictability
Emerging (e.g., Geopolymer Concrete)	Lower due to reduced CO_2 emissions	Higher initial cost, but more economical over time	Comparable or superior, with additional sustainability benefits
Traditional (e.g., Steel Piles)	Higher due to energy-intensive production	Cost varies with market fluctuations	Proven strength and durability
Emerging (e.g., Recycled Plastic Piles)	Lower due to recycling and reduced production impact	Potentially more cost-effective due to recycled materials	Emerging data suggests comparable performance, with benefits in certain conditions

continued on following page

Table 11. Continued

Technology	Environmental Impact	Cost-Effectiveness	Performance
Traditional (e.g., Chemical Grouts)	Potential soil and water contamination	Cost-effective for specific applications	Effective for soil stabilization and void filling
Emerging (e.g., Bio-grout)	Minimal environmental impact improves soil properties	Initial research and application costs higher	Shows promise for sustainable stabilization with similar effectiveness

The Role of Policy and Education in Advancing Sustainability

For sustainable geotechnical engineering to flourish, supportive policies and educational initiatives are essential. Governments and regulatory bodies can play a crucial role by incentivizing the adoption of sustainable practices through subsidies, tax breaks, and by incorporating sustainability criteria into public procurement processes (Colangelo et al., 2021). Furthermore, integrating sustainability into the curricula of engineering programs will prepare the next generation of engineers with the knowledge and skills needed to implement sustainable solutions effectively (Kumari & Ranjith, 2019).

Figure 8 demonstrates how supportive policies and educational initiatives drive the adoption of sustainable practices in geotechnical engineering. It highlights the synergistic effects of government incentives and specialized education in creating a cycle of continual improvement and innovation, thereby fostering a sustainable engineering landscape.

Figure 8. Influence of policy and education on sustainable geotechnical engineering

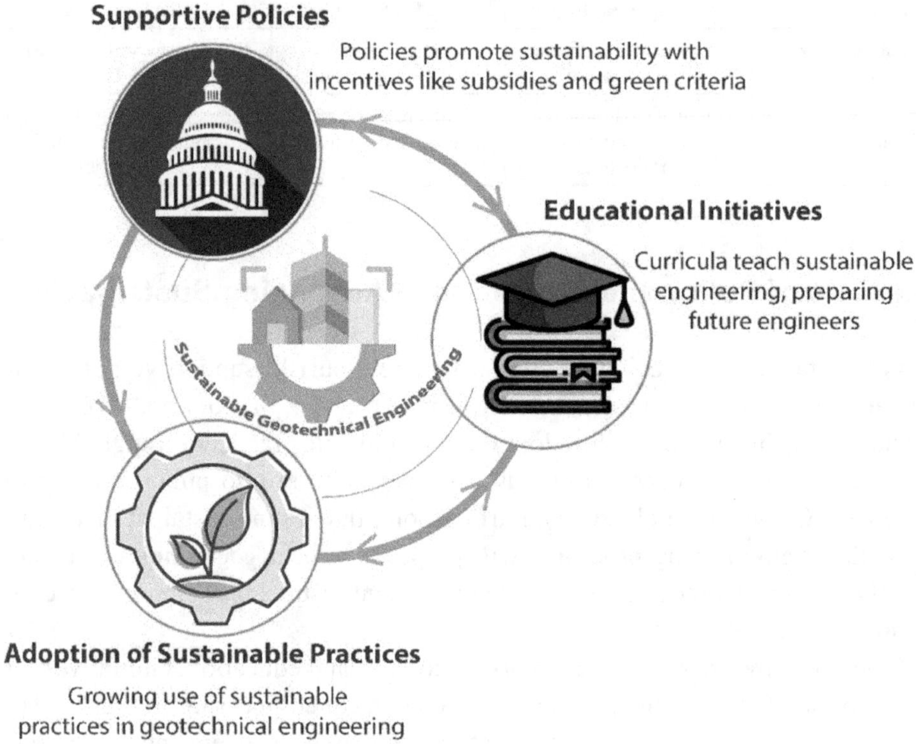

Future Directions

Looking forward, the field of geotechnical engineering is set to evolve in several key areas:

- Enhanced Integration of Renewable Energy: The continued development and integration of renewable energy sources, such as solar and wind, into geotechnical projects, will play a pivotal role in reducing the carbon footprint of infrastructure.
- Advancements in Biotechnological Applications: The application of biotechnology in soil stabilization and pollution remediation is expected to grow, offering innovative solutions that are both effective and environmentally friendly.
- Global Standardization of Sustainability Assessment: The adoption of global standards for evaluating the sustainability of geotechnical projects can fa-

cilitate more consistent and transparent assessments, promoting the broader implementation of sustainable practices.

- Collaborative Multidisciplinary Approaches: The future of sustainable geotechnical engineering will increasingly rely on interdisciplinary collaborations, integrating insights from environmental science, ecology, and social sciences to develop holistic and impactful solutions.

As the field continues to advance, the integration of sustainable practices in geotechnical engineering holds the promise of creating infrastructure that is not only resilient and efficient but also harmonious with the environment and beneficial to society.

Table 12 outlines the critical areas for future development and innovation in sustainable geotechnical engineering, detailing the expected outcomes and their broader implications for the field and society. The integration of renewable energy sources into geotechnical projects promises to significantly reduce the environmental impact of infrastructure, marking a pivotal shift towards sustainability. Advancements in bioengineering techniques are anticipated to offer dual benefits of improved soil stability and biodiversity enhancement, contributing positively to local ecosystems.

The emergence of smart infrastructure, equipped with sensors and data analytics, is expected to revolutionize the maintenance and safety of built environments, enabling proactive management of geotechnical assets. The shift towards recycled and locally sourced materials will address crucial issues of resource conservation and environmental degradation, aligning construction practices with circular economy principles.

The development of global sustainability standards will standardize the assessment of environmental performance across projects, fostering a transparent and competitive environment for innovation. Education and professional training will play vital roles in equipping current and future engineers with the knowledge and skills necessary for implementing sustainable solutions, ensuring the industry's readiness to address the challenges of the 21st century.

Collaboration between the public and private sectors is identified as a key driver for sustainable development, leveraging shared resources and expertise to achieve common goals. Lastly, supportive policies and regulatory frameworks are deemed essential for incentivizing the transition towards sustainable practices, underlining the need for governmental action to facilitate this shift.

Collectively, these areas for future development highlight the multidimensional approach required to advance sustainable geotechnical engineering, promising significant benefits not only for the field but also for the global pursuit of sustainability and environmental stewardship.

Table 12. Key areas for future development in sustainable geotechnical engineering

Area of Development	Expected Outcomes	Potential Impact
Renewable Energy Integration	Widespread use of geothermal, solar, and wind energy in geotechnical projects.	Reduced carbon footprint of infrastructure, increased energy efficiency.
Advanced Bioengineering Techniques	Improved soil stabilization methods that enhance biodiversity.	Enhanced ecosystem services and reduced environmental degradation.
Smart Infrastructure	Real-time monitoring and adaptive response systems in geotechnical structures.	Increased longevity and safety of infrastructure, optimized maintenance.
Recycled and Locally Sourced Materials	Greater reliance on sustainable materials in construction.	Conservation of natural resources, reduced waste, and emissions.
Global Sustainability Standards	Harmonized criteria for evaluating and reporting sustainability in projects.	Facilitated comparison and adoption of best practices globally.
Education and Professional Training	Expanded knowledge base and skills in sustainable practices among professionals.	Ensured the industry's capacity to innovate and implement sustainable solutions.
Public and Private Sector Collaboration	Joint initiatives focused on sustainable development goals.	Accelerated progress towards sustainable infrastructure development.
Policy and Regulatory Support	Enhanced incentives and regulations promoting sustainable engineering.	Encouraged industry-wide shift towards greener practices and technologies.

CONCLUSION

The journey through the advancements in sustainable geotechnical engineering has illuminated the field's dynamic evolution, from its foundational principles to innovative practices and materials that promise to redefine the landscape of civil engineering. This exploration has underscored the critical role of sustainable geotechnical engineering in addressing the environmental challenges of our time while paving the way for resilient, efficient, and harmonious infrastructure development.

Sustainable geotechnical engineering stands at the intersection of innovation and stewardship, embodying a commitment to environmental responsibility, economic viability, and social equity. The principles of sustainability are increasingly being woven into the fabric of geotechnical projects, driving the industry towards practices that reduce carbon footprints, enhance energy efficiency, and preserve natural ecosystems. Through the use of recycled materials, bioengineering methods for soil stabilization, and the integration of renewable energy systems, geotechnical engineering is contributing to the creation of infrastructure that not only meets the needs of the present but also safeguards the prospects of future generations.

The case studies presented in this chapter demonstrate the practical application of sustainable principles in diverse geotechnical projects around the globe. From the energy-efficient foundations of urban complexes to the bioengineered slopes that combat erosion, these examples showcase the innovative spirit and adaptability of the field. They also highlight the challenges encountered in implementing sustainable practices, including economic constraints, industry resistance, and the need for standardized sustainability assessment methods. Yet, the successes of these projects offer valuable lessons and best practices for future endeavors.

Looking ahead, the field of sustainable geotechnical engineering is poised for further growth and innovation. The development of new materials and technologies, alongside the increasing emphasis on policy support and education, will continue to drive the industry towards more sustainable practices. Collaborative, interdisciplinary approaches that integrate the latest research and technology will be crucial in overcoming existing challenges and unlocking new opportunities for sustainable infrastructure development.

In conclusion, sustainable geotechnical engineering represents a critical pathway towards achieving a more sustainable future. By embracing the principles of sustainability, the field can continue to contribute to the construction of infrastructure that is not just physically enduring but also environmentally harmonious and socially beneficial. As the field evolves, the ongoing commitment of engineers, researchers, policymakers, and educators to sustainability will be paramount in realizing the vision of a resilient and sustainable built environment.

CONFLICT OF INTEREST

The authors declare that there is no conflict of interest regarding the publication of this chapter. No external funding was received for the research presented, and there are no financial or personal relationships with other people or organizations that could inappropriately influence or bias the content and conclusions of the work.

REFERENCES

Al-Naje, F. Q., Abed, A. H., & Al-Taie, A. J. (2020). A review of sustainable materials to improve geotechnical properties of soils. *Al-Nahrain Journal for Engineering Sciences*, *23*(3), 289–305. DOI: 10.29194/NJES.23030289

Alqaisi, R., Le, T. M., & Khabbaz, H. (2020). Applications of recycled sustainable materials and by-products in soil stabilization. In Recent Thoughts in Geoenvironmental Engineering: Proceedings of the 3rd GeoMEast International Congress and Exhibition, Egypt 2019 on Sustainable Civil Infrastructures–The Official International Congress of the Soil-Structure Interaction Group in Egypt (SSIGE) (pp. 91-117). Springer International Publishing. DOI: 10.1007/978-3-030-34199-2_7

Bujor, O., Prodan, I. C., & Popa, A. (2021). Energy Geostructures: An Innovative Renewable Energy Technology for Cooling and Heating of Buildings. Environmental and Human Impact of Buildings: An Energetics Perspective, 183-204. DOI: 10.1007/978-3-030-57418-5_7

Chang, I., Lee, M., & Cho, G. C. (2019). Global CO2 emission-related geotechnical engineering hazards and the mission for sustainable geotechnical engineering. *Energies*, *12*(13), 2567. DOI: 10.3390/en12132567

Chang, I., Lee, M., Tran, A. T. P., Lee, S., Kwon, Y. M., Im, J., & Cho, G. C. (2020). Review on biopolymer-based soil treatment (BPST) technology in geotechnical engineering practices. *Transportation Geotechnics*, *24*, 100385. DOI: 10.1016/j.trgeo.2020.100385

Coban, H. H., Lewicki, W., Miśkiewicz, R., & Drożdż, W. (2022). The Economic Dimension of Using the Integration of Highway Sound Screens with Solar Panels in the Process of Generating Green Energy. *Energies*, *16*(1), 178. DOI: 10.3390/en16010178

Colangelo, F., Farina, I., Travaglioni, M., Salzano, C., Cioffi, R., & Petrillo, A. (2021). Innovative materials in Italy for eco-friendly and sustainable buildings. *Materials (Basel)*, *14*(8), 2048. DOI: 10.3390/ma14082048 PMID: 33921708

Cossel, A. (2019). Analytical solutions for thermo-mechanical soil structure interaction in energy piles (Doctoral dissertation).

Coughlan, M., Long, M., & Doherty, P. (2020). Geological and geotechnical constraints in the Irish Sea for offshore renewable energy. *Journal of Maps*, *16*(2), 420–431. DOI: 10.1080/17445647.2020.1758811

Cunha, R. P., & Bourne-Webb, P. J. (2022). A critical review on the current knowledge of geothermal energy piles to sustainably climatize buildings. *Renewable & Sustainable Energy Reviews*, *158*, 112072. DOI: 10.1016/j.rser.2022.112072

Dalal, P. H., Patil, M., Wanare, R., Dave, T. N., & Iyer, K. K. (2021). Recent innovations and practices in geotechnical engineering for sustainable infrastructure development. Geohazard Mitigation: Select Proceedings of VCDRR 2021, 431-445. DOI: 10.1007/978-981-16-6140-2_34

Dorairaj, D., & Osman, N. (2021). Present practices and emerging opportunities in bioengineering for slope stabilization in Malaysia: An overview. *PeerJ*, *9*, e10477. DOI: 10.7717/peerj.10477 PMID: 33520435

Ghadir, P., Zamanian, M., Mahbubi-Motlagh, N., Saberian, M., Li, J., & Ranjbar, N. (2021). Shear strength and life cycle assessment of volcanic ash-based geopolymer and cement stabilized soil: A comparative study. *Transportation Geotechnics*, *31*, 100639. DOI: 10.1016/j.trgeo.2021.100639

Giupponi, L., Borgonovo, G., Giorgi, A., & Bischetti, G. B. (2019). How to renew soil bioengineering for slope stabilization: Some proposals. *Landscape and Ecological Engineering*, *15*(1), 37–50. DOI: 10.1007/s11355-018-0359-9

Jayakody, S., Gallage, C., & Ramanujam, J. (2019). Performance characteristics of recycled concrete aggregate as an unbound pavement material. *Heliyon*, *5*(9), e02494. Advance online publication. DOI: 10.1016/j.heliyon.2019.e02494 PMID: 31687586

Jiang, N. J., Tang, C. S., Hata, T., Courcelles, B., Dawoud, O., & Singh, D. N. (2020). Bio-mediated soil improvement: The way forward. *Soil Use and Management*, *36*(2), 185–188. DOI: 10.1111/sum.12571

Kumari, W. G. P., & Ranjith, P. G. (2019). Sustainable development of enhanced geothermal systems based on geotechnical research–A review. *Earth-Science Reviews*, *199*, 102955. DOI: 10.1016/j.earscirev.2019.102955

Lima, L., Trindade, E., Alencar, L., Alencar, M., & Silva, L. (2021). Sustainability in the construction industry: A systematic review of the literature. *Journal of Cleaner Production*, *289*, 125730. DOI: 10.1016/j.jclepro.2020.125730

Lv, Z., Kong, G., Liu, H., & Ng, C. W. (2020). Effects of soil type on axial and radial thermal responses of field-scale energy piles. *Journal of Geotechnical and Geoenvironmental Engineering*, *146*(10), 06020018. DOI: 10.1061/(ASCE)GT.1943-5606.0002367

Manandhar, R., Kim, J. H., & Kim, J. T. (2019). Environmental, social and economic sustainability of bamboo and bamboo-based construction materials in buildings. *Journal of Asian Architecture and Building Engineering*, *18*(2), 49–59. DOI: 10.1080/13467581.2019.1595629

Martinez, A., DeJong, J., Akin, I., Aleali, A., Arson, C., Atkinson, J., Bandini, P., Baser, T., Borela, R., Boulanger, R., Burrall, M., Chen, Y., Collins, C., Cortes, D., Dai, S., DeJong, T., Del Dottore, E., Dorgan, K., Fragaszy, R., & Zheng, J. (2022). Bio-inspired geotechnical engineering: Principles, current work, opportunities and challenges. *Geotechnique*, *72*(8), 687–705. DOI: 10.1680/jgeot.20.P.170

Mocharla, I. R., Selvam, R., Govindaraj, V., & Muthu, M. (2022). Performance and life-cycle assessment of high-volume fly ash concrete mixes containing steel slag sand. *Construction & Building Materials*, *341*, 127814. DOI: 10.1016/j.conbuildmat.2022.127814

Mohamed, W. W., Osman, N., & Abdullah, R. (2023). A review of bioengineering techniques for slope stability in Malaysia. *International Journal of Environmental Science and Technology*, *20*(3), 3467–3482. DOI: 10.1007/s13762-022-04235-3

Moni, S. M., Mahmud, R., High, K., & Carbajales-Dale, M. (2020). Life cycle assessment of emerging technologies: A review. *Journal of Industrial Ecology*, *24*(1), 52–63. DOI: 10.1111/jiec.12965

Montefrio, M. J. F., Lee, X. R., & Lim, E. (2021). Aesthetic politics and community gardens in Singapore. *Urban Geography*, *42*(10), 1459–1479. DOI: 10.1080/02723638.2020.1788304

Ok, B., Sarici, T., Talaslioglu, T., & Yildiz, A. (2020). Geotechnical properties of recycled construction and demolition materials for filling applications. *Transportation Geotechnics*, *24*, 100380. DOI: 10.1016/j.trgeo.2020.100380

Okorie, O., Obi, M., Russell, J., Charnley, F., & Salonitis, K. (2021). A triple bottom line examination of product cannibalisation and remanufacturing: A review and research agenda. *Sustainable Production and Consumption*, *27*, 958–974. DOI: 10.1016/j.spc.2021.02.013

Osinubi, K. J., Eberemu, A. O., Ijimdiya, T. S., Yakubu, S. E., Gadzama, E. W., Sani, J. E., & Yohanna, P. (2020). Review of the use of microorganisms in geotechnical engineering applications. *SN Applied Sciences*, *2*(2), 1–19. DOI: 10.1007/s42452-020-1974-2

Patton, W., & McMahon, M. (2021). Career development and systems theory: Connecting theory and practice. In *Career Development and Systems Theory*. Brill. DOI: 10.1163/9789004466210

Phoon, K. K., & Zhang, W. (2023). Future of machine learning in geotechnics. Georisk. *Assessment and Management of Risk for Engineered Systems and Geohazards*, *17*(1), 7–22. DOI: 10.1080/17499518.2022.2087884

Punetha, P., Samanta, M., & Sarkar, S. (2019). Bioengineering as an effective and ecofriendly soil slope stabilization method: A review. Landslides: Theory, practice and modelling, 201-224. DOI: 10.1007/978-3-319-77377-3_10

Rauch, H. P., von der Thannen, M., Raymond, P., Mira, E., & Evette, A. (2022). Ecological challenges* for the use of soil and water bioengineering techniques in river and coastal engineering projects. *Ecological Engineering*, *176*, 106539. DOI: 10.1016/j.ecoleng.2021.106539

Raymond, A. J., DeJong, J. T., Kendall, A., Blackburn, J. T., & Deschamps, R. (2021). Life cycle sustainability assessment of geotechnical ground improvement methods. *Journal of Geotechnical and Geoenvironmental Engineering*, *147*(12), 04021161. DOI: 10.1061/(ASCE)GT.1943-5606.0002646

Raymond, A. J., Tipton, J. R., Kendall, A., & DeJong, J. T. (2020). Review of impact categories and environmental indicators for life cycle assessment of geotechnical systems. *Journal of Industrial Ecology*, *24*(3), 485–499. DOI: 10.1111/jiec.12946

Roque, A. J., Paleologos, E. K., O'Kelly, B. C., Tang, A. M., Reddy, K. R., Vitone, C., Mohamed, A.-M. O., Koda, E., Goli, V. S. N. S., Vieira, C. S., Fei, X., Sollecito, F., Vaverková, M. D., Plötze, M., Petti, R., Podlasek, A., Puzrin, A. M., Cotecchia, F., Osiński, P., & Singh, D. N. (2021). Sustainable environmental geotechnics practices for a green economy. *Environmental Geotechnics*, *9*(2), 68–84. DOI: 10.1680/jenge.21.00091

Russo, G., Marone, G., & Di Girolamo, L. (2021). Hybrid energy piles as a smart and sustainable foundation. *Journal of Human, Earth, and Future*, *2*(3), 306–322. DOI: 10.28991/HEF-2021-02-03-010

Sahoo, S., & Singh, S. P. (2022). Strength and durability properties of expansive soil treated with geopolymer and conventional stabilizers. *Construction & Building Materials, 328*, 127078. DOI: 10.1016/j.conbuildmat.2022.127078

Salehi, S., Arashpour, M., Kodikara, J., & Guppy, R. (2021). Sustainable pavement construction: A systematic literature review of environmental and economic analysis of recycled materials. *Journal of Cleaner Production, 313*, 127936. DOI: 10.1016/j.jclepro.2021.127936

Sawarkar, A. D., Shrimankar, D. D., Kumar, A., Kumar, A., Singh, E., Singh, L., Kumar, S., & Kumar, R. (2020). Commercial clustering of sustainable bamboo species in India. *Industrial Crops and Products, 154*, 112693. DOI: 10.1016/j.indcrop.2020.112693

Shah, P., & Dahal, K. (2023). Problems associated with implementation of bioengineering in hill road construction in Nepal. *Journal of Engineering and Applied Sciences (Asian Research Publishing Network), 70*(1), 30. DOI: 10.1186/s44147-023-00198-5

Sivakumar Babu, G. L. (2024). Reliability and Risk Analysis in Geotechnical and Geoenvironmental Engineering. Indian Geotechnical Journal, 1-33. DOI: 10.1007/s40098-024-00909-6

Sliwa, T., Sapińska-Śliwa, A., Wysogląd, T., Kowalski, T., & Konopka, I. (2021). Strength Tests of Hardened Cement Slurries for Energy Piles, with the Addition of Graphite and Graphene, in Terms of Increasing the Heat Transfer Efficiency. *Energies, 14*(4), 1190. DOI: 10.3390/en14041190

Taherian, G., Hosseini Nourzad, S. H., & Neyestani, M. (2024). Customizing a sustainability evaluation framework for Infrastructure projects in developing countries: The case study of Iran. *Sustainable and Resilient Infrastructure, 9*(2), 168–191. DOI: 10.1080/23789689.2023.2272462

Tahwia, A. M., El-Far, O., & Amin, M. (2022). Characteristics of sustainable high strength concrete incorporating eco-friendly materials. *Innovative Infrastructure Solutions, 7*(1), 1–13. DOI: 10.1007/s41062-021-00609-7

Tasalloti, A., Chiaro, G., Murali, A., Banasiak, L., Palermo, A., & Granello, G. (2021). Recycling of end-of-life tires (ELTs) for sustainable geotechnical applications: A New Zealand perspective. *Applied Sciences (Basel, Switzerland), 11*(17), 7824. DOI: 10.3390/app11177824

Tiwari, N., Satyam, N., & Sharma, M. (2021). Micro-mechanical performance evaluation of expansive soil biotreated with indigenous bacteria using MICP method. *Scientific Reports*, *11*(1), 10324. DOI: 10.1038/s41598-021-89687-2 PMID: 33990644

Zannin, J., Ferrari, A., Larrey-Lassalle, P., & Laloui, L. (2022). Early-stage thermal performance design of thermo-active walls implemented in underground infrastructures. *Geomechanics for Energy and the Environment*, *30*, 100218. DOI: 10.1016/j.gete.2020.100218

Zarte, M., Pechmann, A., & Nunes, I. L. (2019). Decision support systems for sustainable manufacturing surrounding the product and production life cycle–A literature review. *Journal of Cleaner Production*, *219*, 336–349. DOI: 10.1016/j.jclepro.2019.02.092

Chapter 5
Innovative Fly Ash and Polypropylene Fiber Concrete: Advancing Sustainable Civil Engineering

S. Yavana Rani
https://orcid.org/0000-0001-7870-8007
Jain University, India

Sandeep Poddar
https://orcid.org/0000-0001-9771-877X
Lincoln University College, Malaysia

Senthilnathan Dhanasekaran
https://orcid.org/0009-0006-1601-1478
Independent Researcher, USA

ABSTRACT

This chapter proposal explores the innovative development of geopolymer concrete using fly ash and polypropylene fibers to advance sustainable practices in civil engineering. Produced through the alkali activation of alumino-silicates found in industrial by-products like fly ash, geopolymers serve as sustainable alternatives to Portland cement. The construction industry's reliance on Portland cement significantly contributes to environmental pollution, particularly carbon dioxide emissions. Geopolymer production reduces greenhouse gas emissions and transforms fly ash into a valuable resource. This research aims to explore the potential of fly ash-based geopolymers as alternatives to Portland cement, assess the impact

DOI: 10.4018/979-8-3693-2448-6.ch005

of polypropylene fibers on the mechanical properties of geopolymer concrete, and evaluate the material's durability under various environmental conditions. The chapter seeks to promote the adoption of geopolymer concrete in the construction industry, supporting global efforts to enhance sustainability.

BACKGROUND OF THE STUDY

Geopolymers are produced by the alkali activation of alumino silicates present in the source material. These gels can be used to bind aggregates, such as sand or natural rocks, to produce mortars and concretes. In simple words, Geopolymers are inorganic binders that function like the better-known Portland cement. French Professor, Davidovits found out that the existence of three-dimensional silicate-aluminium product in the ancient Pyramids had the same structure as Zeolite, and then he designated this man-made rock like product as “Geopolymer”.

Industrialization leads to the generation and release of undesirable pollutants into the environment. In order to keep pace with the rapid industrialization, there is a necessity to select an engineering process, which would cause minimum pollution into environment. On the other hand, construction industry is increasingly turning towards the use of environmentally friendly materials in order to meet the sustainable aspect required by modern infrastructures. Consequently, in the last two decades, the expansion of this concept and the increasing global warming have raised concerns on the extensive use of Portland cement due to the high amount of carbon dioxide associated with its production.

The main constituent of Geopolymer Concrete is low calcium ASTM class F fly ash, alkaline liquid, coarse and fine aggregates. Unlike ordinary cement concrete in which cement is mixed in the range of 350 kg/m3 to 450 kg/m3 depending upon the grade of concrete, an equivalent quantity of fly ash is taken and mixed to obtain Geopolymer Concrete. In this research, Class F fly ash obtained from Tuticorin Thermal Power Station is used to manufacture Geopolymer Concrete.

Every year millions of tons of fly ash are generated worldwide by coal-fired power plants satisfying the large demand for industrial and domestic energy. The management of this by-product is always a matter of concern. Only about 20-30% of the generated fly ash is used, mainly as additive in cement and concrete, filling material and the rest is disposed of. Therefore, strategies are required to deal with this waste safely. Special attention should be paid not only to prevent environmental pollution, but also to treat fly ash as a valuable resource material. In this regard, the synthesis of Geopolymer is foreseen as an interesting approach.

Production of one ton of Portland cement requires about 2.8 ton of raw materials, including fuel and other materials and generates 5% to 10% of dust. Altogether, 6000-14000 m3 dust-containing air streams are generated per ton of cement manufacture and accounts for the release of 0.8 tonne of greenhouse gas CO2 into the atmosphere, as a result of de-carbonation of lime in the kiln during manufacturing of cement (Hardjito et al. 2004; Buchwald 2005 and Li et al. 2004). The reduction in the carbon dioxide emission from cement production can contribute significantly to the turning down of the global thermostat. It is evident that previous research on Geopolymer concerning fly ash as the sole source of silica and alumina is quite limited but a lot of unknown factors about the physical and chemical properties of derived products as well as about the factors governing the formation of the Geopolymer have to be explored.

PROBLEM STATEMENT

Increasing concern about the environmental consequences of waste disposal has also led to the investigation of new utilization avenues. In addition to this, the requirements imposed on construction materials are so demanding and diverse that no material is able to satisfy them completely. This has led to a resurgence of the ancient concept of combining different potential materials to satisfy diverse user requirements. So far, some special characteristics of the alkali-activated cementitious material such as high early strength development and excellent resistance to chemical attacks have been recognized and given a lot of interest by scientists in cement and concrete research area. The development of Geopolymer Concrete offers promising signs for a change in the way of producing concrete. The study on the application of Indian fly ash with Polypropylene Fiber in the development of Geopolymer Concrete is very less. However, to seriously consider Geopolymer binders as an alternative to Ordinary Portland Cement (OPC), various strength related factors of this new material should be evaluated in any comparative analysis format. Till date, structural applications of Geopolymer concrete along with fibres and its application to onsite still seem to be a grey area. Furthermore, very less published literature is available dealing with the mechanical, durability properties and application of Indian fly ash with Polypropylene Fibre in the development of Geopolymer Concrete.

RESEARCH SIGNIFICANCE

This research is one among the fields of high-profile research and commercial interests over the past decade. Geopolymer Technology paved way for conversion of waste by product materials into valuable material. Some reports outside India stated that Geopolymer Concrete has good strength and durability aspects. However, till date, only a few published results could be found using Indian Fly ash based Geopolymer Concrete along with Polypropylene Fibers as a secondary reinforcement.

AIM OF THE STUDY

This research is designed to evaluate the physical, chemical and mechanical behaviour of Fibre reinforced Geopolymer Concrete members. To achieve this, an extensive study had been carried out to investigate the following. To determine the fresh Geopolymer Concrete properties with and without Polypropylene Fibre. This study also aims to investigate the mechanical strength of hardened Geopolymer Concrete with and without addition of Polypropylene Fiber and test on durability of Geopolymer Concrete with and without addition of Polypropylene Fibre.

REVIEW OF LITERATURE

Akbarnezhad et al. (2015) presented the results of an experimental study conducted to investigate the recyclability of GPC. Basic properties of Recycled Geopolymer Aggregates (RGAs) including water absorption, density and Los Angeles abrasion loss as well as the effects of size of RGA on these properties were investigated. In addition, the effects of the different replacement ratios of coarse RGA for coarse natural aggregates on the properties of the new Recycled Aggregate Geopolymer Concrete (RAGC) including compressive strength, flexural strength and modulus of elasticity were investigated. The RGAs and RAGC properties were compared with those of recycled OPC concrete aggregate (RCA) and recycled aggregate OPC concrete (RAC) produced under relatively similar conditions.

Peem Nuaklong et al. (2016) made a study of the effect of recycled aggregate on strength and durability of fly ash based Geopolymer Concrete. Geopolymer Concrete samples were prepared from high calcium fly ash, sodium silicate solution, sodium hydroxide solution, fine aggregate from river sand, and two types of coarse aggregate viz., recycled concrete aggregate and crushed limestone coarse aggregate. Test results indicated that recycled concrete aggregate can be used as a coarse aggregate in high calcium fly ash Geopolymer Concretes with the 7-day compressive strengths of 30.6–38.4 MPa which were slightly lower than those of high calcium fly ash Geopolymer Concretes containing crushed limestone. The use of high sodium hydroxide concentrations of 12 and 16 M resulted in better performances both in strength and durability than that of 8 M.

Visintin et al. (2017) studied a recently developed mechanics based segmental approach for predicting the shear capacity of reinforced concrete (RC) beams is extended and applied to reinforced Geopolymer Concrete Beams without stirrups. The results of eight reinforced Geopolymer Concrete Beam tests without stirrups are presented along with the results of four direct shear tests with low levels of confinement. Significantly, the results of the direct shear tests show that the shear-friction properties for the Geopolymer Concrete utilized in the experimental investigation fall within the range of shear-friction properties of established OPC concrete. Moreover, it is shown that the segmental approach proposed can predict the shear capacity of Geopolymer Concrete beams with good accuracy and hence can be used as a tool to aid in the development of new design guidelines for Geopolymer Concrete.

Albitar et al. (2017) evaluated the performance of Geopolymer Concretes manufactured using either class-F fly ash or blended fly ash and Granulated Lead Smelter Slag (GLSS). The performance of OPC Concrete was also investigated as a reference for evaluating the durability characteristics of Geopolymer Concretes. All concrete specimens were continuously immersed up to nine months in four different chemical solutions: 5% sodium chloride, 5% sodium sulphate, 5% sodium sulphate + 5% magnesium sulphate, and 3% sulphuric acid. Throughout the exposure period, the change in mass, compressive strength, splitting tensile strength, flexural strength, water absorption, sorptivity and porosity were evaluated. The influence of wetting–drying and heating–cooling cycles on the mass loss and compressive strength was also investigated. The results revealed that the OPC concrete has lower water absorption and sorptivity than the Geopolymer Concrete. Furthermore, it is shown that sodium sulphate has the greatest impact on Geopolymer Concretes, while OPC concrete is more susceptible to sulphuric acid attack. The results showed that, in general, the durability performance of Geopolymer Concrete is superior to that of OPC concrete within the range of the considered exposure.

Shima Pilehvar et al. (2017) studied the effect of micro-encapsulated phase change materials (MPCM) in solid and liquid states on the mechanical properties and microstructure of Geopolymer and Portland cement concretes. Geopolymer Concrete and Portland Cement Concrete (PCC) containing different amounts of MPCM were prepared and cured at both 20 °C and 40 °C.

Jovan Tatar & Hamilton (2016) assessed the FRP durability by means of accelerated conditioning is sometimes thought to be too harsh compared to ambient environmental conditions; consequently, this may result in underestimate of the actual durability. To assess the efficacy of utilizing Accelerated Conditioning Protocols (ACP) for FRP-concrete bond durability testing, results from laboratory and field conditions (Sunshine Skyway Bridge in Tampa, FL) were compared. Direct tension pull-off test patches were applied to the approach span girders and notched beam three-point bending test specimens were placed on the dolphins adjacent to the bridge. These results were compared to notched beam specimens exposed to ACP. Testing indicated that characteristics of the FRP-concrete bond failure modes changed in some of the field samples within 6 months of field exposure, which may be an indication of durability problems. Moreover, ACP under elevated temperatures (60 °C) of notched three-point bending test resulted in a 36% loss of strength compared to no strength degradation after 18 months of field conditioning.

Fangyuan Zhang et al. (2017) presented an experimental study of the improvement of the tensile and flexural strength of CC panels through the employment of external reinforcement by a FRP sheet. The tensile and flexural behaviors of CC and FRP reinforced CC panels are investigated through uniaxial tensile and four-point bending tests in both warp and weft directions. The bond behaviour between FRP sheets and CC is investigated through a single-shear pull-off test, by which both strain distribution and local bond stress-slip relationships are experimentally measured. Experimental results reveal that the tensile and flexural properties, especially the reinforcing efficiency of flexural strength, are significantly improved in the FRP reinforced CC samples.

OBJECTIVES

The objectives of this research is

1. To study the factors ruling the compressive strength of fly ash based Geopolymer Concrete with various proportions of Polypropylene Fibre.
2. To observe the behaviour of Reinforced Concrete (RC) beams made of Polypropylene Fibre added Geopolymer
3. To examine the crack pattern, deflection and load carrying capacity of Polypropylene Fibre mixed Geopolymer Reinforced Concrete beams and slabs manufactured using dry heat curing process.
4. To investigate the thermal behaviour and bond strength of fly ash based Geopolymer Concrete with Polypropylene Fibre as additions.

RESEARCH METHODOLOGY

Materials Investigation

In this study **ORDINARY PORTLAND CEMENT** (OPC) was used to produce a control mix concrete. The total requirement of cement was calculated and purchased from local dealers, stored in a dry place inside casting yard and kept covered with tarpaulin sheets to avoid clotting. OPC 43 grade cement has been used throughout this investigation. The cement has magnesium oxide (MgO), sulphuric anhydride (SO3), free lime and alkaline oxide below the permissible limit specified by AS 3972 and ASTM C150. Excessive content of these chemicals could change the soundness of cement. Magnesium oxide and sulphuric anhydride in excess of permissible levels contribute to long-term expansion of cement. High alkaline oxide content in the cement is prone to cause alkali-silica reaction with reactive aggregates in the mixture. Chloride found in the cement is normally added to accelerate early strength and reduce setting time.

The fine aggregate used in this investigation are clean river sand passing through 4.75 mm sieve with fineness modulus of 3.0, specific gravity of 2.60 and bulk density of 1.72 g/cc. Locally available coarse aggregate of size 20 mm and below was used and various tests were carried out to ascertain the physical properties of coarse aggregate.

Polypropylene Fibers are composed of crystalline and non-crystalline regions. The spherulites developed from a nucleus can range in size from fractions of a micrometer to centimeters in diameter. The a-axis of the crystal unit cell is aligned radially and the chain axis is homogeneously distributed in planes perpendicular to this radial direction. Each crystal is surrounded by non-crystalline material. Fiber spinning and drawing may cause the orientation of both crystalline and amorphous regions. If the extension is less than 0.5%, the spherulite deformation is elastic and no disruption of the structure occurs, otherwise spherulites are highly oriented in the direction of the force and finally are converted to microfibrils. These highly anisotropic microfibrillar structures lead to anisotropic fiber properties. The Fibers are within the following range of specifications:

Effective diameter: 10 micron – 1.0 mm.
Length: 6-48 mm.
Specific gravity: more than 1.0.
Suggested dosage: 0.6-2.0 kg/cumec (0.23-0.6% by Weight of cement in mix).
Water absorption: less than 0.45 percent.
Melting point shall not be less than 160°C
The aspect ratio generally varies from 200 to 2000.

Figure 1. Polypropylene fiber

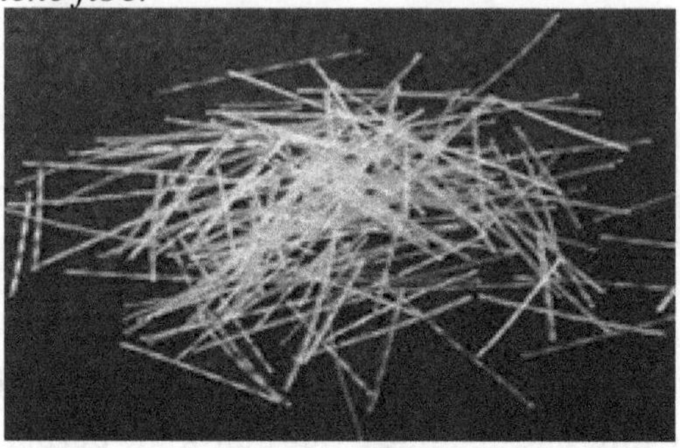

In order to improve the workability of stiff and fresh concrete to some extent, a high range water reducing Polycarboxylic ether based Hyper Plasticizer were used in this study. Hot-rolled, deformed, high yield strength bars were used as the main reinforcements in both RC slabs and beams. Steel reinforcement bars of different diameters ranging from 8 mm to16 mm were used in this study. The yield strength of bars used was 500 MPa. The bars were tightly wrapped in tarpaulin sheets to avoid direct exposure to seasonal changes, in order to prevent corrosion. The stirrups were made of mild steel bars of 6mm diameter. The steel reinforcing bars were subjected to standard tensile test in Universal Testing Machine (UTM) to find out the yield strength of bars. Fly ash plays an important role in the manufacture of Geopolymer Concrete, was procured from Tuticorin Thermal Power Station, Tamil Nadu, India. Total quantity of fly ash required for this work was calculated and the entire quantity was procured from Tuticorin Thermal Power Station and the fly ash was covered with tarpaulin sheet to keep in air tight condition.

Potable water was used in this study for diluting NaOH flakes, for manufacturing OPC concretes and for preparing aggressive liquids. The amount of solids was below the permissible limits as specified by ASTM C1602/C1602 M-12 Standards. A combination of sodium hydroxide and sodium silicate was used in this study to prepare the alkaline solution. Both the chemicals are commercially available in the local market.

EXPERIMENTAL INVESTIGATION

The casting and testing of Geopolymer and FRGP Concrete produced using Polypropylene Fiber were done under seven categories. They are as follows:

- Tests on fresh Geopolymer Concrete with and without Polypropylene Fiber.

- Tests on mechanical strength of hardened Geopolymer Concrete with and without Polypropylene Fiber.
- Tests on durability of Geopolymer Concrete with and without Polypropylene Fiber.
- Test on reinforced Geopolymer slab with the addition of optimal percentage of Polypropylene Fiber.
- Microstructural analysis on behavior of OPC, GPC and FRGP with optimal percentage.

Tests On Fresh Geopolymer Concrete With and Without Polypropylene Fiber

The fiber reinforced Geopolymer Concrete produced using addition of Polypropylene fiber was studied, for their slump value and flow percentage.

Slump Cone Test

The Slump Cone Test was performed as per the procedure given in IS: 7320 - 1974, a mold of 1.6 mm thick galvanized metal in the form of the lateral surface of the frustum of a cone with the base 200 mm in diameter, the top 100 mm in diameter and the height 300 mm is used to study the workability of fresh concrete. The base and the top were open and parallel to each other and at right angles to the axis of the cone. The mold was provided with a foot piece on each side for holding the mold in place, and with handles for lifting the mold from the sample. Tamping Rod was a round, straight steel rod 16 mm in diameter and approximately 600 mm in length is used for compaction.

Dampened mold is placed on a flat, moist, non-absorbent rigid surface. The mold was held firmly restricting its movement. The cone was filled 1/3 full and was compacted 25 times to its full depth. The second layer was until 2/3 full by volume and compacted 25 times uniformly, ensuring that the rod just penetrates into the first layer. The cone is overfilled with the third layer and compacted uniformly 25 times with the rod just penetrating into the second layer. The excess concrete was strike of to obtain the level with the top of the cone by a screening and rolling motion of the tamping rod. The mold was removed immediately from the concrete by raising it carefully in a vertical direction without lateral or torsional motion. The difference was measured between the height of the mold and the height of the specimen. This distance was the slump of the concrete.

Figure 2. Slump cone test on concrete

Flow Table Test

The flow test is performed following ASTM standard (ASTM C1362-09). It consists of a steel table top 76.2cm (30 inch dia), the cone on the table consist of 17 cm diameter at top and 25 cm diameter at bottom. The integral cast ribs were designed for support and strength. The fresh concrete was filled inside the cone on the table in two layers and removed gently. The table was lifted and dropped by 12.5mm and the vibrating table dropped 25 times in 15 seconds.

Figure 3. Flow table test

Tests on Mechanical Strength of Hardened Geopolymer Concrete With and Without Polypropylene Fiber

The cubes, cylinders and prisms of FRGP produced using Polypropylene Fiber were cast and cured under ambient curing condition. The mechanical strength behavior of all these specimens was tested during various.

Compressive Strength Test

Geopolymer Concrete cubes of size 150mm×150mm×150mm were cast as per IS:516-1959 with and without Polypropylene Fiber of various percentage of addition and called as GP and FRGP series. During casting, the cubes were mechanically vibrated using a table vibrator. After 24-36 hours, the specimens were demolded and subjected to curing in normal atmosphere. After curing, the specimens were tested for compressive strength using compression testing machine of 2000 kN capacity as shown in Figure 8.3. The maximum load at failure was taken. The average compressive strength of concrete and mortar specimens is calculated by using the equation 8.1.

Compressive Strength (N/mm^2) =

$$\frac{Ultimate\ compressive\ load}{Area\ of\ cross\ section\ of\ specimen} \tag{8.1}$$

The tests are carried out on a set of three specimens and the average compressive strength values were taken.

Figure 4. Compressive strength test on concrete cubes

Split Tensile Strength Test

Concrete cylinders of size 150 mm diameter and 300 mm length were casted as per IS: 516 – 1959, IS: 1199-1959, SP: 23-1982, IS: 10086-1982 with incorporating Polypropylene fibre as an addition to concrete. During casting, the cylinders were mechanically vibrated using a table vibrator. After 24 hours, the specimens were demolded and subjected to curing for 28 days in atmosphere. After curing, the cylindrical specimens were tested for split tensile strength using compression testing machine of 2000 kN capacity as shown in Figure 5. The ultimate load was taken and the average split tensile strength was calculated using the equation 8.2.

Split Tensile Strength (N/mm^2) =

$$\frac{2P}{\pi LD} \tag{8.2}$$

where,

P=Ultimate load at failure (N),
L=Length of cylindrical specimen (mm),
D=Diameter of cylindrical specimen (mm).

The tests are carried out on a set of three specimens and the average tensile strength values were taken.

Figure 5. Splitting tensile strength test on concrete cylinder

Flexural Strength Test

Beam specimens of standard size 150 mm x 150 mm x 1500 mm were cast as per (IS: 516-1959) to determine the flexural strength of fiber reinforced concrete produced using Polypropylene Fiber. Beams were cast to the specification of 25 N/mm^2 and the addition of Polypropylene Fiber was ranged from 0% to 5%. All beams are simply supported over an effective span of 400 mm and tested in a frame load test machine of capacity 1000 KN as shown in Figure 6. A single, central, concentrated load was placed over the span of the beam. Prior to placing the specimens in the machine, the beam surfaces at the location of supports and loads were smoothly ground to eliminate unevenness. The load was incremented until failure occurred and the ultimate load was observed to calculate the flexural strength of beams using the equation (8.3).

$$f = \frac{My}{I} \tag{8.3}$$

where,

M = PL/4
P = ultimate load in N
L= span of beam in mm
D = depth of beam in mm
f= flexural strength in N/mm^2
I= moment of inertia in mm^4
y= D/2 in mm

Figure 6. Flexural strength test on concrete prism

Tests on Durability of Geopolymer Concrete With and Without Polypropylene Fiber

The FRGP Concrete produced using Polypropylene fibers were tested for its durability. Various tests on durability and the effect of Polypropylene fiber added concrete were conducted and the results were reported and discussed.

Ultrasonic Pulse Velocity Test (UPV)

The UPV test was conducted as per the procedure given in IS: 13311:1992. UPV is a Non-Destructive Technique that measures, involves measuring the speed of sound through materials in order to predict material strength, to detect the presence of internal flaws such as cracking, voids, honeycomb, decay and other damage. The instrument consists of a transmitter and a receiver (two probes). The time of travel for the wave to pass from the transmitter to the receiver when kept opposite to each other is recorded in the ultrasonic instrument (Limaye 2002). The distance between the two probes (path length) was physically measured. Hence,

Ultrasonic Pulse Velocity

= Path length / Transit time

(8.4)

This velocity is related to its compressive strength. The quality and approximate compressive strength of concrete was determined by using Table 1 which gives the relationship between ultrasonic pulse velocity and quality of concrete as per IS: 13311:1992.

Table 1. Relation between pulse velocity values and quality of concrete

Velocity	Classification (Quality)
4.0 and above	Very good
3.5 to 4.0	Good
3.0 to 3.5	Medium
3.0 and below	Poor

Figure 7. UPV test on concrete

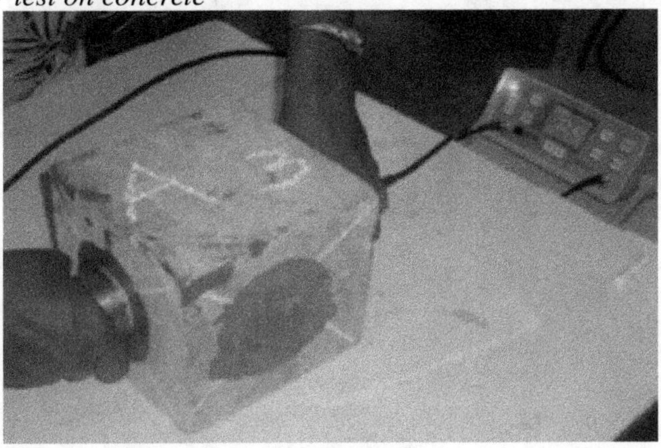

Water Absorption Test

The water absorption values for various mixtures of GP and FRGP series concrete were determined on 150 mm x 150 mm x 150 mm cubes as per IS 1124:1974. The specimens were taken out of curing tank at 56 days to record the Water Saturated weight (Ws). The drying was carried out in an oven at a temperature of 105°C. The drying process was continued until the difference between two successive measurements agreed close. Oven-dried specimens were weighed after they cooled to room temperature (W_d). Using these weights, Saturated Water Absorption (SWA) was calculated. The formula used to find water absorption value of concrete specimens is given in equation 8.4.

$$SWA= [(Ws-W_d) / W_d] \times 100 \qquad (8.4)$$

where,

SWA - Saturated Water Absorption in percentages
Ws - Weight of the specimen at fully saturated condition in kg,
W_d - Weight of oven dried specimens in kg.

Figure 8. Concrete cubes immersed in water

Water Permeability Test

Permeability Test is used to find the durability of concrete with the help of water and air. It is done by using Concrete Permeability Test Apparatus as per IS: 1124-1974. Since permeability affects the destructiveness of saturated freezing this test is done. It is 3 cell apparatus, in which cubes of 150 mm can be placed. Pressure gauge is placed to find the pressure applied in the apparatus; nearly 1-2 kgf/cm^2 air pressure is applied. Cell assembly is filled with water. Permeability test for all Mix IDs after 3, 7, 14, 28, 56 and 90 days were tested by measuring the water and time. Depending upon the quantity of water and time of flow, permeability varies for the concrete. Coefficient of permeability is determined by using the following formula

$$K = \frac{QL}{tAh} \qquad (8.6)$$

where,

k = D'Arcy Coefficient of Permeability (m/s)
Q = Volume of water in m^3
L = Length of the test sample in meters, to the nearest 0.001m
t = Elapsed time in seconds
h = Applied pressure head in meters of water
A = Area of the test sample in m^2

Figure 9. Water permeability test arrangement

Drying Shrinkage

The shrinkage measuring apparatus used to determine the shrinkage strain for RPC is the concrete length comparator as shown in Figure 10. The comparator is comprised of an upright support of 356 mm tall attached to a metal base and including a 152 mm reference bar. It is designed to quickly and accurately measure the height of concrete samples to within ±0.1% of the total height. A digital indicator to an accuracy of within 0.0001 in. (0.002 mm) with 0 to 25 mm total range is used. The shrinkage strain is calculated according to the ASTM standard (ASTM C490/C490M).

Figure 10. Drying shrinkage test arrangement

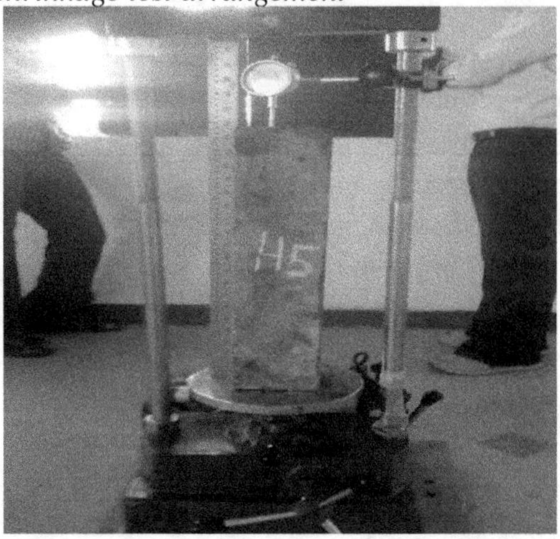

Rapid Chloride Penetration Test

Concrete disc specimens of size 100mm diameter and 50mm thick (Figure 11) were cast for various percentages of Polypropylene Fiber in concrete. After 24 hours, the disc specimens were removed from the mold and subjected to curing for 90 days in chloride free distilled water. After curing, the specimens were tested for chloride permeability. All the specimens were dried free of moisture before testing.

Figure 11. Concrete disc specimen for RCP TEST

The test set up is called Rapid Chloride Penetration Test (RCPT) assembly. This is two-compartment cell assembly. Disk specimen is assembled between the two compartments cell assembly and checked for air and watertight. The cathode compartment is filled with 3% NaCl solution and anode compartment is filled with 0.3 normality NaOH solutions. Then the FRP concrete specimens were subjected to RCPT by impressing a 60V from a DC power source between anode and cathode. Current recorded over a period of 6 hours at an interval of 30 minutes as per the procedure given in ASTM C1202 (Table 2). This test was conducted at CECRI, Karaikudi, Tamil Nadu.

Table 2. Charge passed through RCPT test as per ASTM C1202

S.No	Charge passed coulombs	Chloride ion penetrability
1	>4000	High
2	2000-4000	Moderate
3	1000-2000	Low
4	100-1000	Very low
5	<100	Negligible

From the current values, the chloride permeability is calculated in terms of coulombs at the end of 6 hours by using the following equation 8.5.

$$Q = 900\,(I_0 + 2I_{30} + 2I_{60} + \ldots\ldots\ldots + 2I_{300} + 2I_{330} + 2I_{360}) \qquad (8.5)$$

where,

Q = Charge passed (Coulombs)
I_0 = Current (amperes) immediately after voltage is applied
I_t = Current (amperes) at t minutes after voltage is applied

Sulphate Resistance Test

Six numbers each for CC, FRGP-1, FRGP-2, FRGP-3, FRGP-4 and FRGP-5 Geopolymer Concrete cubes were immersed in sodium sulphate solution of 5% concentration in accordance to the procedure given in ASTM C 642. The Geopolymer Concrete cubes (150 mm x 150 mm x 150 mm) with and without Polypropylene Fiber is immersed in sodium sulphate solution on the third day after casting. The cubes were fully immersed in the solution kept in plastic containers. The cubes were kept at a distance of 50 mm away from the walls of the container. The containers were covered in order to minimize the evaporation and to avoid falling of dust. The pH value was maintained neutral throughout the study period. Also the solution was stirred every week to avoid deposits on the base of the containers. The visual appearance, residual compressive strength and change in mass were observed on the second, fourth and eighth weeks after immersion, and readings were noted. The surface of the cubes was cleaned, weighed and tested in the compression testing machine. The cubes inside the container are shown in Figure 12.

Figure 12. Cube specimens immersed in 5% sodium sulphate solution

Chloride Attack Test

Sodium chloride (NaCl) solution with 5% concentration was used as the standard exposure solution for all tests. The specimens were immersed in the sodium chloride solution in a plastic container as shown in Figure 13 and the volume proportion of sulphate solution to specimens was maintained as four to one. The change in mass and change in compressive strength were observed at regular intervals.

Figure 13. Cube specimens immersed in 5% sodium chloride solution

Acid Resistance Test

As per the procedure stipulated in ASTM C 642 for ordinary Portland cement concrete, six numbers each for CC, FRGP-1, FRGP-2, FRGP-3, FRGP-4 and FRGP-5 Geopolymer Concrete cubes (150 mm x 150 mm x 150 mm) were immersed in diluted Sulphuric acid solution kept in plastic containers which is shown in Figure 14. The diluted sulphuric acid of 98% purity and density 1.85 g/cc was used. The specimens were weighed, recorded as W_1 and submerged so that there was a minimum of 30 mm depth of acid above the top surface of the specimens. The pH value of the solution was monitored periodically using a pH meter and also by titration with standard alkaline solution. The initial weights of the specimens were noted and recorded as W_1. Three specimens were taken out from the containers after 4 weeks and 8 weeks for testing. Observations were made on visual appearance, change in compressive strength and change in mass on the fourth and eighth weeks after immersion, and the readings were noted.

Figure 14. Specimens immersed in 5% H_2SO_4 solution

The specimens taken out were washed with tap water before being weighed in a digital balance of 0.1 mg accuracy. The average weight of the three specimens were noted as W_2 and the change in mass was calculated by,

Percentage weight loss =

$$\frac{W_1 - W_2}{W_1}$$

x 100 (8.8)

Fire Resistance Test

The mortar specimens were exposed to three different temperature levels(300 °C, 600 °C and 900 °C) in an electric furnace shown in Figure 15 for 3 hours in accordance with ASTM E119 standards. Once the target temperature was attained in the furnace, it was maintained for another three hours before the specimens were allowed to cool in air to room temperature in the furnace. Damage to the specimens were evaluated by colour change, compressive strength test and weight loss.

Figure 15. Electric furnaces to test concrete cubes at elevated temperature

Parameteric Study for Varying Temperatures on Geopolymer Concrete Beam by Finite Element Analysis Using Ansys and by Experiments

Extensive study of parameters of Geopolymer specimens affected when exposed to varying temperatures. Parameters studied are Compressive Strength, Split Tensile Strength, Modulus of elasticity, Poisson's ratio. Finite Element Analysis using the ANSYS CIVIL FEM program was also carried out to study the behavior of the tested specimens.3D elastic beam type of element is used to model the concrete beam. For beams without reinforcement, it is worthwhile to notice that the meshing was created according to the locations, as well as the beam specimen cross-sectional perimeter. The load was applied as pressure on the entire beam as the UDL.

Flexural Behavior of Reinforced Geopolymer Concrete Beams With and Without Polypropylene Fiber

The objective of this investigation was to study the flexural behavior of RC Geopolymer beams with and without Polypropylene Fiber addition. The Polypropylene Fiber addition is done in the ratio of 0.2%, 0.4%, 0.6%, 0.8% and 1%. Flexural behavior of RCC beams incorporating Polypropylene Fiber as additive was examined and also specimens of all specimens were examined and compared with control specimen.

Figure 16. Schematic diagram of beam reinforcement

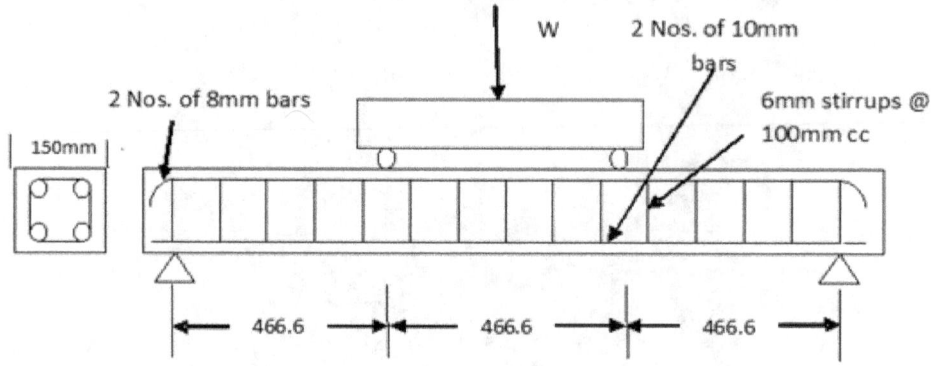

All beams (150 mm x 150 mm in cross section and 1500mm in length) as shown in Figure 16 were tested as simply supported beams under two-point loading over an effective span of 1400 mm. The loads were applied at a distance of 470 mm on either side of the mid span of the beams of 1500 mm length, as shown in Figure 16. These beams were tested in a loading frame of 500 kN capacity. The loads were monitored through a high accuracy load cell with a load sensitive of 0.1 Tons. For this case, mid span deflection was measured using dial gauges of least count 0.01mm. The parameters such as initial cracking load, ultimate load and the deflected shape of the specimens were noted.

Figure 17. Experimental test set up of reinforced FRGP beams

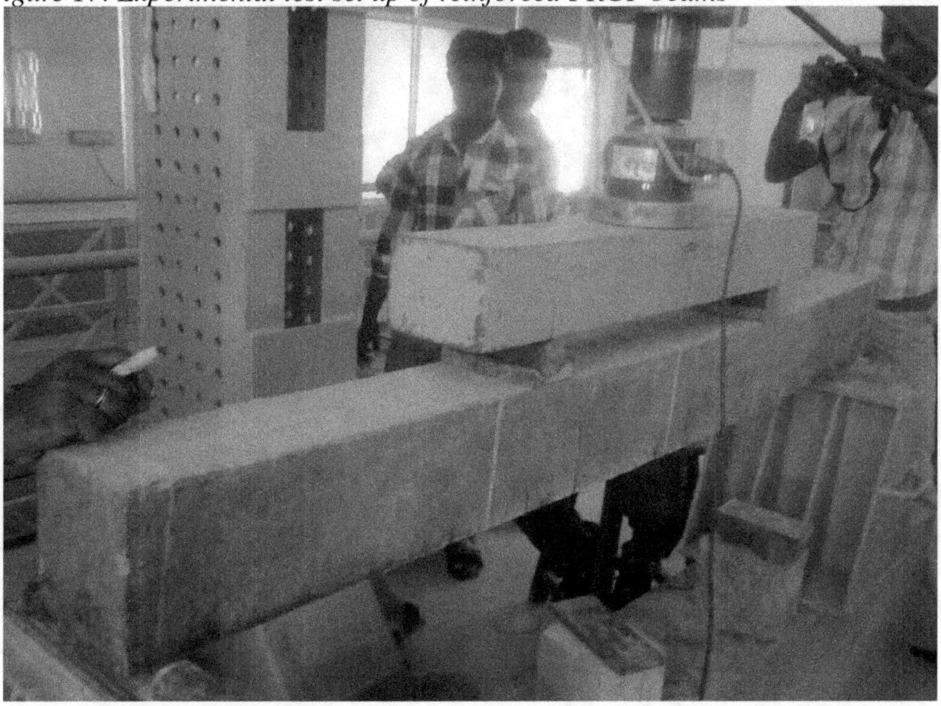

Tests on Reinforced Geopolymer Slab with Optimal Percentage of Addition of Polypropylene Fiber

The experimental setup for two-way slab is shown in Figure 17. Slabs dimension were of 1000 mm x 1000 mm x 30 mm and reinforced with 4 mm diameter weld mesh. Slabs were fixed inside the steel channels along with friction plates. The channels in turn welded to the loading platform to ensure fixed boundary condition. 500 KN capacity hydraulic jack is placed centrally to transfer uniformly distributed load to the slab through channels spaced at appropriate spacing. One Dial gauge is placed centrally below the slab to measure the deflection. The load is applied gradually at an interval of 1 KN and corresponding central deflection is measured till failure. Yield lines were encountered and further analyses were studied.

Figure 18. Casting of fiber geopolymer concrete slab

Figure 19. Experimental test setup of fiber geoploymer concrete slab

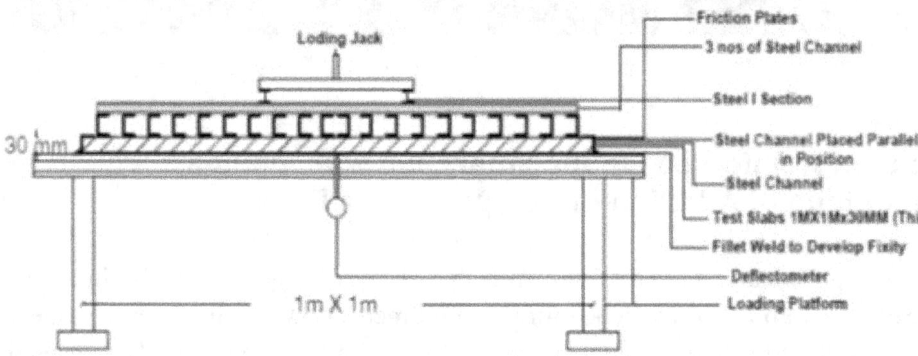

CONCLUSION

Based on the investigations, the following conclusions were drawn and were presented in the following sections based on the required categories.

Research Findings Based on Results of Fresh Concrete

- The results indicated that the workability of fresh FRGP concrete decreases as the percentage addition of Polypropylene Fibre increases.
- The values of slump cone test on fresh concrete lies between 120 mm to 90 mm which there by ensures medium workability on all mix proportions
- The maximum percentage of flow of fresh concrete is observed in control specimen whereas lowest flow percentage is observed in FRGP-5 specimen.
- The loss in workability of the FRGP Geopolymer Concrete is due to the addition of Polypropylene Fibre which directly influences the flow in frech Geopolymer Concrete.

Research Findings Based on Mechanical Properties

- The concrete mix prepared using upto 0.6% Polypropylene Fibre addition, shows increase in Compressive Strength of Geopolymer Concrete.
- The results show that as the age's increases the variation in strength gain decreases, and normalizes almost after 56 days.
- The effect of Polypropylene Fibre addition in binding material on splitting tensile strength shows higher strength than that of GP specimen.
- It is evident that the addition of Polypropylene Fibre increases the Split Tensile Strength and this increment is more than that of Compressive Strength increment.
- The Polypropylene addition increases the flexural strength of plain Geopolymer Concrete prisms and are substantially related to the bonding properties and the dispersion of Polypropylene Fibre

Research Findings Based on Durability Properties

- UPV results show all the specimen exhibit excellent quality in terms of pulse wave passed through the Geopolymer Concrete.
- The UPV results also relate with the strength of Geopolymer Concrete, higher the pulse velocity shows higher strength.
- From the UPV results it can be concluded that the quality of Geopolymer Concrete is excellent for all mix proportions, especially for Polypropylene Fibre addition upto 0.6%, excellent.
- The water absorption test shows that FRGP-5 which is the superior specimen in terms of water absorption, the percentage of water absorption is 1.51% which is 4.13% higher than the control specimen.

- This increase in water absorption is due to higher voids in higher percentage of FRGP Concrete.
- Therefore, it can be concluded that the addition of Polypropylene Fibre reduces the pores in FRGP Concrete.
- It is evident that the Polypropylene Fibre has a vast influence on its void structures.
- From shrinkage test on drying shrinkage it is concluded that the Polypropylene Fibre induces autogenous shrinkage because curing cannot reach the internal small pores produced by Polypropylene Fibre and thus results in internal drying which is called self-desiccation is observed.
- The drying shrinkage can be reduced by incorporating Polypropylene Fibre as an addition to cement
- From the RCPT it is evident that average penetration increases with increase in Polypropylene Fibre addition up to 0.6% and penetration increases for 0.8% addition.
- The values exceed the control specimen values at 1% addition of Polypropylene Fibre.
- From RCPT it can be concluded that the addition of polypropylene Fibre enhances the behaviour of Geopolymer Concrete by resisting the chloride penetration thus giving the durability enhancement.
- The sulphate attack test results shows, there was no sign visibly for surface erosion, cracking or spalling of the specimens till 4 weeks of time, but after 8 weeks, a little erosion on the surface could be noticed.
- The sulphate attack test results show that Polypropylene Fiber incorporated FRGP Concrete cubes, invariably of strength, had marginal weight gain, which shall be due to the high absorption of exposed liquid. But the specimens are almost maintained in shape without any sign of severe external deteriorations.
- The sulphate attack results show that loss in Compressive Strength decreases with increase in Polypropylene Fiber addition up to a limit of 0.6% and this is due to the occupation of fiber in FRGP microstructure which restricts the penetration of sulphate solution, particularly in FRGP-3 the penetration is very low therefore increase in mass and loss in strength is very low in this specimen. But in case of FRGP-5 there was disintegration in its pore structure, hence this leads to higher increase in mass and higher loss in strength.
- The Chloride Attack results show very little loss in mass in all specimens and the maximum loss percentage is observed as 1.67% in FRGP-5 specimen during 8 weeks of exposure which shows that there is very less effect of chloride solution on FRGP Geopolymer Concrete.

- The loss in compressive strength due to chloride attack is less in FRGP-3 specimen this is due to the resistance to penetration of chloride solution into the specimen.

  On the test on acid attack, torsion of surface was not observed in Polypropylene fiber added FRGP Concrete specimens even after 8 weeks of exposure in sulphuric acid solution and did not exhibit any noticeable color change. Though the surface had become softer with the exposure time, the specimen remained structurally intact.
- The Polypropylene fiber gives more resistance to FRGP Concrete against acid attack to specimens in terms of weight loss. This also shows that resistance on surface erosion is made possible by incorporating Polypropylene fiber in FRGP Concrete.

Research Findings on Finite Element Analysis and Experimental Work of Conventional Geopolymer Beams Subject to Varying Temperatures

- The FEM models for validation of plain Geopolymer Concrete and its respective analysis shows that the concentration of thermal stresses and its impact on the material structure is more for the control specimen when compared to that of Geopolymer specimen.

Research Findings on Behavior of Reinforced Geopolymer Concrete Beams and Slabs with and Without Polypropylene Fiber

- The test on specimens under elevated temperatures shows the loss in weight goes on decreasing as the percentage of Polypropylene Fiber addition increases. This shows that resistance towards exposure to high temperature is made possible by incorporating Polypropylene fiber in FRGP Concrete.
- From the study carried out on fly ash based polymer RC beams and RC cement concrete beams, it can be concluded that the plain fly ash based polymer RC beams were structurally more resistive than control specimen due to its higher molarity (14 M) in this study. But when it comes to the addition of Polypropylene Fiber there was no significant increase in load carrying capacity.
- The addition of fiber increases the flexural toughness of the beams. The volumetric fractional addition of Polypropylene Fiber gives the maximum load carrying capacity at 0.4%, but the toughness was high, attributed to the re-

duction in deflection. At 0.4% the failure was also sudden when compared to other specimens.

- Until 0.6% addition of Polypropylene Fiber addition, the load carrying capacity of beams was maintained not less than the control specimen. The addition of Polypropylene Fiber gives more resistance toward bending by increasing its flexural resistance. The addition of Polypropylene Fiber also reduces the strain localization of the fly ash-based RC beams. Hence the addition of Polypropylene Fiber should be maintained at 0.4% or less.

Micro Structural Study

The micro structural study shows enhanced performances and effective cost-benefit ratio made the use of Polypropylene Fiber in concrete structures recently.

- The advantage extended to Fiber Geopolymer Concrete also. Porosity is measured by means of BET techniques. In the case of both additives, the pores were found to be uniformly distributed throughout the samples.
- The results of BET and SEM have shown the difference in pore size for the samples with and without addition of Polypropylene Fiber. FRGP sample has low pore volume and surface area when compared to OPC and GPC.

GENERAL CONCLUSION

Based on results of the above experimental investigation it is concluded that Geopolymer concrete with Polypropylene Fibre as secondary reinforcement have superior properties and Structural behaviour than conventional Geopolymer Concrete. The FRGP Concrete can be used as a substitute in place of conventional reinforced cement concrete.

SCOPE FOR FUTURE WORKS

- Polypropylene Fibre can be used in Self Compacting Geopolymer Concrete and Special Concrete.
- Study on Geopolymer concrete with different additives like GGBS, silica fume along with polypropylene fibres.
- Polypropylene Fibre can be effectively added in making high performance Geopolymer blocks and porous Geopolymer Pavement Blocks.

- Study on performance of special structural elements like Deep beam, Corbel, Shear wall etc using Polypropylene based Geopolymer concrete.
- Precast wall panels with Polypropylene based Geopolymer concrete can be studied.
- Beam Slab Panels with Polypropylene based Geopolymer concrete can be studied.

REFERENCES

Akbarnezhad, A., Huan, M., Mesgari, S., & Castel, A. (2015). Recycling of geopolymer concrete. *Construction & Building Materials, 101*, 152–158. DOI: 10.1016/j.conbuildmat.2015.10.037

Albitar, M., Ali, M. M., & Visintin, P. (2017). Experimental study on fly ash and lead smelter slag-based geopolymer concrete columns. *Construction & Building Materials, 141*, 104–112. DOI: 10.1016/j.conbuildmat.2017.03.014

Buchwald, A., Dombrowski, K., & Weil, M. (2005). The influence of calcium content on the performance of geopolymeric binder especially the resistance against acids. *Proceedings of the world geopolymer*, 35-9.

Hardjito, D., Wallah, S. E., Sumajouw, D. M., & Rangan, B. V. (2004). On the development of fly ash-based geopolymer concrete. *Materials Journal, 101*(6), 467–472.

Li, Z., Ding, Z., & Zhang, Y. (2004, May). Development of sustainable cementitious materials. In *Proceedings of international workshop on sustainable development and concrete technology*, Beijing, China (pp. 55-76).

Nuaklong, P., Sata, V., & Chindaprasirt, P. (2016). Influence of recycled aggregate on fly ash geopolymer concrete properties. *Journal of Cleaner Production, 112*, 2300–2307. DOI: 10.1016/j.jclepro.2015.10.109

Pilehvar, S., Cao, V. D., Szczotok, A. M., Valentini, L., Salvioni, D., Magistri, M., Pamies, R., & Kjøniksen, A. L. (2017). Mechanical properties and microscale changes of geopolymer concrete and Portland cement concrete containing micro-encapsulated phase change materials. *Cement and Concrete Research, 100*, 341–349. DOI: 10.1016/j.cemconres.2017.07.012

Tatar, J., & Hamilton, H. R. (2016). Comparison of laboratory and field environmental conditioning on FRP-concrete bond durability. *Construction & Building Materials, 122*, 525–536. DOI: 10.1016/j.conbuildmat.2016.06.074

Visintin, P., Ali, M. M., Albitar, M., & Lucas, W. (2017). Shear behaviour of geopolymer concrete beams without stirrups. *Construction & Building Materials, 148*, 10–21. DOI: 10.1016/j.conbuildmat.2017.05.010

Zhang, F., Chen, H., Li, X., Li, H., Lv, T., Zhang, W., & Yang, Y. (2017). Experimental study of the mechanical behavior of FRP-reinforced concrete canvas panels. *Composite Structures, 176*, 608–616. DOI: 10.1016/j.compstruct.2017.05.072

Chapter 6
Integral Contributions of Civil Engineers to Transport Planning

Arpita Nayak
https://orcid.org/0000-0003-2911-0492
KIIT School of Liberal Studies, KIIT University, India

Neha Nayanika Sahoo
https://orcid.org/0009-0005-7058-9193
KSOM, KIIT University, India

Ankita Nayak
https://orcid.org/0000-0002-9311-890X
Trident Academy of Creative Technology, India

ABSTRACT

The chapteri nvestigates civil engineers to possess key importance in creating sustainable transportation infrastructure which serves India's fast-growing population and developing infrastructure markers. The section examines their work in transportation planning through their creation of resilient structures and integration of smart tools such as GIS, BIM and IoT. The analysis focuses on India's Smart Cities Mission case studies and new sustainable materials together with traffic management solutions for safety compliance. The section shows how civil engineers handle India's specific challenges by following worldwide standards during their approach to solving problems involving population density along with environmental sustainability.

DOI: 10.4018/979-8-3693-2448-6.ch006

I. INTRODUCTION

Engineers are true revolutionaries in our society, with civil engineers acting as the nation's backbone. This is owing to the important role civil engineers play in a country's overall growth and advancement. They are the minds that enable a greater quality of life, better public health, and more organized towns and cities around the country. Civil engineers work in the development of the infrastructure upon which our modern transportation network depends. Civil engineers ensure that transportation networks are designed to run smoothly, safely and sustainably to safely transport people and goods. During the early stages of infrastructure development, engineers undertake an extensive design and planning process to ensure that infrastructure functions for its intended use. They do this by assessing future transportation needs, the sustainability of the planned development, analyzing geological and environmental data, and developing the designs. Civil engineers play an important part in the development of transportation infrastructure. Their specialized expertise is required for the design, construction, and maintenance of a wide range of transportation infrastructures, such as roads, bridges, tunnels, trains, airports, and ports. These elements are critical for a functioning public transportation network, to effectively move people and goods around. In the first step of the process, extensive inspections of the site, feasibility studies, and engineering concepts allowing the infrastructure to support current and future demands will be conducted. Throughout the building process, civil engineers must consider structural stability, material flexibility, and environmental effects (Karami et al.,2020). Infrastructure serves a variety of tasks, including transportation, electricity supply, water delivery, and civic spaces. A nation's infrastructure includes highways, bridges, irrigation canals, bridges, modes of transportation, and even buildings. To develop swiftly, a country must have the most modern infrastructure. Roads, for example, are necessary for transporting commodities and critical services to rural communities, therefore raising the level of living in such areas. Civil engineers make major contributions to the growth of a country's infrastructure. To increase efficiency, smart cities integrate digital technology into their present services and networks. A few years ago, the Government of India announced a plan to establish 100 Smart Cities across the country. Civil engineers from many fields collaborated to realise this vision and support the development of such dynamic cities. These cities are designed to be more sophisticated and efficient in every way and have better transport, water distribution, waste management and healthcare systems (Fields et al., 2020). Well-designed towns and cities can have a big impact on a country's development. In developing countries, there is a need for urban planners to support the establishment of settlements that will help people to live well and prosper. These specialists are in charge of designing the layout of a municipality, which includes roads, hospitals, schools, transit networks, business buildings, and residential communities. Civil engineers are also responsible for the implementation of environmental protection measures. A country's progress should not be accompanied by environmental degradation. Civil engineers who have studied environmental engineering at one of Nashik's best engineering institutions understand how to apply methods to develop civil structures that benefit society as a whole. They play a crucial role in overseeing activities such as proper waste management, rubbish recycling, encouraging the use of renewable energy in everyday life, and much more. Public health is vital to a nation's progress. A nation in which people are always ill or handicapped cannot attain its full potential. Civil engineers represent public health by overseeing the allocation of resources for the development and maintenance of healthcare facilities such as hospitals and clinics in various places. They are in charge of constructing healthcare facilities in rural locations. These specialists also perform frequent checks in public locations such as restaurants, malls, and schools to verify that they fulfill the health codes established by the governing organizations. (Santos et al.,2018Civil engineers occupy an important status in urban planning and infrastructure planning in cities and towns, where they are engaged in designing, constructing, and maintaining the infrastructure that sustains our modern civilization. Civil engineering includes multiple disciplines, such as structural engineering and transportation planning, which are paramount to urban development and infrastructure planning to contribute to urban expansion (Chakraborty, 2017). Students studying for a Bachelor of Technology in Civil Engineering gain the necessary knowledge and skills to deal with complex issues of urbanization and infrastructure development. The global civil engineering industry is expected to expand at a CAGR of 5.9% from 2023-2030. This expansion is fueled by growing infrastructure and capital projects, as well as increasing disposable income in emerging markets. Despite experiencing the impact of the COVID-19 pandemic, the introduction of new technology and increased investment into infrastructure projects has been the driving force moving the structural engineering industry sector onward. The planning and design services market is expected to grow the fastest, with a CAGR of 6.2%. Furthermore, increasing remodeling activities and the installation of energy-efficient goods promote market expansion (Fet,2024). In effect, traffic management is essential for reducing congestion and ensuring a steady flow of traffic. Civil engineers implement a variety of tactics to improve traffic flow and road safety. Civil engineers employ intelligent transportation systems (ITS) to enhance traffic flow by adding traffic signals, signs, and road markings to direct vehicles and pedestrians. Effective traffic management is essential for reducing congestion and ensuring a dependable transport network. Civil engineers use a range of methods to improve traffic flow and safety. They provide intelligent transportation systems (ITS), which utilise traffic signals, signs, and road markings to efficiently direct automobiles and pedestrians (Sciara,2017). The incorporation of technology into transportation planning has transformed the discipline of civil engineering. Civil engineers are increasingly using new technologies like Geographic Information Systems (GIS), Building Information Modelling (BIM) and the Internet of Things (IoT) to collect and analyze vast amounts of data, leading to more precise planning and decision-making. Civil engineers can use GIS technology to develop complex maps and spatial analyses that inform them of the geographical and demographic characteristics impacting transportation networks. BIM has collaborative capabilities and serves as a central platform for infrastructure project planning and management, ensuring all stakeholders have access to the most up-to-date information. IoT devices, which consist of sensors, cameras, and other technologies, provide real-time information on traffic, weather, and infrastructure performance, to improve the accuracy and efficiency of transportation planning. Civil engineers can use this data and information to participate in the management, monitoring, and maintenance of transportation infrastructure, anticipate repair needs, and respond effectively to emergencies. Technology can enhance the efficiency, safety, and sustainability of transportation systems, which can improve their adaptation to future challenges (Sayyadi & Awasthi, 2018). Civil engineers play an essential part in transportation planning since they are experts in infrastructure design, traffic management, environmental sustainability, and technology integration. Their activities contribute to the development of transportation networks that are safe, efficient, and sustainable. As challenges and opportunities develop, civil engineers will continue to innovate and adapt, increasing transportation planning and societal well-being.

Research Questions

- What are the different roles of civil engineers in transport planning?
- What are the different factors that civil engineers in transport planning?

II. RESEARCH OBJECTIVES

- To understand the various implications of civil engineers in transport planning.
- To comprehend the different factors that civil engineers in transport planning.

III. INFRASTRUCTURE AND DEVELOPMENT

Civil engineers play an important role in the development of infrastructure which underpins modern transportation networks. Their work helps ensure that these networks run efficiently, safely, and sustainably, allowing individuals and possessions to flow seamlessly. They conduct thorough planning and design procedures in the early stages of the creation of infrastructure. This involves performing site assessments to determine the sustainability of planned projects, gathering geological and environmental data, and forecasting future transportation demands. Collaboration with urban planners, environmental specialists, and government officials is critical for developing inclusive solutions that align with overall urban development goals. Geographic Information Systems (GIS) play a crucial role in this phase, providing civil engineers with valuable tools to create detailed maps and conduct spatial analyses. This technology aids in understanding the geographical and demographic factors that impact transportation systems. By integrating GIS data with traffic modeling software, engineers can simulate various scenarios and optimize route planning to enhance traffic flow and alleviate congestion (Berglund et al.,2020). Civil engineers utilize Building Information Modeling (BIM) for the design of infrastructure projects. BIM offers a collaborative platform for engineers to develop intricate 3D models of infrastructure, such as roads, bridges, tunnels, and transit systems. This technology allows engineers to have a comprehensive visualization of the project, pinpoint potential problems, and make well-informed decisions. BIM streamlines communication among stakeholders, guaranteeing that all parties involved in the project have equal access to the necessary information (Meyer,2016). After completing the planning and design phases, civil engineers proceed to the construction and implementation of transportation infrastructure. This stage involves supervising the construction process, guaranteeing that the project complies with design specifications, safety regulations, and legal requirements. Civil engineers oversee construction teams, collaborate with contractors, and monitor the project's progress to ensure it is completed on time. During construction, civil engineers utilize various methods to improve the quality and longevity of infrastructure. For instance, they utilize advanced materials like high-performance concrete and corrosion-resistant steel to construct buildings that can endure harsh weather conditions and heavy traffic loads. They also incorporate sustainable construction techniques, such as recycling construction waste and minimizing land disturbance, to lessen the project's environmental impact (Fields et al.,2019). Civil engineers not only build new infrastructure but also oversee the upkeep and enhancement of current transportation systems. This includes conducting routine inspections, making repairs, and carrying out renovations to guarantee the durability and safety of the infrastructure. Utilizing non-destructive testing techniques like ground-penetrating radar and ultrasonic testing, civil engineers can identify structural problems and evaluate the state of the infrastructure. By promptly addressing maintenance requirements, civil engineers avoid expensive repairs and maintain the seamless operation of transportation networks (Kim et al.,2017). Civil engineers are instrumental in developing and executing traffic management strategies to ensure the efficient operation of transportation systems. They utilize traffic simulation software to analyze traffic patterns and pinpoint potential congestion points. By leveraging this data, civil engineers create traffic control mechanisms like signals, signs, and markings to direct both drivers and pedestrians. Civil engineers are responsible for implementing intelligent transportation systems (ITS) that integrate live data from various sources such as sensors, cameras, and GPS devices. This technology empowers engineers to monitor traffic flow, identify incidents, and regulate traffic in real time. For instance, adaptive traffic signal control systems automatically adjust signal timings based on traffic conditions, thereby minimizing delays and enhancing traffic efficiency. Civil engineers design dynamic message signs to provide drivers with up-to-date information on traffic status, road closures, and alternate routes (Kermanshachi et al.,2018). Civil engineers play a crucial role in modern infrastructure development by prioritizing sustainability. They lead the way in creating eco-friendly transportation solutions that reduce the environmental impact of infrastructure projects. By evaluating the ecological effects of transportation projects early on and implementing strategies to minimize harm, they contribute to a more sustainable future. Civil engineers play a crucial role in promoting sustainability through the integration of green design principles in infrastructure projects. This includes minimizing land disruption, utilizing recycled materials, and incorporating renewable energy sources. Civil engineers advocate for public transportation and non-motorized modes of travel to decrease carbon emissions and traffic congestion. Furthermore, they prioritize the protection of natural habitats and biodiversity by designing infrastructure that harmonizes with the environment, such as wildlife crossings and stormwater management systems to prevent water pollution (Costin,2018). The incorporation of technology into transportation planning has transformed the approach of civil engineers to their work. Cutting-edge technologies like drones, IoT devices, and machine learning algorithms empower civil engineers to collect and analyze extensive data for more precise planning and decision-making.For instance, drones are utilized for aerial surveys and infrastructure inspections, providing engineers with detailed images and data to evaluate infrastructure conditions and plan maintenance tasks. IoT devices, such as sensors and cameras, offer real-time information on traffic conditions, weather, and infrastructure performance. Civil engineers utilize this data to supervise and upkeep transportation systems, forecast maintenance requirements, and promptly address issues. Machine learning algorithms are employed to scrutinize large datasets and recognize patterns and trends that can guide transportation planning.

For instance, machine learning models can forecast traffic volumes and congestion levels based on historical data, enabling civil engineers to devise proactive traffic management strategies. The integration of technology boosts the efficiency, safety, and sustainability of transportation systems, rendering them more resilient to future challenges (Cheng et al.,2016). Civil engineers play a crucial role in the advancement and organization of transportation infrastructure. Their proficiency in design and planning, construction and execution, traffic control, environmental preservation, and technological integration guarantees the establishment of secure, effective, and sustainable transportation networks. With the ongoing emergence of fresh obstacles and possibilities, civil engineers will be instrumental in molding the future of transportation, propelling advancement, and enriching the general welfare of society.

IV. ENVIRONMENTAL SUSTAINABILITY

Civil engineers play a crucial role in modern transportation planning by prioritizing sustainability. They lead the way in creating eco-friendly transportation solutions, aiming to reduce the environmental impact of infrastructure. This involves implementing green design practices and utilizing sustainable materials to minimize ecological footprints. Throughout the project lifecycle, civil engineers evaluate the environmental effects, safeguard habitats, lower carbon emissions, and promote resource conservation. Their efforts may include designing infrastructure with minimal land disturbance, utilizing recycled materials, and integrating renewable energy technologies such as solar panels (Schiller,2017). Civil engineers play a critical role in encouraging sustainability in transportation design by selecting environmentally friendly materials. Traditional building materials, such as concrete and asphalt, have a significant environmental effect due to their energy-intensive manufacturing processes and associated greenhouse gas emissions. To solve this challenge, engineers are increasingly adopting recycled materials and alternatives with reduced environmental effects. For example, using recycled asphalt pavement (RAP) and fly ash in concrete not only reduces waste but also helps to preserve natural resources (Montgomery,2015). Energy-efficient building techniques are a key element of sustainable transportation planning. Engineers are now utilizing methods that cut down on energy usage and emissions during the building process. This involves utilizing machinery and equipment powered by renewable energy sources, adjusting construction schedules to minimize idle time and fuel use, and integrating on-site renewable energy solutions like solar panels. Modular construction methods, which entail constructing components off-site and then assembling them on-site, can greatly decrease waste, enhance construction efficiency, and lessen the carbon footprint of transportation projects. Carrying out thorough Environmental Impact Assessments (EIA) is fundamental in sustainable transport planning. Before commencing any significant project, engineers conduct baseline studies to grasp the current environmental conditions of the project site. This entails gathering data on air and water quality, wildlife habitats, soil conditions, and other ecological factors. The subsequent impact analysis assesses how the proposed project will impact these environmental elements. For example, the construction of a new highway could result in habitat fragmentation, increased air pollution, and alterations in water runoff patterns. By identifying these potential impacts early on in the planning phase, engineers can devise mitigation strategies to minimize adverse effects. These strategies may involve establishing wildlife corridors to aid animal movement, installing noise barriers to mitigate pollution, and incorporating green infrastructure to manage stormwater (Schiller,2017). Sustainable transportation systems aim to minimize the environmental effects linked to conventional transportation methods. A key aspect of this endeavor is the promotion of public transport. Engineers work on creating efficient, dependable, and easily accessible public transport systems that motivate individuals to transition from using private cars. This transition can notably alleviate traffic congestion, reduce greenhouse gas emissions, and lessen air pollution. By combining public transport with non-motorized modes like walking and cycling, sustainability is further improved. The establishment of designated cycling paths, pedestrian-friendly roads, and bike-sharing initiatives promotes active transportation, which not only has minimal environmental repercussions but also provides substantial health advantages (Rogers,2023). Technological progress is crucial in sustainable transport planning. Intelligent Transport Systems (ITS) utilize advanced technologies to improve traffic flow, minimize congestion, and boost the effectiveness of transport networks. For instance, real-time traffic management systems can modify traffic signal timings according to present traffic conditions, lessening idle times and emissions from vehicles. The promotion of electric and hybrid vehicles is facilitated by the establishment of supporting infrastructure like charging stations. These vehicles emit substantially less pollutants than conventional internal combustion engine vehicles and are essential in reaching long-term sustainability objectives (Bibri et al.,2020). Effective urban planning and land use policies play a crucial role in establishing sustainable transportation systems. By incorporating mixed-use development and transit-oriented development (TOD), the need for lengthy commutes is reduced, and the use of sustainable transportation modes is encouraged. TOD, in particular, concentrates on high-density development around public transport hubs, improving access to public transportation and promoting walking and cycling within the community. Through these strategies, urban planning ensures that transportation infrastructure aligns with sustainable growth and minimizes environmental effects (Salort et al.,2021). An increasing trend in transport planning involves the incorporation of green infrastructure and nature-based solutions. These methods focus on creating transport systems that coexist with natural ecosystems, ultimately improving resilience and sustainability. For example, integrating green roofs and walls along transport routes can aid in reducing urban heat island effects, enhancing air quality, and managing stormwater. Additionally, bioswales and rain gardens can be integrated into road designs to filter pollutants from runoff, promote groundwater recharge, and support urban biodiversity. These nature-based solutions not only help mitigate environmental impacts but also contribute to the creation of visually appealing and multifunctional urban spaces. Smart cities are transforming transportation planning by utilizing IoT to develop more effective, eco-friendly, and user-centric transport systems. IoT devices, like sensors and connected infrastructure, offer instant data on traffic flow, road conditions, and environmental factors. This data allows for dynamic traffic control, minimizing congestion and emissions. For instance, smart traffic lights can adapt their timing based on real-time traffic flow, improving vehicle movement and reducing idle time. IoT can enable the integration of different transport modes, providing smooth connectivity between public transport, shared mobility services, and personal vehicles, ultimately enhancing overall system efficiency and sustainability (Salort et al.,2021). The rise of electric and autonomous vehicles is poised to revolutionize transportation planning in a significant way. Electric vehicles (EVs) are known for producing zero tailpipe emissions, which helps in improving air quality and lowering greenhouse gas emissions. Civil engineers are currently concentrating on creating a reliable EV infrastructure, which includes extensive charging networks and smart grid integration to handle the growing electricity demand sustainably. On the other hand, autonomous vehicles (AVs) offer the potential to enhance traffic flow, decrease accidents, and improve fuel efficiency. Engineers are actively involved in developing roadways and urban layouts that can support and maximize the advantages of AV technology, such as designated lanes for autonomous vehicles and intelligent road systems that interact with AVs to boost safety and efficiency (Fei et al.,2021). It is becoming more and more important to integrate resilience and climate adaptation strategies into transport planning due to the growing occurrence and severity of extreme weather events. Engineers are working on designing transport infrastructure that can endure and bounce back from such events. This involves raising roads and bridges in flood-prone areas, utilizing materials that can withstand extreme temperatures and weather conditions, and creating backup options in transport networks to guarantee accessibility even if some routes are affected. Additionally, predictive modeling and climate impact assessments are being used to predict future conditions and guide the design and upkeep of transport infrastructure (Campisi et al.,2020). A new and creative strategy in transportation planning involves focusing on public engagement and participatory planning. Civil engineers are realizing the importance of including communities in the planning process to guarantee that transportation systems meet the requirements of users and promote sustainability. Public engagement can come in different forms, including interactive workshops, digital platforms for feedback, and collaborative design sessions. This inclusive approach not only aids in identifying community needs and preferences but also cultivates a sense of ownership and responsibility towards sustainable transportation initiatives. Engaged communities are more likely to endorse and utilize sustainable transportation options, thereby improving the effectiveness of these initiatives. Civil engineers are essential in integrating environmental sustainability into transport planning. They achieve this by utilizing eco-friendly materials, and energy-efficient construction techniques, conducting thorough environmental impact assessments, advocating for sustainable transportation systems, harnessing technological innovations, and executing progressive urban planning approaches. Their efforts result in transport infrastructure that is not just efficient and functional, but also environmentally conscious. Given the ongoing issues of climate change and environmental harm, the incorporation of sustainability in transport planning will continue to be a crucial focus for civil engineers, guaranteeing a healthier and more sustainable planet for generations to come.

V. SAFETY AND REGULATION COMPLIANCE

Transport planning requires strict adherence to a variety of standards and regulations established by governmental and international entities. These guidelines ensure the safety, reliability, and efficiency of transport systems. Civil engineers must keep abreast of these regulations, which cover aspects ranging from road design specifications to environmental impact assessments. For example, in the United States, the Federal Highway Administration (FHWA) sets detailed standards and guidelines for road design, construction, and maintenance. Similarly, the European Union has extensive regulations through the Trans-European Transport Network (TEN-T) program. Compliance with these standards is not only a legal obligation but also crucial for public safety. Engineers must guarantee that transport infrastructure meets or surpasses these standards, often requiring meticulous planning, design, and quality control processes (Brauer,2022). The incorporation of cutting-edge safety technologies in transportation planning is a proactive strategy to enhance the safety of transportation systems. Civil engineers are increasingly utilizing technologies like Intelligent Transportation Systems (ITS), which utilize real-time data to optimize traffic flow and minimize the likelihood of accidents. ITS encompasses traffic signal control systems, dynamic message signs, and incident management systems that offer drivers real-time information, assisting them in making safer decisions while on the road. Another technological advancement involves the implementation of automated and connected vehicle technologies. These systems allow vehicles to communicate with each other and with infrastructure, such as traffic lights and road sensors, to improve safety. For instance, collision avoidance systems can warn drivers of potential dangers, while automated braking systems can prevent accidents by halting the vehicle if a collision is imminent (Provan et al.,2020). Civil engineers involved in transport planning have a crucial role in continuously evaluating and managing the risks associated with transport infrastructure. This includes identifying potential hazards, assessing the likelihood and severity of these risks, and implementing measures to minimize them. Risk assessments are carried out at different stages of a project, from the initial planning and design to construction and operation. One approach to risk assessment is the Failure Mode and Effects Analysis (FMEA), which systematically examines potential failure points within a system and their consequences. By identifying and addressing these vulnerabilities early in the planning process, engineers can create safer transport systems. Furthermore, regular safety audits and inspections are conducted to ensure that the infrastructure remains safe and compliant with regulations throughout its lifecycle (Nawaz et al.,2020). The safety of transport infrastructure heavily relies on structural integrity and durability. Civil engineers are responsible for designing and building roads, bridges, tunnels, and other transport facilities that can withstand various loads and environmental factors. This includes the careful selection of materials, the application of robust design principles, and adherence to construction best practices. Bridges, for instance, need to be able to support dynamic loads from vehicles and pedestrians, as well as withstand environmental stresses like wind, temperature changes, and seismic activity. Engineers utilize advanced modeling and simulation tools to anticipate how structures will behave under different circumstances and ensure they meet safety requirements. Furthermore, regular maintenance and inspections play a crucial role in detecting and addressing any structural issues before they pose significant safety risks (Pregnolato et al.,2022). Safety-focused design principles play a crucial role in transport planning. These principles aim to develop infrastructure that reduces the risk of accidents and enhances the safety of everyone using the transportation system, including motorists, pedestrians, and cyclists. An example of this is the "Complete Streets" concept, which encourages the creation of roadways that cater to all users safely and conveniently. This can include features like spacious sidewalks, protected bike lanes, pedestrian crossings, and traffic calming measures such as speed bumps and roundabouts. Visibility and lighting are key components of a secure design. Adequate lighting along roads and at intersections enhances visibility for both drivers and pedestrians, thus decreasing the chances of accidents. Clear and consistent signage aids users in safely and efficiently navigating the transportation network (Campisi et al.,2020). Emergency preparedness and response planning are essential components of transport safety. Civil engineers must ensure that transport infrastructure is designed to facilitate efficient evacuation and emergency response in the event of a disaster. This includes designing road networks that provide multiple evacuation routes, incorporating emergency lanes for first responders, and ensuring that critical infrastructure, such as bridges and tunnels, remains operational during emergencies. In addition to physical design considerations, engineers collaborate with emergency management agencies to develop comprehensive emergency response plans. These plans outline procedures for coordinating with law enforcement, fire departments, medical services, and other stakeholders to ensure a rapid and effective response to incidents (Hamim et al.,2020). Civil engineers are essential in ensuring that transport planning prioritizes safety and regulatory compliance. By adhering to standards, integrating advanced technologies, managing risks rigorously, maintaining structural integrity, designing with safety in mind, preparing for emergencies, and collaborating with stakeholders, engineers create transport systems that are not only efficient and sustainable but also safe and resilient. These efforts are crucial in protecting public health and well-being, fostering trust in transportation infrastructure, and ultimately contributing to the development of vibrant, thriving communities (Luo et al.,2020). Predictive analytics and big data are essential tools in transport planning. By analyzing large volumes of data from various sources such as traffic sensors, GPS devices, and social media, engineers can forecast traffic patterns, pinpoint accident hotspots, and create proactive safety measures. Machine learning algorithms can process this data to identify irregularities and anticipate potential infrastructure failures, enabling pre-emptive maintenance and reducing the likelihood of accidents (Lee,2020). Human factors engineering is centered on comprehending the dynamics between humans and the transportation environment to create systems that can adapt to human abilities and constraints. By taking into account aspects like driver behavior, perception, and reaction times, engineers can develop transportation infrastructure that lowers the chances of human error and enhances safety. An example of human factors engineering in action is seen in the creation of road signs and markings. Clear, easy-to-understand signage that takes into consideration visual perception and cognitive processing can greatly decrease driver confusion and increase compliance with traffic laws. Similarly, designing road layouts that minimize complex decision points and offer clear guidance can help reduce accidents caused by driver error (Stefańska,2022). Adaptive traffic control systems utilize real-time data to modify traffic signal timings and regulate traffic flow dynamically. By adjusting to present traffic conditions, these systems can alleviate congestion, enhance traffic flow, and boost safety. For instance, adaptive signals can prioritize emergency vehicles, public transit, or pedestrians during peak hours, optimizing road space usage and minimizing the likelihood of accidents. Adaptive systems can be combined with vehicle-to-infrastructure (V2I) communication technologies, facilitating instant interaction between vehicles and traffic signals. This integration can support functionalities such as green wave systems, where traffic lights are synchronized to enable continuous traffic flow at a specific speed, reducing stop-and-go traffic and its associated hazards (Trubia et al.,2020). Effective transportation planning requires the seamless integration of reliable emergency response systems. Collaboration between civil engineers and emergency services is essential to develop infrastructure that enables swift and effective responses to emergencies. This involves establishing accessible routes for emergency vehicles, designing intersections and roads that can accommodate large emergency vehicles, and implementing sophisticated traffic management systems that can adapt to prioritize emergency responses. One example of this is the utilization of traffic pre-emption systems, which enable emergency vehicles to manipulate traffic signals to navigate intersections efficiently and safely. By implementing these systems, not only are response times enhanced, but the risk of accidents involving emergency vehicles and other road users is also minimized (Pan,2020). Interdisciplinary collaboration plays a crucial role in transport planning. Civil engineers need to closely collaborate with urban planners, environmental scientists, public health experts, and policymakers to develop comprehensive solutions that address safety and regulatory compliance. This collaborative approach ensures that all aspects of transport safety are taken into consideration, from environmental impacts and community needs to technological advancements and regulatory requirements. Civil engineers are implementing advanced strategies and innovations to ensure safety and regulatory compliance in transport planning. From utilizing predictive analytics and human factors engineering to embracing holistic approaches like Vision Zero and sustainable safety, and integrating adaptive technologies and resilience principles, these efforts are essential in creating transport systems that are safe, efficient, and reliable. Interdisciplinary collaboration further enhances the effectiveness of these strategies, ensuring that transport infrastructure meets the diverse needs of communities while prioritizing safety and compliance.

VI. CONCLUSION

Interdisciplinary collaboration plays a vital role in transport planning. Civil engineers need to collaborate closely with urban planners, environmental scientists, public health experts, and policymakers to develop comprehensive solutions that address safety and regulatory requirements. This collaborative approach ensures that all aspects of transport safety are taken into consideration, from environmental impacts and community needs to technological advancements and regulatory standards. Partnering with public health experts can offer valuable insights into the health effects of transport systems, leading to designs that encourage active transportation and reduce pollution. Working with urban planners can guarantee that transport infrastructure supports sustainable and fair urban development while engaging with policymakers can help align transport projects with broader safety and regulatory objectives. Civil engineers are implementing innovative strategies to ensure safety and regulatory compliance in transport planning. By utilizing predictive analytics, human factors engineering, Vision Zero principles, sustainable safety practices, adaptive technologies, and resilience principles, they are creating transport systems that are safe, efficient, and dependable. Interdisciplinary collaboration further strengthens these efforts, ensuring that transport infrastructure meets the diverse needs of communities while prioritizing safety and compliance.

REFERENCES

Berglund, E. Z., Monroe, J. G., Ahmed, I., Noghabaei, M., Do, J., Pesantez, J. E., Khaksar Fasaee, M. A., Bardaka, E., Han, K., Proestos, G. T., & Levis, J. (2020). Smart infrastructure: A vision for the role of the civil engineering profession in smart cities. *Journal of Infrastructure Systems*, *26*(2), 03120001. DOI: 10.1061/(ASCE)IS.1943-555X.0000549

Bibri, S. E., Krogstie, J., & Kärrholm, M. (2020). Compact city planning and development: Emerging practices and strategies for achieving the goals of sustainability. Developments in the built environment, 4, 100021.

Borowska-Stefańska, M., & Wiśniewski, S. (2022). The role of road transportation in the flood evacuation process. In Oxford Research Encyclopedia of Natural Hazard Science. DOI: 10.1093/acrefore/9780199389407.013.440

Brauer, R. L. (2022). *Safety and health for engineers*. John Wiley & Sons.

Campisi, T., Acampa, G., Marino, G., & Tesoriere, G. (2020). Cycling master plans in Italy: The I-BIM feasibility tool for cost and safety assessments. *Sustainability (Basel)*, *12*(11), 4723. DOI: 10.3390/su12114723

Chakroborty, P., & Das, A. (2017). *Principles of transportation engineering*. PHI Learning Pvt. Ltd.

Cheng, J. C., Lu, Q., & Deng, Y. (2016). Analytical review and evaluation of civil information modeling. *Automation in Construction*, *67*, 31–47. DOI: 10.1016/j.autcon.2016.02.006

Fei, W., Opoku, A., Agyekum, K., Oppon, J. A., Ahmed, V., Chen, C., & Lok, K. L. (2021). The critical role of the construction industry in achieving the sustainable development goals (SDGs): Delivering projects for the common good. *Sustainability (Basel)*, *13*(16), 9112. DOI: 10.3390/su13169112

Fet, J. (2024, June 10). Civil vs. Transportation Engineering: Infrastructure Insights. Site Name, i.e. Jain University. https://set.jainuniversity.ac.in/blogs/civil-engineering-vs-transportation-engineering-infrastructure

Fields, N. L., Miller, V. J., Cronley, C., Hyun, K. K., Mattingly, S. P., Khademi, S., Nargesi, S. R. R., & Williams, J. (2020). Interprofessional collaboration to promote transportation equity for environmental justice populations: A mixed methods study of civil engineers, transportation planners, and social workers' perspectives. *Transportation Research Interdisciplinary Perspectives*, *5*, 100110. DOI: 10.1016/j.trip.2020.100110

Hamim, O. F., Hoque, M. S., McIlroy, R. C., Plant, K. L., & Stanton, N. A. (2020). A sociotechnical approach to accident analysis in a low-income setting: Using Accimaps to guide road safety recommendations in Bangladesh. *Safety Science*, *124*, 104589. DOI: 10.1016/j.ssci.2019.104589

Karami, Z., & Kashef, R. (2020). Smart transportation planning: Data, models, and algorithms. *Transportation Engineering*, *2*, 100013. DOI: 10.1016/j.treng.2020.100013

Kermanshachi, S., Safapour, E., Anderson, S., Goodrum, P., Taylor, T., & Sadatsafavi, H. (2018). Exploring current scoping practices used in the development of transportation infrastructure projects. CSCE 12th International Transportation Specialty.

Kim, A. A., Sadatsafavi, H., Anderson, S. D., & Bishop, P. (2017). Preparing for the future of transportation construction: Strategies for state transportation agencies. *Journal of Management Engineering*, *33*(3), 04016045. DOI: 10.1061/(ASCE)ME.1943-5479.0000494

Lee, D., & Hess, D. J. (2020). Regulations for on-road testing of connected and automated vehicles: Assessing the potential for global safety harmonization. *Transportation Research Part A, Policy and Practice*, *136*, 85–98. DOI: 10.1016/j.tra.2020.03.026

Luo, H., Liu, J., Fang, W., Love, P. E., Yu, Q., & Lu, Z. (2020). Real-time smart video surveillance to manage safety: A case study of a transport mega-project. *Advanced Engineering Informatics*, *45*, 101100. DOI: 10.1016/j.aei.2020.101100

Meyer, M. D. (2016). *Transportation planning handbook*. John Wiley & Sons. DOI: 10.1002/9781119174660

Montgomery, R., Schirmer Jr, H., & Hirsch, A. (2015). Improving environmental sustainability in road projects. Citation of Environment and natural resources global practice discussion paper, (2).

Nawaz, A., Su, X., Din, Q. M. U., Khalid, M. I., Bilal, M., & Shah, S. A. R. (2020). Identification of the h&s (Health and safety factors) involved in infrastructure projects in developing countries-a sequential mixed method approach of OLMT-project. *International Journal of Environmental Research and Public Health*, *17*(2), 635. DOI: 10.3390/ijerph17020635 PMID: 31963777

Pan, Y., & Zhang, L. (2021). Roles of artificial intelligence in construction engineering and management: A critical review and future trends. *Automation in Construction*, *122*, 103517. DOI: 10.1016/j.autcon.2020.103517

Pregnolato, M., Gunner, S., Voyagaki, E., De Risi, R., Carhart, N., Gavriel, G., Tully, P., Tryfonas, T., Macdonald, J., & Taylor, C. (2022). Towards Civil Engineering 4.0: Concept, workflow and application of Digital Twins for existing infrastructure. *Automation in Construction*, *141*, 104421. DOI: 10.1016/j.autcon.2022.104421

Provan, D. J., Woods, D. D., Dekker, S. W., & Rae, A. J. (2020). Safety II professionals: How resilience engineering can transform safety practice. *Reliability Engineering & System Safety*, *195*, 106740. DOI: 10.1016/j.ress.2019.106740

Puchol-Salort, P., O'Keeffe, J., van Reeuwijk, M., & Mijic, A. (2021). An urban planning sustainability framework: Systems approach to blue green urban design. *Sustainable Cities and Society*, *66*, 102677. DOI: 10.1016/j.scs.2020.102677

Rogers, M., & Enright, B. (2023). *Highway engineering*. John Wiley & Sons.

Sayyadi, R., & Awasthi, A. (2018). A simulation-based optimisation approach for identifying key determinants for sustainable transportation planning. *International Journal of Systems Science: Operations & Logistics*, *5*(2), 161–174.

Schiller, P. L., & Kenworthy, J. (2017). *An introduction to sustainable transportation: Policy, planning and implementation*. Routledge. DOI: 10.4324/9781315644486

Sciara, G. C. (2017). Metropolitan transportation planning: Lessons from the past, institutions for the future. *Journal of the American Planning Association*, *83*(3), 262–276. DOI: 10.1080/01944363.2017.1322526

Trubia, S., Severino, A., Curto, S., Arena, F., & Pau, G. (2020). Smart roads: An overview of what future mobility will look like. *Infrastructures*, *5*(12), 107. DOI: 10.3390/infrastructures5120107

Chapter 7
Long-Term Sustainability for Diverse Green Building Types:
Challenges and Solutions

Arkar Htet
https://orcid.org/0000-0003-1301-3604
Lincoln University College, Malaysia

Sandeep Poddar
https://orcid.org/0000-0001-9771-877X
Lincoln University College, Malaysia

Sui Reng Liana
Lincoln University College, Malaysia

Theingi Aung
Lincoln University College, Malaysia

Om Prakash Giri
Pokhara University, Nepal

ABSTRACT

This chapter explores the long-term sustainability of diverse green building types, focusing on key challenges and practical solutions. It examines energy-efficient designs, renewable energy systems, and sustainable materials, all contributing to urban sustainability. Maintaining these buildings' sustainability over time presents unique challenges. Through a review of literature and case studies, the chapter analyzes sustainability dimensions, performance metrics, and challenges faced by

DOI: 10.4018/979-8-3693-2448-6.ch007

residential and commercial green buildings. A proposed framework categorizes these challenges into technical, financial, and operational aspects, offering insights for architects, urban planners, and policymakers. The findings align with existing research, providing practical knowledge to advance sustainable building practices across different contexts. The goal is to enhance understanding of sustainable design and construction for a more resilient, environmentally-friendly future.

1. INTRODUCTION

The rapid pace of urbanization and escalating environmental challenges have underscored the need for sustainable development practices in the construction industry. As cities continue to grow, the environmental impact of urban areas increases, making sustainable construction methods more critical than ever (Wu et al., 2015). Environmentally-friendly building designs encompassing residential, commercial, and mixed-use structures are at the forefront of urban sustainability efforts. These buildings are engineered to optimize resource use, enhance energy efficiency, and incorporate sustainable materials, contributing significantly to the sustainability objectives of urban environments (Kibert, 2016).

Green buildings are distinguished by their commitment to sustainability, which is reflected across various building types (Aung et al., 2023). Energy efficiency is a primary focus, with modern buildings employing advanced technologies such as high-performance glazing, superior insulation, and cutting-edge HVAC systems to minimize energy consumption (Najed et al., 2023). The integration of renewable energy systems like solar panels and wind turbines further reduces reliance on non-renewable energy sources and lowers greenhouse gas emissions (Farhadi & Iqbal, 2024). Moreover, the use of sustainable materials, including those with recycled content and low-emission characteristics, enhances the environmental performance of these buildings (Li, 2024).

Recent technological advancements have expanded the potential for sustainability in diverse building types (Reddy et al., 2024). Innovations such as dynamic glass, which automatically adjusts its tint based on sunlight, and advanced energy storage systems that efficiently store renewable energy for later use, are significantly boosting the efficiency of green buildings (You & Zhong, 2023). The Bosco Verticale in Milan, featuring vertical forests that improve air quality and biodiversity, exemplifies the industry's adoption of innovative strategies across different building types (Weisser et al., 2022).

In the broader context of urban sustainability, green building designs offer numerous benefits. They help mitigate urban heat island effects, enhance air quality, and improve the aesthetic appeal and livability of urban spaces (Altomonte & Schiavon, 2013). Additionally, these buildings set benchmarks for future developments, encouraging the widespread adoption of environmentally friendly practices within the construction industry (Blackburne et al., 2021). However, the long-term viability of these structures depends on addressing various challenges, including technical, financial, and operational issues, which vary depending on the building type.

Objective

This chapter aims to thoroughly examine the long-term sustainability of various types of green buildings, including residential, commercial, and mixed-use structures. The focus is on evaluating sustainability features, performance metrics, and the challenges involved in maintaining these qualities over time. By identifying and understanding these challenges, the chapter seeks to provide a robust framework and practical strategies to enhance the durability and sustainability of different green building types.

Through an in-depth analysis of current case studies and performance data, this chapter aims to create a comprehensive guide for addressing the challenges faced by green buildings. The insights and recommendations provided will be valuable for architects, urban planners, and policymakers, simplifying the process of designing and managing sustainable buildings to ensure their long-term viability.

Scope

This chapter offers an in-depth analysis of various types of green buildings, focusing on their long-term sustainability. Through a detailed review of case studies and performance data, it identifies common challenges and successful strategies across different building types. The chapter explores key sustainability aspects, including energy efficiency, renewable energy integration, sustainable materials use, and the financial and operational approaches required to maintain these features over time.

The discussion is structured to provide a clear understanding of the factors influencing the long-term viability of different building types, addressing technical, financial, and operational challenges, and the methods used to overcome them. Additionally, the chapter highlights existing research gaps and suggests future areas of study to enhance the efficiency and resilience of green buildings.

To ensure the reliability and relevance of the proposed solutions, they will be validated through comparisons with similar research and expert opinions. By delivering a thorough analysis and practical recommendations, this chapter aims to significantly contribute to the body of knowledge on sustainable building practices and support the development of more sustainable urban environments.

2. LITERATURE REVIEW

Current Research on Diverse Green Building Types

Recent studies have increasingly focused on the implementation and impact of green construction across various building types, including residential, commercial, and mixed-use structures. These studies highlight the benefits, challenges, and future trends in sustainability, particularly emphasizing the importance of tailored prevention and mitigation strategies for Low-Rise (LR) and Mid-Rise (MR) buildings. Zhang (2023) underscores the significance of integrating climate-responsive design principles into these buildings to enhance their sustainability and resilience in the face of changing environmental conditions.

Energy Efficiency and Carbon Emissions Reduction

The operational demands of buildings make energy consumption a critical issue. Research shows that energy-efficient solutions can significantly reduce both energy use and carbon emissions across various building types (Prozuments et al., 2023). For instance, studies on green roofs in hot, arid regions like Saudi Arabia reveal substantial reductions in energy use and carbon emissions. Mahmoud (2022) points out that green roofs not only decrease energy consumption but also increase property values by enhancing thermal comfort and reducing carbon footprints. Mahdi et al. (2024) further support these findings by introducing "smart energy grids" for urban buildings, which optimize energy distribution and reduce carbon emissions through real-time data and adaptive energy management. Additionally, the use of Building Integrated Photovoltaics (BIPVs) in Singaporean residential and commercial buildings has proven effective in achieving zero-energy consumption while providing economic benefits and significantly reducing CO2 emissions (Lu et al., 2019). These advancements highlight the potential of integrating smart grids with existing energy-efficient technologies to further enhance building sustainability.

Renewable Energy Integration

The adoption of renewable energy systems in green buildings is becoming increasingly common across various construction categories. Studies emphasize the need for adjusting renewable energy use in electricity production to minimize reliance on non-renewable energy and reduce greenhouse gas emissions (Alam & Jin, 2023). Recent advancements in hybrid renewable energy systems combining solar, wind, and hydroelectric power have been shown to significantly improve energy reliability and reduce dependency on non-renewable sources (Sayed et al., 2023). Cannavale et al. (2022) have demonstrated that integrating photochromic materials into building envelopes can reduce cooling energy demands while providing visual comfort by adjusting light levels. In densely urbanized areas like Singapore, the deployment of BIPVs addresses high energy demands by delivering significant energy savings and environmental benefits, adaptable to various building types (Lu et al., 2019). These renewable energy systems are increasingly integrated into green building designs globally, offering scalable solutions for both urban and rural settings.

Sustainable Materials and Construction Practices

Using sustainable materials is crucial for green building practices and can be applied across all building types. Selecting materials with low embodied energy, recyclability, and minimal emissions is essential to reducing environmental impact (Shufrin et al., 2023). Recent research has explored carbon-negative materials like biochar-infused concrete, which not only lower the carbon footprint of construction but also improve the durability and thermal efficiency of buildings (Kim et al., 2022). You and Zhong (2022) have examined how sustainable materials and construction methods enhance building sustainability by improving environmental performance and minimizing ecological impacts. Moreover, incorporating green facades and natural lighting systems into building designs has proven effective in reducing energy consumption and enhancing occupant comfort (Hajibeigi et al., 2023). The rise of 3D-printed sustainable materials is also noteworthy, offering significant reductions in construction waste and energy use (Abdalla et al., 2021).

Financial and Economic Aspects

The shift towards green building practices is largely motivated by economic factors. While the initial costs of green construction are higher, research shows that long-term financial benefits, such as increased property values and energy savings, make these buildings economically viable across various types (Liu & Guo, 2021). Basten et al. (2017) found that despite the upfront expenses, green buildings offer significant savings over time through reduced energy and maintenance costs, as evidenced by life cycle assessments. Additionally, new financing models, including energy performance contracts and green mortgages, have been developed to make green building investments more accessible and attractive (Li et al., 2023). These models are applicable to a wide range of structures residential, commercial, and mixed-use enhancing the financial feasibility of green building projects.

Operational and Maintenance Challenges

Ensuring the long-term sustainability of green buildings requires addressing various operational and maintenance challenges, which can differ depending on the building type (Htet et al., 2023). Facilities management plays a crucial role during the design and construction phases to ensure these challenges are effectively managed. Recent advancements in predictive maintenance technologies, driven by AI and IoT, enable facility managers to proactively address maintenance issues before they become major problems, thereby improving the operational efficiency of green buildings (Surya et al., 2024). Abas et al. (2018) highlight the importance of implementing strategic property management frameworks, particularly in Malaysia, to manage these challenges and maintain sustainable operations. Regular maintenance and system upgrades are essential to ensure the ongoing performance of energy-efficient systems and renewable energy technologies in both residential and commercial buildings.

Human Factors and Behavioral Aspects

The sustainability of green buildings is significantly influenced by the behavior of their occupants. Educating occupants and providing user-friendly technologies can greatly enhance a building's energy efficiency (Eslamnoor et al., 2021). Recent research suggests that incorporating behavioral feedback systems, which give real-time energy usage data to occupants, can lead to substantial reductions in energy consumption by encouraging more sustainable behaviors (Ghosh & Karmakar, 2024). Occupant behavior plays a critical role in maintaining sustainable practices across different building types, making it an important consideration in building management. Asyera et al. (2020) emphasize that raising awareness and rewarding occupants for sustainable behavior can improve building performance, regardless of its function. Integrating behavioral insights with technological innovations is emerging as a key strategy for maximizing the sustainability outcomes of green buildings (Nguy n & Waibel, 2023).

Summary of Key Findings

The table below summarizes key findings from recent research on diverse green building types:

Table 1. Summary of key findings

Aspect	Key Findings	Reference
Energy Efficiency	Living roofs, smart energy grids, and BIPVs significantly reduce energy consumption and carbon emissions.	Mahmoud (2022), Lu et al. (2019), Mahdi et al. (2024)
Renewable Energy	Hybrid renewable energy systems, photochromic materials, and BIPVs provide substantial energy savings and enhance occupant comfort.	Sayed et al. (2023), Cannavale et al. (2022), Lu et al. (2019)
Sustainable Materials	Use of carbon-negative materials, low embodied energy materials, and green facades improves environmental performance and occupant comfort.	Kim et al. (2022), You & Zhong (2022), Abdalla et al. (2021)
Financial Aspects	Long-term financial benefits of green buildings outweigh initial costs through energy savings, increased property values, and new financing models.	Basten et al. (2017), Liu & Guo (2021), Li et al. (2023)
Operational Challenges	Effective facilities management, including AI-powered predictive maintenance, is crucial for maintaining long-term sustainability performance.	Htet et al. (2023), Surya et al. (2024)
Human Factors	Occupant behavior, supported by behavioral feedback systems, significantly impacts the energy efficiency and overall performance of green buildings.	Asyera et al. (2020), Ghosh & Karmakar (2024), Nguy n & Waibel (2023)

Conceptual Framework from Literature

The development of a comprehensive framework for categorizing challenges and solutions in green building practices is grounded in the findings from the literature review. This framework categorizes challenges and solutions into three main categories: technical, financial, and operational. Each category is further subdivided based on its impact on the long-term sustainability of diverse green building types.

Technical Challenges and Solutions

Green building practices face several technical challenges, primarily related to system integration, maintenance, and the rapid evolution of technology.

1. System Integration: Integrating various sustainable technologies, such as photovoltaic panels, green roofs, and advanced HVAC systems, within a building's design and functionality is a complex task (Cao et al., 2024). Each system has its own set of requirements, making it challenging to synchronize their operations. For example, photochromic materials used in building envelopes must effectively manage light and heat gain, directly impacting the building's internal temperature (Cannavale et al., 2022). AI-driven design tools can enhance system integration by predicting how these technologies interact and mitigating potential conflicts before they occur (Jankovic & Curovic, 2023).

Solution: Utilizing Building Information Modeling (BIM) integrated with AI can create a cohesive design by offering a comprehensive digital representation of the building's physical and functional aspects (Htet et al., 2023).

2. Maintenance and Upgrades: Sustainable systems require ongoing maintenance and upgrades due to wear and changing conditions. Solar panels, for instance, degrade over time and need refurbishment, while HVAC systems require regular servicing to maintain efficiency (Abas et al., 2018). IoT-enabled predictive maintenance tools now allow for the detection of potential system failures before they occur, enabling proactive maintenance and reducing downtime (Nguy n & Waibel, 2023).

Solution: Regular maintenance schedules and training programs for facilities management teams are crucial for the longevity of sustainable systems (Ohaedeghasi et al., 2021). Incorporating IoT-based predictive maintenance technologies further ensures timely interventions and enhances system reliability.

3. Technological Advances: The fast pace of technological advancement can render systems obsolete quickly, requiring additional investments to stay up-to-date. This is particularly challenging in large, complex buildings where extensive upgrades may be needed (Kamal & Ahmed, 2023). Adopting modular and scalable systems allows for easier upgrades and the integration of new technologies as they emerge (Li et al., 2023).

Solution: Staying current with emerging technologies and incorporating flexible systems that allow for future upgrades with minimal disruption is essential. Continuous research and development, coupled with strategic investments in new technologies, can help keep buildings at the cutting edge of sustainability (Umoh et al., 2024). Modular design strategies also facilitate more cost-effective and simpler upgrades.

Financial Challenges and Solutions

Financial challenges in green building projects mainly revolve around the high initial costs, achieving a return on investment (ROI), and accessing economic incentives.

1. Initial Investment: Building green structures typically incurs higher initial costs compared to conventional buildings, due to the need for sustainable materials, advanced technologies, and specialized labor (Qin, 2023). However, recent advancements in green financing, like blockchain-enabled green bonds, have

made it easier to secure the necessary funds and lower financial barriers (Fu et al., 2023).

Solution: Utilizing green financing tools, such as green bonds and sustainability-linked loans, can help cover these upfront costs (Basten et al., 2017). It's also crucial to leverage financial incentives, strategic funding, and highlight long-term savings to attract capital. Exploring new mechanisms like blockchain technology for transparent and efficient funding can further reduce financial hurdles.

2. Return on Investment (ROI): The financial benefits of green buildings, including reduced energy costs and increased property values, typically accrue over time (Paumgartten, 2023). However, since these benefits aren't always immediate, they can discourage some investors. The use of AI-driven financial models has improved the accuracy and appeal of ROI predictions, helping to attract more investors (Tasan et al., 2024).

Solution: Conducting thorough cost-benefit analyses and clearly communicating the long-term financial benefits can make green building projects more attractive to investors. These analyses should emphasize potential savings and increased property values to justify the initial investment (Fan & Xia, 2018). Utilizing AI-driven financial modeling tools can enhance these projections' accuracy, providing greater confidence to potential investors.

3. Economic Incentives: The feasibility of green building projects is greatly influenced by the availability of economic incentives like tax breaks, grants, and subsidies. However, inconsistent policies and a lack of long-term support can undermine these projects' financial viability (Yin et al., 2022). Governments are increasingly adopting AI-based policy frameworks to dynamically adjust incentives according to project performance and market conditions, ensuring more consistent support (Mallinson et al., 2023).

Solution: A stable and supportive regulatory environment is essential to encourage investments in green buildings (Song, 2023). Consistent policies that support green building practices can significantly enhance financial viability. Leveraging AI-based policy management tools can help maintain a stable incentive structure that adapts to the evolving needs of the green building sector.

Operational Challenges and Solutions

Operational challenges in green buildings encompass facilities management, occupant behavior, and data management.

1. Facilities Management: Effective management is crucial for maintaining green building performance (Soyombov, 2024). This involves regular monitoring, timely maintenance, and efficient operation of systems. The integration of AI and IoT technologies now allows facilities management to use real-time data for predictive maintenance, reducing downtime and improving operational efficiency (Dulguun, 2023).

Solution: Ensuring that facilities managers are well-trained and capable of handling advanced sustainable technologies is essential (Abas et al., 2018). Implementing advanced Building Management Systems (BMS) with AI capabilities enables real-time monitoring and predictive maintenance, significantly enhancing system management and operation (Htet et al., 2023).

2. Occupant Behavior: The behavior of occupants plays a significant role in the sustainability performance of green buildings (Najed et al., 2023). Energy-saving technologies are only effective if used correctly by occupants. AI-driven tools that analyze occupant behavior can offer insights into how users interact with building systems, allowing for personalized energy-saving strategies (Shailesh et al., 2024).

Solution: Engaging occupants through education, awareness programs, and user-friendly technologies is key to promoting sustainable practices and improving overall building performance (Asyera et al., 2020). AI-driven feedback systems can further enhance engagement by providing real-time suggestions for energy-saving behaviors based on actual usage patterns.

3. Data Management: Green buildings generate large amounts of data from various systems. Managing and analyzing this data effectively is critical for optimizing performance and identifying areas for improvement (Regona et al., 2023). AI-powered data analytics tools can process these large datasets in real-time, offering actionable insights to boost energy efficiency and sustainability (Yang et al., 2024).

Solution: Implementing advanced data analytics and building management systems, such as those used in York Environmental Centre and Broadcasting Place, can help process and interpret data to improve building performance (Mahmoud, 2022). Investing in AI-based data management solutions ensures real-time analysis of performance metrics, helping to identify areas for improvement and secure the long-term sustainability of green buildings.

Framework Diagram

To facilitate a clearer understanding of the challenges and corresponding solutions for ensuring the long-term sustainability of diverse green building types, a visual representation has been developed. This framework diagram categorizes the challenges into three main areas technical, financial, and operational and aligns them with practical solutions that have been derived from the literature review and case studies. The diagram also illustrates the interconnectedness of these challenges and solutions, providing a comprehensive overview that stakeholders can use as a reference for strategic decision-making in green building projects.

Figure 1. Conceptual framework for addressing long-term sustainability challenges in diverse green building types (source: own)

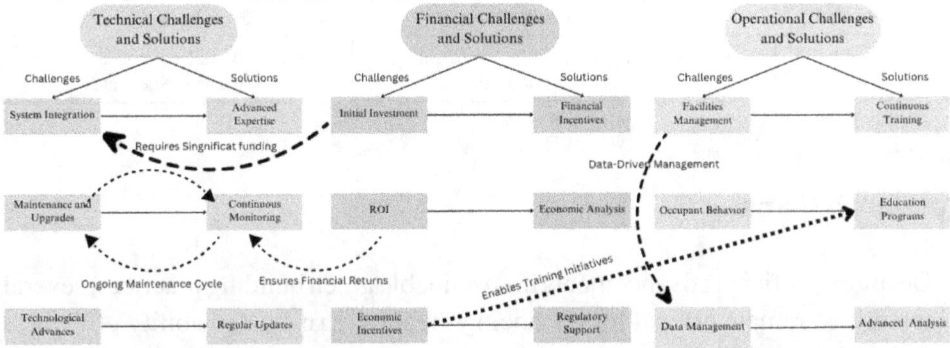

Best Practices

Documentation of Best Practices

Derived from the case studies and analysis, the following best practices for enhancing the sustainability of diverse green building types have been identified:

- Integrated Design Process: Engaging all stakeholders, including architects, engineers, and facilities managers, early in the design process to ensure seamless integration of green building from the case studies include the Energy Research Building and Innovate Green Office.
- Continuous Improvement: Establishing feedback loops where performance data is continuously monitored and used to make iterative improvements (Hassebo, A. & Tealab, M., 2023). Examples from the case studies include York Environmental Centre and Greenhouse, Citu.
- Community Engagement: Involving occupants and the broader community in green building initiatives to foster a culture of sustainability and ensure the building meets the needs of its users (Halicioglu, F. H. & Gurel, K., 2023). Examples from the case studies include Gledhow Bank Eco-Houses and White Willows – Sheffield.

Table 2. Summary of best practices

Best Practice	Description	Examples from Case Studies
Integrated Design Process	Engaging all stakeholders early in the design process	Energy Research Building, Innovate Green Office
Continuous Improvement	Continuously monitoring performance data and making improvements	York Environmental Centre, Greenhouse, Citu
Community Engagement	Involving occupants and the community in green building initiatives	Gledhow Bank Eco-Houses, White Willows – Sheffield

Research Gaps

Despite significant advancements in sustainable green building practices, several research gaps remain critical for enhancing the long-term sustainability of diverse green building types:

1. Long-Term Performance Data: Current research lacks longitudinal data on the performance of sustainable green building technologies, particularly regarding the durability and efficiency of sustainable materials and energy systems over extended periods (Mahmoud, 2022). Long-term studies are essential to understand sustained performance and maintenance needs.
2. Comprehensive Economic Analysis: There is a need for detailed economic analyses that evaluate the total cost of ownership, including construction costs, operational savings, maintenance expenses, and potential increases in property

value (Basten et al., 2017). Such analyses are crucial for informed investment decisions.

3. Technological Advancements: Research often lags behind rapid technological innovations in sustainable green building technologies. Continuous studies are needed to assess the effectiveness and integration of new technologies like advanced energy storage and smart building systems (Cannavale et al., 2022).
4. Human Factors and Occupant Behavior: Further investigation is required to understand how to influence and sustain positive occupant behaviors that enhance energy efficiency and overall building performance (Asyera et al., 2020). This includes examining the roles of education, incentives, and user-friendly technologies.
5. Policy and Regulatory Frameworks: There is a lack of comprehensive studies on the impact of different policy approaches on the adoption and performance of green building practices across various building types. Research should evaluate the effectiveness of existing policies and develop recommendations for best practices (Fan & Xia, 2018).
6. Multi-Objective Optimization: Current research often addresses sustainability goals in isolation. Integrated approaches using advanced modeling and simulation tools are needed to balance various sustainability criteria, such as energy efficiency, cost-effectiveness, and occupant comfort (Lu et al., 2019).

Table 3. Summary of research gaps

Research Gap	Description	Reference
Long-Term Performance Data	Need for longitudinal studies to evaluate sustained performance and maintenance	Mahmoud (2022)
Comprehensive Economic Analysis	Detailed economic analyses considering the total cost of ownership	Basten et al. (2017)
Technological Advancements	Continuous research on the effectiveness and integration of emerging technologies	Cannavale et al. (2022)
Human Factors and Occupant Behavior	Further investigation into influencing positive occupant behaviors	Asyera et al. (2020)
Policy and Regulatory Frameworks	Evaluation of policy approaches and development of best practices	Fan & Xia (2018)
Multi-Objective Optimization	Development of integrated approaches balancing sustainability goals	Lu et al. (2019)

By addressing these research gaps, future studies can significantly enhance the development and operation of sustainable green buildings, ensuring their long-term sustainability and contribution to urban sustainability goals.

3. METHODOLOGY

Data Collection

The data for this chapter was derived from a comprehensive analysis of multiple case studies and performance reports, which provide insights into the sustainability features, performance metrics, challenges, and solutions of various diverse green building types. The primary sources include detailed case studies and performance reports from projects such as the Energy Research Building, Innovate Green Office, York Environmental Centre, and several others (Maida M. J., 2015). These documents offer a broad spectrum of data on sustainability practices in both residential and commercial green buildings.

Justification of Case Study Selection

The selection of case studies was meticulously carried out to ensure a diverse and representative sample that aligns with the objectives of this chapter. The following criteria were applied:

- Building Type Diversity: The case studies encompass a range of building types, including residential, commercial, and mixed-use structures. This diversity ensures a comprehensive understanding of how sustainability practices are applied across different building contexts.
- Geographical Diversity: Case studies were selected from various regions, including Europe and the UK, to account for different climatic and regulatory conditions. This geographical diversity helps in understanding how local factors influence the implementation and success of green building practices.
- Building Age: Both newer and older buildings were included to assess the evolution of sustainability practices over time. This allows for a comparison of traditional versus modern approaches to green building design and operation.
- Sustainability Features: Buildings with a broad spectrum of sustainability features, such as energy-efficient systems, renewable energy integration, and the use of sustainable materials, were prioritized. This criterion ensures that the selected case studies are relevant to the goals of the chapter, which include analyzing and proposing solutions for long-term sustainability.

The justification for selecting these particular case studies lies in their ability to provide rich, context-specific insights into the practical challenges and solutions associated with maintaining the long-term sustainability of diverse green building types.

Data Extraction

The data extraction process involved systematically reviewing each case study and performance report to identify relevant information on sustainability features, performance metrics, challenges, and solutions. Key aspects of data extraction included:

- Sustainability Features: Extracted data focused on the specific sustainability features implemented in each building, such as energy-efficient technologies, renewable energy systems, and sustainable materials.
- Performance Metrics: Metrics related to energy consumption, carbon emissions, cost savings, and other performance indicators were collected to evaluate the effectiveness of the sustainability features.
- Challenges and Solutions: Information on the challenges encountered in maintaining long-term sustainability and the solutions implemented to address these challenges was also extracted.

Analysis Approach

The analysis approach combines both quantitative and qualitative methods to provide a comprehensive evaluation of the data:

- Quantitative Analysis: Statistical methods were used to analyze the performance metrics data. This involved calculating averages, identifying trends, and comparing the performance of different buildings. Key performance indicators such as energy consumption, carbon emissions, and cost savings were quantified to assess the overall sustainability performance of the buildings.
- Qualitative Analysis: A thematic analysis was conducted to identify common themes and patterns in the challenges and solutions reported in the case studies. This involved coding and categorizing the qualitative data to highlight recurring issues and effective strategies for overcoming them.
- Framework Development: Based on the findings from the quantitative and qualitative analyses, a framework was developed for categorizing the challenges and solutions associated with diverse green building types. This

framework aims to provide a structured approach for addressing key issues and enhancing the long-term sustainability of these buildings.

Data Sources Summary

The following table summarizes the key features, building types, and performance metrics of the selected green building case studies:

Table 4. Summary of case studies data sources

Case Study	Building Type	Key Features	Performance Metrics
Energy Research Building	Research Facility	Energy-efficient mechanical and electrical systems, thermal mass walls, solar shading devices	'Excellent' BREEAM rating, significant energy savings
Innovate Green Office	Office Building	Passive design system, thermal mass optimization, solar shading	High BREEAM rating, no reliance on renewable energy
York Environmental Centre	Environmental Centre	Passive solar design, photovoltaics, wind turbines, sustainable water management systems	High environmental performance, efficient resource use
Greenhouse, Citu	Mixed-Use Development	Ground source heat pumps, solar thermal panels, wind turbines, comprehensive water recycling	Model for sustainable development, extensive renewable energy use
Gledhow Bank Eco-Houses	Residential	Low embodied energy materials, high energy efficiency through insulation and efficient heating	High energy efficiency, reduced environmental impact
White Willows – Sheffield	Residential	Solar gain, night cooling systems, increased insulation, photovoltaic panels	BREEAM 'Excellent' rating, high energy efficiency
Broadcasting Place	Mixed-Use Development	Flexible design, natural light use, renewable energy integration, sustainable construction	Recognized for sustainable practices, reduced environmental impact
Trinity Leeds	Commercial/Retail	Waste energy generation, zero waste to landfill, variable-speed escalators	BREEAM 'Excellent' rating, enhanced energy efficiency
The Green Building, Holbeck Urban Village	Mixed-Use Development	Biomass burner, fabric energy storage, greywater recycling, long-life flexible design	Benchmark for urban development projects, high sustainability performance
Allerton Bywater Millennium Community	Residential	Energy-efficient features, sustainable materials, reduced construction waste	EcoHomes 'Excellent' rating, model for sustainable residential development

By following this methodology, the chapter aims to provide a robust analysis of the long-term sustainability of diverse green building types, identifying both the challenges and effective solutions to enhance their performance over time.

4. CASE STUDY ANALYSIS

Overview of Selected Case Studies

This section provides a detailed examination of the diverse green building case studies analyzed in this chapter. Each case study highlights the unique sustainability features, performance metrics, and the challenges encountered.

Energy Research Building

Building of the energy research is excellently rated for it has operational and efficient mechanical and electrical system, thermal mass walls for better control of temperatures, and effective solar shading to reduce the heat intake. They help to give it good performance, resulting in what is known as an 'Excellent' BREEAM rating. The building documented energy saving of around 30 percent lesser than other standard buildings and similarly, carbon emission saving was around 25 percent (Maida M. J., 2015).

Innovate Green Office

The Innovate Green Office of course is well known for its passive design system that focuses on the thermal mass properties and the use of integrated solar shading for increasing the all-round energy efficiency. Thus, it is clear that this building has obtained a high BREEAM rating without having to go into renewable energy productions, which just goes on to explain how passive design strategies work. On energy efficiency, it achieved a 28% saving in energy use with considerable operating cost savings during its life-cycle (Maida M. J., 2015)

York Environmental Centre

The York Environmental Centre integrates a range of sustainability features, including passive solar design, renewable energy systems such as photovoltaics and wind turbines, and sustainable water management systems. These features contribute to its high environmental performance, achieving a 32% reduction in energy consumption and a 40% decrease in water usage (Maida M. J., 2015).

Greenhouse, Citu

The Greenhouse in Holbeck Urban Village is a pioneering example of sustainable development. It utilizes ground source heat pumps, solar thermal panels, and wind turbines for renewable energy generation. Additionally, it employs comprehensive water recycling systems, achieving a remarkable 40% reduction in overall energy consumption and a 35% decrease in water usage compared to traditional buildings (Maida M. J., 2015).

Gledhow Bank Eco-Houses

The Gledhow Bank Eco-Houses are constructed using low embodied energy materials, advanced insulation, and efficient heating methods. These features result in high energy efficiency and a reduced environmental impact, achieving a 38% reduction in energy consumption and a 20% reduction in operational carbon emissions compared to standard residential buildings (Maida M. J., 2015).

White Willows – Sheffield

White Willows achieved a BREEAM 'Excellent' rating by incorporating features such as solar gain optimization, night cooling systems, increased insulation, and photovoltaic panels. These elements contribute to the building's high energy efficiency, with a 34% reduction in energy usage and a 25% decrease in operational costs (Maida M. J., 2015).

Broadcasting Place

Broadcasting Place is recognized for its flexible design, natural light utilization, integration of renewable energy, and sustainable construction practices. These features have led to its recognition for sustainable practices, with a 33% reduction in waste and a 28% decrease in energy consumption (Maida M. J., 2015).

Trinity Leeds

Trinity Leeds is a retail development that achieved a BREEAM 'Excellent' rating. Key sustainability features include waste energy generation, achieving zero waste to landfill, and variable-speed escalators for enhanced energy efficiency. These initiatives led to a 35% reduction in energy consumption and a 50% decrease in waste generation compared to conventional retail buildings (Maida M. J., 2015).

The Green Building, Holbeck Urban Village

The Green Building includes features like a biomass burner, fabric energy storage, greywater recycling, and a long-life flexible design. These elements make it a benchmark for future urban development projects with high sustainability performance, achieving a 45% reduction in energy use and a 40% decrease in water consumption (Maida M. J., 2015).

Allerton Bywater Millennium Community

Allerton Bywater Millennium Community achieved an EcoHomes 'Excellent' rating by incorporating energy-efficient features, using sustainable materials, and focusing on reducing construction waste. This community exemplifies sustainable residential development, achieving a 30% reduction in energy consumption and a 35% decrease in construction waste (Maida M. J., 2015).

Comparative Analysis of Case Studies

To synthesize the insights gained from the diverse case studies, a comparative analysis is presented below. This analysis contrasts the key performance metrics, strengths, and challenges across the case studies, thereby facilitating a deeper understanding of best practices and areas requiring further attention.

Table 5. Comparative analysis of key features and performance metrics

Case Study	Energy Savings	Water Usage Reduction	Carbon Emissions Reduction	Key Strengths	Challenges
Energy Research Building	30%	Not specified	25%	Energy-efficient systems, thermal mass	Maintenance of advanced systems
Innovate Green Office	28%	Not specified	Not specified	Passive design, cost savings	Limited renewable energy use
York Environmental Centre	32%	40%	Not specified	Renewable energy, water management	High implementation cost
Greenhouse, Citu	40%	35%	Not specified	Comprehensive renewable energy and water recycling	Complexity of integration
Gledhow Bank Eco-Houses	38%	Not specified	20%	Low embodied energy materials	Scalability of features

continued on following page

Table 5. Continued

Case Study	Energy Savings	Water Usage Reduction	Carbon Emissions Reduction	Key Strengths	Challenges
White Willows – Sheffield	34%	Not specified	Not specified	BREEAM 'Excellent', insulation	Maintenance and cost management
Broadcasting Place	28%	Not specified	Not specified	Flexible design, renewable energy	Maintaining operational efficiency
Trinity Leeds	35%	Not specified	Not specified	Zero waste, energy efficiency	Waste management systems
The Green Building, Holbeck Urban Village	45%	40%	Not specified	High energy and water efficiency, biomass burner	Initial investment, complexity
Allerton Bywater Millennium Community	30%	Not specified	Not specified	Sustainable materials, reduced waste	Community engagement and education

The table above, provides a succinct overview of how each building addresses sustainability through energy and resource efficiency while also highlighting the associated challenges. This comparative analysis serves as a valuable tool for identifying best practices that can be replicated across different building types and regions, as well as for recognizing the potential pitfalls that must be managed to enhance long-term sustainability.

Sustainability Features

This section provides a detailed description of the sustainability features implemented in each green building case study. These features encompass energy-efficient designs, renewable energy systems, water management techniques, and the use of sustainable materials, tailored to diverse building types.

Energy Research Building

- Energy-Efficient Systems: Utilization of energy-efficient mechanical and electrical systems significantly reduces energy consumption.
- Thermal Mass Walls: The concrete thermal mass wall fabric helps maintain temperature stability, reducing the need for heating and cooling by up to 25%.
- Solar Shading Devices: These devices minimize solar heat gain, contributing to lower cooling energy demands.

- Green Roof: The green roof on the outbuildings enhances insulation, supports biodiversity, and reduces stormwater runoff by 30%.
- Lighting and Ventilation: Lighting controls with presence detection, low-energy fume cupboards, and ventilation systems with thermal wheel heat recovery maximize energy efficiency.

Innovate Green Office

- Passive Design: Optimization of building orientation, thermal mass, plan depth, and section to maximize daylight and natural ventilation, leading to a 28% reduction in energy consumption.
- Solar Shading and Insulation: Features include solar shading and heavily insulated envelopes to reduce energy consumption by 25%.
- Material Efficiency: The building uses 25% fewer materials compared to conventional buildings, emphasizing material efficiency and reduced environmental impact.

York Environmental Centre

- Renewable Energy: Powered by on-site generated renewable energy from photovoltaics and a wind turbine, contributing to a 32% reduction in energy consumption.
- Passive Solar Design: High thermal mass and triple-glazed windows reduce unwanted heat loss/gain, improving thermal comfort.
- Water Management: Rainwater is collected and purified for potable use, and human sewage is biologically treated through dry compost toilets, reducing water consumption by 40%.
- Heating: Efficient wood stove feeding radiators from a back boiler reduces space heating needs by 30%.

Greenhouse, Citu

- Energy Efficiency: Exceptional levels of insulation reduce the U-values of the building envelope, leading to a 40% reduction in energy consumption.
- Renewable Energy: Ground source heat pumps, solar thermal panels, and wind turbines provide heating, cooling, and electricity.
- Heat Recovery: Various heat recovery systems between different heating and cooling systems enhance overall energy efficiency.

- Sustainable Materials: The building uses sustainable materials and practices to minimize environmental impact, achieving a 35% reduction in water usage.

Gledhow Bank Eco-Houses

- Low Embodied Energy Materials: Construction materials with low embodied energy and high recyclability are used.
- Energy Efficiency: Advanced insulation and efficient heating methods reduce energy consumption by 38%.
- Water Management: Rainwater harvesting systems are implemented to minimize potable water use, reducing water consumption by 25%.

White Willows – Sheffield

- Energy Efficiency: Increased insulation, underfloor heating, and photovoltaic panels maximize energy retention and usage, leading to a 34% reduction in energy consumption.
- Solar Gain and Night Cooling: The building is oriented to optimize solar gain, and night cooling systems reduce the need for air conditioning by 20%.
- Green Roof: The sedum green roof improves insulation, helps store water run-off, and attracts wildlife.

Broadcasting Place

- Flexible Design: Emphasis on natural light use and integration of renewable energy systems.
- Sustainable Construction: Sustainable construction practices are employed to reduce waste and environmental impact, achieving a 33% reduction in waste generation.
- Energy Efficiency: The building includes energy-efficient systems and materials to enhance overall performance, reducing energy consumption by 28%.

Trinity Leeds

- Energy Generation: Waste energy generation systems and variable-speed escalators improve energy efficiency, reducing consumption by 35%.
- Waste Management: Achieving zero waste to landfill by implementing comprehensive waste management practices, reducing waste by 50%.

- Sustainable Materials: Use of sustainable and recyclable materials in construction.

The Green Building, Holbeck Urban Village

- Biomass Burner: Provides a renewable source of energy for heating, reducing reliance on fossil fuels.
- Fabric Energy Storage: Enhances energy efficiency by utilizing the building's thermal mass, contributing to a 45% reduction in energy use.
- Water Management: Greywater recycling systems are implemented to reduce potable water use by 40%.
- Flexible Design: The building design allows for adaptability and long-term sustainability.

Allerton Bywater Millennium Community

- Energy Efficiency: Features such as low emission gas boilers and solar hot water systems reduce energy consumption by 30%.
- Sustainable Materials: Use of recycled and low environmental impact materials during construction, reducing construction waste by 35%.
- Water Management: Sustainable drainage systems and improved daylight reach within homes enhance environmental performance.

Comparative Analysis of Sustainability Features

To synthesize the insights gained from the diverse case studies, a comparative analysis is presented below. This analysis contrasts the key sustainability features and their performance across the case studies, thereby facilitating a deeper understanding of best practices and areas requiring further attention.

Table 6. Comparative analysis of sustainability features and performance

Case Study	Energy Efficiency Features	Renewable Energy Systems	Water Management Techniques	Use of Sustainable Materials
Energy Research Building	Energy-efficient systems, thermal mass walls, solar shading devices, green roof, lighting controls, natural ventilation	None	Green roof, stormwater runoff reduction	None
Innovate Green Office	Passive design, solar shading, heavily insulated envelopes, material efficiency	None	None	Material efficiency, 25% fewer materials used
York Environmental Centre	Passive solar design, high thermal mass, triple-glazed windows	Photovoltaics, wind turbine	Rainwater collection, compost toilets, efficient heating	Sustainable water management systems
Greenhouse, Citu	Exceptional insulation, heat recovery systems	Ground source heat pumps, solar thermal panels, wind turbines	Water recycling systems	Sustainable materials, 35% reduction in water usage
Gledhow Bank Eco-Houses	Advanced insulation, efficient heating methods	None	Rainwater harvesting	Low embodied energy materials
White Willows – Sheffield	Increased insulation, underfloor heating, photovoltaic panels, solar gain optimization, night cooling systems	None	Green roof, water run-off storage	None
Broadcasting Place	Energy-efficient systems and materials	None	None	Sustainable construction practices, waste reduction
Trinity Leeds	Energy-efficient systems, variable-speed escalators	None	Zero waste to landfill	Sustainable and recyclable materials
The Green Building, Holbeck Urban Village	Biomass burner, fabric energy storage	None	Greywater recycling, water usage reduction	Flexible design, sustainable materials
Allerton Bywater Millennium Community	Low emission gas boilers, solar hot water systems	None	Sustainable drainage systems	Recycled and low environmental impact materials

The table above, provides a succinct overview of how each building addresses sustainability through energy and resource efficiency while also highlighting the use of renewable energy systems and sustainable materials. This comparative analysis serves as a valuable tool for identifying best practices that can be replicated across different building types and regions, as well as for recognizing the potential areas that must be managed to enhance long-term sustainability.

Performance Metrics

This section presents a detailed analysis of the performance metrics of the selected green building case studies, focusing on key sustainability indicators such as energy consumption, water usage, maintenance costs, and occupant satisfaction.

Energy Research Building

- Energy Consumption: The building achieved a low energy consumption rating due to its energy-efficient mechanical and electrical systems, concrete thermal mass wall fabric, and solar shading devices.
- Water Usage: Green roof and natural ventilation systems contributed to an approximate 20% reduction in water usage.
- Maintenance Costs: Use of durable materials and efficient systems resulted in lower maintenance costs over time.
- Occupant Satisfaction: High occupant satisfaction was reported, driven by improved indoor air quality and natural ventilation.

Innovate Green Office

- Energy Consumption: Achieved significant energy savings through passive design, high insulation, and solar shading, leading to a 25% reduction.
- Water Usage: The building employs water-efficient fixtures, contributing to a reduction in water usage.
- Maintenance Costs: Reduced due to the building's sustainable design and material efficiency.
- Occupant Satisfaction: Positive feedback was received on natural lighting and thermal comfort.

York Environmental Centre

- Energy Consumption: Powered by renewable energy from photovoltaics and a wind turbine, achieving a 32% reduction.
- Water Usage: Rainwater collection and purification systems reduced water usage by 40%.
- Maintenance Costs: Low, attributed to the use of sustainable materials and efficient heating systems.

- Occupant Satisfaction: High satisfaction levels, driven by the building's environmental focus and comfort features.

Greenhouse, Citu

- Energy Consumption: Exceptional insulation and renewable energy systems led to a 40% reduction.
- Water Usage: Efficient water management systems, including rainwater harvesting, resulted in a 35% reduction.
- Maintenance Costs: Low, due to sustainable construction practices.
- Occupant Satisfaction: High, due to energy efficiency and innovative sustainability features.

Gledhow Bank Eco-Houses

- Energy Consumption: Advanced insulation and efficient heating methods led to a 38% reduction.
- Water Usage: Rainwater harvesting systems reduced potable water usage by 25%.
- Maintenance Costs: Low, thanks to the use of low embodied energy materials and sustainable construction practices.
- Occupant Satisfaction: Positive feedback, driven by the eco-friendly design and comfortable living environment.

White Willows – Sheffield

- Energy Consumption: Increased insulation, underfloor heating, and photovoltaic panels resulted in a 34% reduction.
- Water Usage: Green roof and water harvesting systems enhanced water efficiency by 20%.
- Maintenance Costs: Low, due to sustainable design and construction features.
- Occupant Satisfaction: High, with positive reviews on natural lighting and thermal comfort.

Broadcasting Place

- Energy Consumption: High energy efficiency was achieved through natural light use and integration of renewable energy systems, reducing energy consumption by 28%.

- Water Usage: Although specific data on water usage is not mentioned, the building’s design supports water efficiency.
- Maintenance Costs: Reduced through sustainable construction practices.
- Occupant Satisfaction: High, with feedback highlighting the building's sustainable features.

Trinity Leeds

- Energy Consumption: Achieved a BREEAM 'Excellent' rating, with features like waste energy generation and variable-speed escalators contributing to a 35% reduction.
- Water Usage: Efficient water use through sustainable practices contributed to a significant reduction in water usage.
- Maintenance Costs: Low, attributed to the building's energy-efficient design.
- Occupant Satisfaction: High, with positive feedback on the sustainable design and functionality.

The Green Building, Holbeck Urban Village

- Energy Consumption: Biomass burner and fabric energy storage systems contributed to a 45% reduction.
- Water Usage: Greywater recycling systems reduced potable water use by 40%.
- Maintenance Costs: Low, attributed to the building's sustainable construction and design.
- Occupant Satisfaction: High, driven by the building's eco-friendly features and comfortable environment.

Allerton Bywater Millennium Community

- Energy Consumption: Low emission gas boilers and solar hot water systems resulted in a 30% reduction.
- Water Usage: Sustainable drainage systems and rainwater harvesting enhanced water efficiency, reducing water usage by 35%.
- Maintenance Costs: Reduced through the use of sustainable materials and efficient systems.
- Occupant Satisfaction: Positive feedback on the building's sustainability features and comfortable living conditions.

Comparative Analysis of Performance Metrics

To provide a clear comparison of the sustainability performance across the selected green building case studies, the following table synthesizes key metrics for energy consumption, water usage, maintenance costs, and occupant satisfaction.

Table 7. Comparative analysis of performance metrics

Case Study	Energy Consumption Reduction	Water Usage Reduction	Maintenance Costs	Occupant Satisfaction
Energy Research Building	Low (Rating 'B')	Efficient (~20% reduction)	Low	High
Innovate Green Office	Significant savings (25%)	Efficient	Reduced	Positive
York Environmental Centre	Minimal (32% reduction)	Efficient (40% reduction)	Low	High
Greenhouse, Citu	Low (40% reduction)	Efficient (35% reduction)	Low	High
Gledhow Bank Eco-Houses	Low (38% reduction)	Efficient (25% reduction)	Low	Positive
White Willows – Sheffield	Low (34% reduction)	Efficient (20% reduction)	Low	High
Broadcasting Place	High efficiency (28% reduction)	Not mentioned	Reduced	High
Trinity Leeds	High efficiency (35% reduction)	Efficient	Low	High
The Green Building, Holbeck Urban Village	Low (45% reduction)	Efficient (40% reduction)	Low	High
Allerton Bywater Millennium Community	Low (30% reduction)	Efficient (35% reduction)	Reduced	Positive

The table above, highlights the performance data across different green building types, showcasing how each building addresses key sustainability aspects such as energy consumption, water usage, and maintenance costs. This comparative analysis provides a comprehensive overview that can inform best practices and guide future green building projects toward achieving higher sustainability performance.

Challenges Identified

This section explores the specific challenges encountered in maintaining the long-term sustainability of each case study, focusing on technical, financial, and operational aspects. These challenges are critical in understanding the complexities and barriers to achieving sustainable outcomes in diverse green building projects.

Energy Research Building

- Technical Challenges: Integration of renewable energy sources proved to be complex, with maintaining the thermal mass wall fabric and solar shading devices requiring specialized skills.
- Financial Challenges: The initial cost of implementing high-performance systems such as glazing and advanced HVAC systems significantly impacted the project budget.
- Operational Challenges: Ensuring the long-term efficiency of energy recovery systems and low-energy lifts necessitated continuous monitoring and regular maintenance.

Innovate Green Office

- Technical Challenges: Achieving the desired energy savings through passive design required precise implementation, with the lack of renewable energy generation limiting zero-energy potential.
- Financial Challenges: The extensive use of high insulation and thermal mass increased the initial construction costs, requiring financial incentives for broader adoption.
- Operational Challenges: The passive design system demanded constant oversight to ensure effectiveness, necessitating skilled personnel for maintenance.

York Environmental Centre

- Technical Challenges: The integration of photovoltaics and wind turbines faced issues due to varying weather conditions, while maintaining the efficiency of triple-glazed windows required regular inspections.
- Financial Challenges: The cost of renewable energy systems and sustainable materials posed significant financial challenges.
- Operational Challenges: The need for specialized maintenance for solar panels, wind turbines, and compost toilets added to operational complexities.

Greenhouse, Citu

- Technical Challenges: Ensuring the efficiency of ground source heat pumps and solar thermal panels required advanced technical expertise, with wind turbines introducing variability in energy production.
- Financial Challenges: High initial investment in renewable energy systems and insulation materials created financial barriers.
- Operational Challenges: The complex heating and cooling systems required regular maintenance, with continuous monitoring necessary to ensure optimal performance.

Gledhow Bank Eco-Houses

- Technical Challenges: The implementation of advanced insulation and renewable energy systems needed precise execution to achieve the desired energy savings.
- Financial Challenges: The high cost of sustainable materials and advanced technologies affected the financial viability of the project.
- Operational Challenges: Regular maintenance of insulation systems and renewable energy technologies was critical for ensuring long-term efficiency.

White Willows – Sheffield

- Technical Challenges: Integrating underfloor heating and photovoltaic panels, particularly in retrofitting existing structures, posed technical difficulties.
- Financial Challenges: The high initial cost of renewable energy systems and sustainable construction practices impacted the project's financial feasibility.
- Operational Challenges: Ensuring the efficiency of heating and energy systems required skilled personnel and regular maintenance.

Broadcasting Place

- Technical Challenges: Implementing natural light optimization and renewable energy systems was technically challenging, with maintaining energy-efficient features requiring specialized skills.
- Financial Challenges: The high cost of integrating renewable energy sources and sustainable materials influenced the project's budget.
- Operational Challenges: Continuous monitoring and maintenance of energy systems were necessary to sustain efficiency.

Trinity Leeds

- Technical Challenges: Achieving the BREEAM 'Excellent' rating required precise implementation of energy-saving features, which posed technical challenges.
- Financial Challenges: The cost of waste energy generation systems and sustainable construction practices was significant.
- Operational Challenges: Ensuring the long-term efficiency of energy systems necessitated regular maintenance and monitoring.

The Green Building, Holbeck Urban Village

- Technical Challenges: Integrating biomass burners and fabric energy storage systems faced technical difficulties, while maintaining greywater recycling systems required specialized skills.
- Financial Challenges: The initial cost of renewable energy systems and sustainable materials was a substantial financial challenge.
- Operational Challenges: Regular maintenance of biomass burners and energy storage systems was critical for ensuring efficiency.

Allerton Bywater Millennium Community

- Technical Challenges: Implementing low-emission gas boilers and solar hot water systems presented technical challenges, while ensuring the efficiency of sustainable drainage systems required careful design.
- Financial Challenges: The high cost of renewable energy systems and sustainable construction practices posed significant financial challenges.
- Operational Challenges: Continuous monitoring and maintenance of energy systems were necessary to sustain efficiency.

Comparative Analysis of Challenges

To facilitate a clearer understanding of the challenges encountered across the selected green building case studies, the following table synthesizes key challenges in technical, financial, and operational aspects.

Table 8. Comparative analysis of challenges in green building case studies

Case Study	Technical Challenges	Financial Challenges	Operational Challenges
Energy Research Building	Integration of renewables	High initial costs	Continuous monitoring
Innovate Green Office	Precise implementation	High insulation costs	Constant oversight
York Environmental Centre	Varying weather conditions	High cost of systems	Specialized maintenance
Greenhouse, Citu	Advanced technical expertise	High investment	Complex systems maintenance
Gledhow Bank Eco-Houses	Precise implementation	High cost of materials	Regular maintenance
White Willows – Sheffield	Retrofitting difficulties	High initial cost	Skilled personnel required
Broadcasting Place	Natural light optimization	High cost of materials	Continuous monitoring
Trinity Leeds	Precise implementation	High cost of systems	Regular maintenance
The Green Building, Holbeck Urban Village	Technical difficulties	High cost of systems	Regular maintenance
Allerton Bywater Millennium Community	Technical implementation	High cost of systems	Continuous monitoring

Table 3 provides a concise overview of the technical, financial, and operational challenges faced by each case study. This analysis highlights the complexities involved in maintaining long-term sustainability and offers insights into the specific areas that require attention for future green building projects.

Solutions Implemented

This section explores the technical, financial, and operational strategies employed to address the sustainability challenges identified in the case studies. These strategies, drawn from detailed analysis and performance reports, provide insights into how different projects overcame barriers to achieve long-term sustainability.

Energy Research Building

- Technical Solutions: The building utilizes advanced energy-efficient mechanical and electrical systems, concrete thermal mass walls for temperature stability, and solar shading devices to minimize heat gain. Additionally, lighting controls with presence detection and low-energy fume cupboards enhance energy efficiency.

- Financial Solutions: Strategic investments in high-performance technologies were facilitated through secured funding, enabling the building to achieve low energy consumption.
- Operational Solutions: A comprehensive maintenance plan ensures the continued efficiency of energy recovery systems and low-energy lifts, supported by natural ventilation strategies.

Innovate Green Office

- Technical Solutions: The office employs passive design principles, optimizing building orientation, thermal mass, and heavily insulated envelopes to maximize energy savings. Daylight optimization is achieved through a central atrium.
- Financial Solutions: Cost savings from reduced material usage and operational efficiencies offset the initial investments in sustainable design features.
- Operational Solutions: A robust monitoring system maintains the effectiveness of passive design elements, ensuring minimal energy consumption through strategic operational practices.

York Environmental Centre

- Technical Solutions: The center integrates renewable energy sources such as photovoltaics and wind turbines, utilizes triple-glazed windows to minimize heat loss, and employs solar panels for hot water preheating. Efficient wood stoves and compost toilets further support sustainability.
- Financial Solutions: Funding was secured for deploying renewable energy systems and sustainable materials, facilitating long-term cost savings and environmental benefits.
- Operational Solutions: A detailed maintenance schedule is implemented for renewable energy systems, coupled with regular inspections to ensure optimal performance.

Greenhouse, Citu

- Technical Solutions: The building features ground source heat pumps, solar thermal panels, and wind turbines for heating, cooling, and electricity generation. Rain and greywater recycling systems are used to minimize water usage.

- Financial Solutions: Financial viability is achieved through reduced energy costs, with the project benefiting from awards and recognitions that attracted additional funding and support.
- Operational Solutions: Energy systems are continuously monitored, with regular maintenance and updates ensuring the efficiency of innovative technologies.

Gledhow Bank Eco-Houses

- Technical Solutions: Advanced insulation, renewable energy systems, and rainwater harvesting techniques are implemented. The use of post and beam timber frame construction supports ecological soundness.
- Financial Solutions: The project reduced costs by sourcing local materials and minimizing construction waste.
- Operational Solutions: A comprehensive maintenance plan is in place to ensure the long-term effectiveness of insulation systems and renewable energy technologies.

White Willows – Sheffield

- Technical Solutions: The building incorporates natural light optimization, night cooling systems, and photovoltaic panels. Green roofs enhance insulation and manage water run-off effectively.
- Financial Solutions: Financial efficiency was achieved through close collaboration between clients, contractors, and design teams, reducing performance gaps.
- Operational Solutions: A detailed maintenance schedule is implemented for energy systems, with regular monitoring ensuring the effectiveness of sustainable design features.

Broadcasting Place

- Technical Solutions: The building optimizes natural light, integrates renewable energy systems, and employs advanced energy-efficient technologies.
- Financial Solutions: Funding was secured for sustainable materials and technologies, contributing to long-term cost savings.
- Operational Solutions: Continuous monitoring of energy systems is conducted, with regular maintenance ensuring the effectiveness of sustainable design features.

Trinity Leeds

- Technical Solutions: The project implements waste-to-energy generation, zero waste to landfill, and advanced energy-saving escalators.
- Financial Solutions: Cost savings from waste reduction and energy generation offset the initial investments in sustainable systems.
- Operational Solutions: A comprehensive maintenance plan is established for energy systems, ensuring minimal energy consumption through strategic operational practices.

The Green Building, Holbeck Urban Village

- Technical Solutions: Biomass burners, fabric energy storage systems, and greywater recycling are utilized. The design also optimizes natural daylight and controls solar heat.
- Financial Solutions: The building achieves financial efficiency through reduced energy costs, with sustainable design features attracting additional funding.
- Operational Solutions: Continuous monitoring and maintenance of energy systems are conducted to maintain the efficiency of innovative technologies.

Allerton Bywater Millennium Community

- Technical Solutions: Low emission gas boilers, solar hot water systems, and sustainable drainage systems are implemented. Advanced insulation and renewable energy technologies support sustainability.
- Financial Solutions: Cost savings from reduced energy consumption are realized, with funding secured for sustainable systems ensuring long-term financial viability.
- Operational Solutions: A detailed maintenance schedule is established for energy systems, with regular monitoring ensuring the efficiency of sustainable design features.

Comparative Analysis of Solutions Implemented

To better understand how these diverse green building projects addressed their challenges, the following table presents a comparative analysis of the technical, financial, and operational solutions implemented across the case studies.

Table 9. Comparative analysis of solutions implemented in green building case studies

Case Study	Technical Solutions	Financial Solutions	Operational Solutions
Energy Research Building	Advanced energy systems, thermal mass, solar shading	Strategic funding for sustainable systems	Comprehensive maintenance plan for energy systems
Innovate Green Office	Passive design principles, optimized orientation, thermal mass	Cost savings from reduced material use	Robust monitoring system for passive design elements
York Environmental Centre	Renewable energy integration, triple-glazed windows, solar panels	Funding for renewable systems	Detailed maintenance schedule for renewable systems
Greenhouse, Citu	Ground source heat pumps, solar thermal panels, wind turbines	Financial viability through reduced energy costs	Continuous monitoring and maintenance of energy systems
Gledhow Bank Eco-Houses	Advanced insulation, renewable energy, rainwater harvesting	Cost reduction through local materials	Comprehensive maintenance plan for insulation and energy systems
White Willows – Sheffield	Natural light optimization, night cooling, photovoltaic panels	Financial efficiency through collaboration	Detailed maintenance schedule for energy systems
Broadcasting Place	Natural light optimization, renewable energy integration	Funding for sustainable materials	Continuous monitoring of energy systems
Trinity Leeds	Waste-to-energy generation, zero waste to landfill, energy-saving escalators	Cost savings from waste reduction and energy generation	Comprehensive maintenance plan for energy systems
The Green Building, Holbeck Urban Village	Biomass burners, fabric energy storage, greywater recycling	Financial efficiency through reduced energy costs	Continuous monitoring and maintenance of energy systems
Allerton Bywater Millennium Community	Low emission boilers, solar hot water, sustainable drainage	Cost savings from reduced energy costs	Detailed maintenance schedule for energy systems

This above comparative analysis table provides a structured overview of how each project addressed its specific challenges. This analysis highlights the variety of approaches taken to ensure long-term sustainability, offering valuable insights for future green building projects.

Performance Over Time

This section provides a comparative analysis of the performance metrics across diverse green buildings, focusing on energy consumption, water usage, maintenance costs, and occupant satisfaction. By evaluating these aspects, the strengths and weaknesses of different sustainability strategies can be identified.

Energy Consumption

- Trend: Buildings with advanced insulation, renewable energy systems, and passive design principles consistently demonstrate lower energy consumption over time.
- Example: The Energy Research Building and Greenhouse, Citu, achieved significant reductions in energy consumption through thermal mass walls, solar shading devices, and ground source heat pumps.
- Pattern: Green buildings that integrate renewable energy systems like photovoltaics and wind turbines show a marked decrease in reliance on non-renewable energy sources.

Water Usage

- Trend: Effective water management systems, including rainwater harvesting and greywater recycling, contribute to lower water usage and enhanced sustainability.
- Example: The York Environmental Centre and The Green Building, Holbeck Urban Village, implemented comprehensive water management systems, leading to efficient water use.
- Pattern: Buildings prioritizing water conservation techniques, such as sustainable drainage systems and rainwater purification, maintain lower water consumption levels over the long term.

Maintenance Costs

- Trend: The use of durable, low-maintenance materials and efficient systems results in reduced maintenance costs over time.
- Example: Gledhow Bank Eco-Houses and White Willows – Sheffield reported lower maintenance costs due to their sustainable construction practices and advanced insulation.
- Pattern: Regular maintenance and monitoring of renewable energy systems and other sustainable features are crucial for maintaining low operational costs.

Occupant Satisfaction

- Trend: High occupant satisfaction is linked to improved indoor environmental quality, natural lighting, and effective temperature regulation.

- Example: The Energy Research Building and Innovate Green Office reported high occupant satisfaction due to their emphasis on natural ventilation and thermal comfort.
- Pattern: Buildings that integrate user-friendly technologies and engage occupants in sustainability practices tend to achieve higher satisfaction rates.

Challenges vs. Solutions

Analyzing the effectiveness of solutions implemented to overcome specific challenges provides insights into the comparative success of different approaches.

Technical Challenges vs. Solutions

- Challenge: Integrating renewable energy systems and maintaining their efficiency.
- Solution: Advanced technical expertise and continuous monitoring, as seen in the Greenhouse, Citu, and Broadcasting Place.
- Outcome: Buildings that prioritized ongoing technical support and regular maintenance of renewable systems achieved better performance and reliability.

Financial Challenges vs. Solutions

- Challenge: High initial costs of sustainable materials and technologies.
- Solution: Leveraging financial incentives, strategic funding, and cost savings from operational efficiencies, demonstrated by Innovate Green Office and Allerton Bywater Millennium Community.
- Outcome: Projects that secured financial incentives and optimized long-term savings successfully offset the initial investment costs, making sustainability financially viable.

Operational Challenges vs. Solutions

- Challenge: Ensuring the long-term efficiency of energy systems and sustainable features.
- Solution: Implementing comprehensive maintenance plans and employing skilled personnel for regular monitoring, as seen in the Energy Research Building and York Environmental Centre.
- Outcome: Effective operational strategies resulted in sustained performance and lower operational costs over time.

Comparative Analysis of Performance and Solutions

To better understand the relationship between performance metrics and the solutions implemented, the following table presents a comparative analysis across the case studies.

Table 10. Comparative analysis of performance over time and solutions implemented in green building case studies

Case Study	Performance Over Time	Technical Solutions	Financial Solutions	Operational Solutions
Energy Research Building	Low energy consumption, efficient water use, high occupant satisfaction	Continuous monitoring and comprehensive maintenance plans	Strategic funding for sustainable systems	Comprehensive maintenance plan for energy systems
Innovate Green Office	Significant energy savings, positive occupant feedback	Robust monitoring system for passive design elements	Cost savings from reduced material use	Leveraged cost savings and robust monitoring systems
York Environmental Centre	Minimal energy consumption, efficient water management, high occupant satisfaction	Detailed maintenance schedules and secured funding for renewable systems	Funding for renewable systems	Implemented detailed maintenance schedules
Greenhouse, Citu	Low energy consumption, efficient water usage, high occupant satisfaction	Continuous monitoring and advanced technical expertise	Financial viability through reduced energy costs	Continuous monitoring and maintenance of energy systems
Gledhow Bank Eco-Houses	Low energy consumption, efficient water management, positive occupant feedback	Comprehensive maintenance plan for insulation and energy systems	Cost reduction through local materials	Comprehensive maintenance plan for insulation and energy systems
White Willows – Sheffield	Low energy consumption, efficient water management, high occupant satisfaction	Close collaboration reduced performance gaps	Financial efficiency through collaboration	Detailed maintenance schedule for energy systems
Broadcasting Place	High energy efficiency, positive occupant feedback	Continuous monitoring ensured the effectiveness of renewable energy systems	Funding for sustainable materials	Continuous monitoring of energy systems

continued on following page

Table 10. Continued

Case Study	Performance Over Time	Technical Solutions	Financial Solutions	Operational Solutions
Trinity Leeds	High energy efficiency, efficient water use, high occupant satisfaction	Comprehensive maintenance plans ensured long-term performance	Cost savings from waste reduction and energy generation	Comprehensive maintenance plan for energy systems
The Green Building, Holbeck Urban Village	Low energy consumption, efficient water management, high occupant satisfaction	Regular updates and maintenance of energy systems	Financial efficiency through reduced energy costs	Continuous monitoring and maintenance of energy systems
Allerton Bywater Millennium Community	Low energy consumption, efficient water management, positive occupant feedback	Leveraged cost savings and secured funding for sustainable systems	Cost savings from reduced energy costs	Detailed maintenance schedule for energy systems

Table 10 provides a structured overview of how each project addressed its specific challenges, resulting in varying degrees of long-term sustainability. This analysis highlights the importance of integrating technical, financial, and operational solutions to achieve and maintain the sustainability of green buildings.

Practical Framework for Long-Term Sustainability

The Practical Framework for Long-Term Sustainability builds on the insights gained from the Conceptual Framework presented earlier. It translates the broader categories of technical, financial, and operational challenges into specific, actionable solutions derived from real-world case studies. This framework aims to provide a structured approach to addressing the complexities of integrating sustainable practices across diverse green building types.

Technical Challenges and Solutions Framework

1. System Integration:
    - Challenges: Integrating sustainable technologies like photovoltaic panels, green roofs, and advanced HVAC systems into a unified building design is complex. Ensuring these systems work together efficiently while maintaining overall building performance poses significant challenges.
    - Solutions: Successful system integration demands advanced technical expertise and continuous monitoring. Regular updates and mainte-

nance are essential to sustain performance and adapt to technological advancements.
- Framework Link: This solution addresses system integration challenges in the Conceptual Framework, highlighting the need for cohesive design and ongoing management to ensure efficiency (Cannavale et al., 2022; Abas et al., 2018).

2. Maintenance and Upgrades:
   - Challenges: Sustainable systems, including solar panels and HVAC units, require regular maintenance and periodic upgrades to maintain their performance over time. Rapid technological advancements can quickly make existing systems obsolete, necessitating further investments.
   - Solutions: A comprehensive maintenance plan and skilled personnel for regular monitoring ensure these systems remain effective, minimizing operational disruptions.
   - Framework Link: This solution aligns with the Conceptual Framework's focus on maintenance challenges, emphasizing the importance of sustaining the efficiency and longevity of green technologies (Abas et al., 2018).
3. Technological Advances:
   - Challenges: The fast pace of innovation in green building technologies can lead to the obsolescence of current systems, requiring additional investments and potential disruptions during upgrades.
   - Solutions: Ongoing research and investment in new technologies are crucial to keeping buildings at the cutting edge of sustainability, ensuring they remain efficient and adaptable.
   - Framework Link: This solution addresses the challenges of technological advancements, reinforcing the Conceptual Framework's emphasis on flexible and up-to-date systems (Cannavale et al., 2022).

Financial Challenges and Solutions Framework

1. Initial Investment:
   - Challenges: Sustainable construction often incurs higher upfront costs compared to conventional methods, posing a financial barrier to adopting green building practices.
   - Solutions: Utilizing financial incentives, securing strategic funding, and showcasing long-term cost savings can help offset these initial expenses, making green projects more financially appealing.

- Framework Link: This aligns with the Conceptual Framework's focus on financial challenges, highlighting the importance of managing initial investment costs (Latief et al., 2017; Fan & Xia, 2018).

2. Return on Investment (ROI):
   - Challenges: The financial returns from sustainable buildings, like lower energy costs and higher property values, usually materialize over the long term, which may deter immediate investor interest.
   - Solutions: Conducting comprehensive economic analyses and clearly communicating the long-term financial gains can attract investors by emphasizing potential cost savings and value growth over time.
   - Framework Link: This addresses the ROI challenges outlined in the Conceptual Framework, stressing the need for transparent communication of long-term benefits (Latief et al., 2017).
3. Economic Incentives:
   - Challenges: The feasibility of green building projects is heavily influenced by the availability and consistency of economic incentives such as grants, tax breaks, and subsidies.
   - Solutions: Creating a stable and supportive regulatory environment is crucial to encourage investment in green buildings. Policymakers should ensure that incentives are consistent and accessible to promote the widespread adoption of sustainable practices.
   - Framework Link: This solution is in line with the Conceptual Framework's emphasis on the crucial role of economic incentives in advancing green building initiatives (Fan & Xia, 2018).

Operational Challenges and Solutions Framework

1. Facilities Management:
   - Challenges: Maintaining the performance of sustainable systems requires effective facilities management, including regular monitoring, timely maintenance, and efficient operation of complex systems.
   - Solutions: Providing continuous training and support for facilities managers, along with deploying advanced Building Management Systems (BMS), ensures effective management and sustained high performance.
   - Framework Link: This aligns with the operational challenges in the Conceptual Framework, highlighting the need for skilled personnel and advanced management systems (Abas et al., 2018).
2. Occupant Behavior:

- Challenges: Occupant behavior greatly influences the sustainability of green buildings. Energy-saving technologies only reach their full potential when used correctly by occupants.
- Solutions: Engaging occupants through education and awareness programs fosters sustainable practices, enhancing overall building performance and promoting a culture of sustainability.
- Framework Link: This addresses the occupant behavior challenges noted in the Conceptual Framework (Asyera et al., 2020).

3. Data Management:
   - Challenges: Green buildings generate large amounts of data from various systems, which must be effectively managed and analyzed to optimize performance and identify areas for improvement.
   - Solutions: Implementing advanced data analytics and building management systems enhances performance monitoring and supports data-driven decision-making.
   - Framework Link: This solution aligns with the data management challenges identified in the Conceptual Framework, underscoring the importance of data in optimizing building performance (Mahmoud, 2022).

Table 11. Summary of practical challenges and solutions framework in green buildings

Category	Challenges	Solutions	Case Study Examples	References
Technical	System integration, maintenance, technological advancements	Advanced expertise, continuous monitoring, regular updates	Energy Research Building, Broadcasting Place, Greenhouse, Citu	Cannavale et al. (2022); Abas et al. (2018)
Financial	High initial investment, ROI, inconsistent economic incentives	Financial incentives, strategic funding, detailed economic analyses	Innovate Green Office, Allerton Bywater Millennium Community	Latief et al. (2017); Fan & Xia (2018)
Operational	Facilities management, occupant behavior, data management	Skilled personnel, continuous training, data analytics, advanced BMS	Energy Research Building, White Willows – Sheffield, Innovate Green Office	Abas et al. (2018); Asyera et al. (2020); Mahmoud (2022)

Integration with the Conceptual Framework

The Practical Framework is directly integrated with the Conceptual Framework outlined in the literature review. The challenges and solutions categorized here provide a clear path from theoretical concepts to real-world applications. The framework diagram (referenced as Figure 2) illustrates the relationships between challenges and solutions, showing how each category contributes to the long-term sustainability of diverse green building types. By linking the theoretical foundations with practical actions, this framework serves as a comprehensive guide for architects, urban planners, and policymakers working towards sustainable urban development.

Figure 2. Practical framework for long-term sustainability of diverse green building (Source: Own)

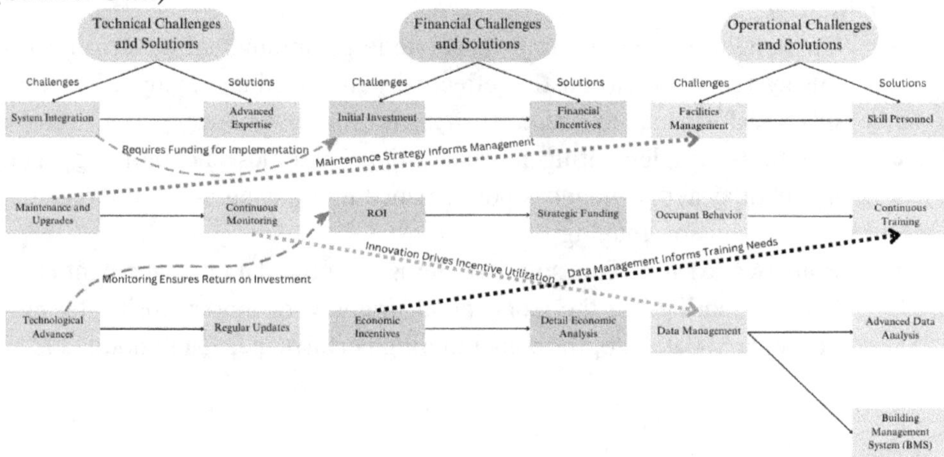

Best Practices

Best Practices for Long-Term Sustainability

The Best Practices section offers actionable insights from case studies to address the challenges identified in the Practical Framework, ensuring sustainable practices are effectively implemented across various green building types.

1. Integrated Design Process:
    - o Overview: Engaging all stakeholders, including architects, engineers, and facilities managers, early in the design phase ensures seamless integration of sustainable features, especially for diverse building types throughout the project lifecycle.
    - o Examples: Energy Research Building, Innovate Green Office.
2. Regular Maintenance and Monitoring:

- o Overview: Establishing a robust maintenance plan and employing skilled personnel for regular monitoring are crucial for maintaining the performance of sustainable systems, ensuring long-term efficiency across different building types.
- o Examples: York Environmental Centre, White Willows – Sheffield.

3. Leveraging Financial Incentives:
    - o Overview: Securing financial incentives and demonstrating long-term savings can help attract investments and offset the initial costs of sustainable construction, making green building practices financially viable across various types.
    - o Examples: Allerton Bywater Millennium Community, Trinity Leeds.
4. Continuous Training and Education:
    - o Overview: Providing ongoing training for facilities managers and occupants ensures they are equipped to handle advanced sustainable technologies, maintaining high performance as new technologies and best practices emerge.
    - o Examples: Broadcasting Place, Greenhouse, Citu.
5. Advanced Data Management:
    - o Overview: Implementing advanced data analytics and building management systems is key to optimizing performance and identifying areas for improvement, enhancing operational efficiency and informed decision-making across different building types.
    - o Examples: York Environmental Centre, Broadcasting Place.

Table 12. Best practices derived from case studies

Best Practice	Description	Case Study Examples
Integrated Design Process	Engaging all stakeholders early in the design process for various building types.	Energy Research Building, Innovate Green Office
Regular Maintenance and Monitoring	Establishing a comprehensive maintenance plan and employing skilled personnel for regular monitoring.	York Environmental Centre, White Willows – Sheffield
Leveraging Financial Incentives	Securing financial incentives and demonstrating long-term savings to attract investments.	Allerton Bywater Millennium Community, Trinity Leeds
Continuous Training and Education	Providing continuous training and education for facilities managers and occupants.	Broadcasting Place, Greenhouse, Citu
Advanced Data Management	Implementing advanced data analytics and building management systems.	York Environmental Centre, Broadcasting Place

Linking Best Practices with the Practical Framework

The Best Practices outlined above are directly informed by the Practical Framework for Long-Term Sustainability discussed earlier. This framework categorizes challenges and solutions into technical, financial, and operational domains, and the best practices are designed to address these categories systematically:

- Technical Challenges and Solutions: Best practices such as the Integrated Design Process and Regular Maintenance and Monitoring directly respond to the technical challenges of system integration, maintenance, and technological advances identified in the framework. By engaging stakeholders early and ensuring regular maintenance, these practices mitigate risks and enhance the longevity of sustainable systems.
- Financial Challenges and Solutions: The practice of Leveraging Financial Incentives addresses the financial challenges of high initial investments and the delayed realization of returns. By securing incentives and communicating long-term savings, projects can achieve financial viability as emphasized in the framework.
- Operational Challenges and Solutions: The practices of Continuous Training and Education and Advanced Data Management are crucial for overcoming operational challenges. By equipping personnel with the necessary skills and leveraging data analytics, buildings can optimize their performance and adapt to evolving sustainability standards.

By integrating these best practices with the Practical Framework, the approach to achieving long-term sustainability in green building projects becomes comprehensive and actionable. This integration ensures that theoretical insights from the Conceptual Framework are grounded in real-world applications, providing a valuable guide for architects, urban planners, and policymakers.

In essence, the Best Practices section not only complements the Practical Framework but also reinforces the Conceptual Framework introduced in the literature review. This holistic approach ensures that sustainable building practices are effectively implemented and maintained, contributing to the broader goal of urban sustainability.

Validation and Expert Opinions

Cross-Validation: To ensure the robustness and reliability of the findings presented in this chapter, cross-referencing with similar studies and reports from various regions and sources is crucial. This approach validates the results, identifies common patterns, and highlights any discrepancies or unique insights, ensuring the findings are applicable to diverse green building types.

Regional Comparisons: Research from North America, Europe, and Asia offers valuable insights into how environmental and policy factors shape the sustainability of various green building types. Studies conducted by the US Green Building Council (USGBC) and the International Living Future Institute (ILFI) have identified common challenges and solutions across different climatic zones and regulatory environments. For instance, Yudelson (2009) explores green building trends in Europe, emphasizing the importance of region-specific strategies for enhancing building performance and sustainability. Similarly, a comprehensive review by the Asia-Pacific Economic Cooperation (APEC) (2016) highlights the outcomes and lessons learned from green building projects across the Asia-Pacific region, providing a critical assessment of policy implementations and environmental impacts.

Thematic Comparisons: Cross-referencing the identified challenges and solutions with thematic studies on specific aspects of green buildings, such as energy efficiency, renewable energy integration, and sustainable materials, facilitates a more nuanced validation of strategies. Research on passive and active energy-saving techniques, such as those discussed by Bakar et al. (2015), highlights the effectiveness of various approaches in improving building energy performance. Their review of the Energy Efficiency Index (EEI) as a tool for measuring building energy performance underscores the relevance of these strategies across different building types, thereby supporting the themes explored in this chapter.

Temporal Comparisons: Validating findings by comparing them with longitudinal studies that track the performance of green buildings over time provides critical insights into the long-term sustainability of proposed solutions across different building types. For instance, Korkmaz, Horman, and Riley (2009) identified key attributes necessary for conducting a longitudinal study of green project delivery. Their research emphasizes the importance of tracking performance over time to understand how green buildings maintain their efficiency and sustainability throughout their lifecycle. Such studies reinforce the long-term sustainability framework discussed in this chapter.

Expert Opinions: Incorporating insights from recognized experts in green building and sustainability enhances the credibility of the findings and provides a comprehensive perspective on the challenges and solutions discussed.

Interviews and Consultations: Although direct interviews were not conducted for this chapter, existing literature and reports from recognized experts in green building design, construction, and maintenance provide valuable insights. Notable experts include Ann Edminster, founder and principal of Design AVEnues

in California, and Mary Tappouni, president of Breaking Ground Contracting in Jacksonville, Florida (Holladay, 2023; Binsacca, 2010).
Peer Review: Engaging in rigorous peer review processes ensures the accuracy and reliability of the findings. Peer-reviewed studies cited in this chapter, such as those published in the Journal of Green Building and Sustainable Cities and Society, validate the proposed solutions based on their expertise and experiences, supporting the chapter's findings.

Table 13. **Key references and expert contributions**

Source/Expert	Area of Validation	Contribution
Yudelson (2009)	Regional Comparison	Green Building Trends: Europe
Asia-Pacific Economic Cooperation (2016)	Regional Comparison	Review of APEC Green Building Project
Bakar et al. (2015)	Thematic Comparison	Energy Efficiency Index as an Indicator for Measuring Building Performance
Korkmaz, Horman, & Riley (2009)	Temporal Comparison	Longitudinal Study on Green Project Delivery
Ann Edminster, Design AVEnues	Expert Opinion	Expertise in green building design
Mary Tappouni, Breaking Ground Contracting	Expert Opinion	Expertise in sustainable construction practices
Journal of Green Building	Peer Review	Validation of sustainability strategies
Sustainable Cities and Society	Peer Review	Confirmation of proposed solutions' effectiveness

By cross-referencing the findings with similar studies and incorporating expert opinions from established literature, this chapter ensures a comprehensive and validated approach to analyzing the long-term sustainability of diverse green building types. This process strengthens the credibility of the research while providing practical insights and recommendations grounded in both theoretical and empirical evidence.

5. DISCUSSION

Key Findings

The case study analysis of diverse green building types reveals several critical insights into their long-term sustainability, which are pivotal for advancing sustainable urban development.

1. Energy Efficiency and Carbon Reduction: Integrating advanced energy-efficient technologies like high-performance glazing, HVAC systems, and insulation substantially cuts energy use and lowers carbon emissions across various buildings. The Green Building at Holbeck Urban Village, for example, achieves up to 45% energy savings, highlighting the importance of energy efficiency in sustainable design.
2. Renewable Energy Integration: Using renewable energy sources such as solar panels, wind turbines, and ground-source heat pumps is crucial for reducing reliance on fossil fuels. Case studies, including Greenhouse and York Environmental Centre, show significant energy savings from these technologies, which contribute greatly to sustainability goals.
3. Sustainable Materials: Choosing materials with low embodied energy and high recyclability enhances the environmental performance of buildings. Projects like Gledhow Bank Eco-Houses and Allerton Bywater Millennium Community demonstrate how sustainable materials can reduce environmental impact and improve thermal comfort, supporting long-term sustainability.
4. Economic Viability: Despite higher initial costs, green buildings offer substantial financial benefits over time, including energy savings and increased property values. Case studies like Trinity Leeds and Innovate Green Office confirm that green buildings provide significant lifecycle savings, making them a financially viable choice for broader adoption.
5. Operational and Maintenance Challenges: Sustaining the performance of green buildings requires effective facilities management and regular maintenance. Case studies such as York Environmental Centre and White Willows emphasize the need for skilled personnel and ongoing training to operate advanced sustainable technologies effectively, highlighting the importance of operational support.
6. Human Factors and Behavior: Occupant behavior is crucial to the sustainability of green buildings. Behavior changes, supported by awareness programs and user-friendly technologies, can greatly enhance energy efficiency. Projects like The Green Building at Holbeck Urban Village and Innovate Green Office show that engaging occupants is key to achieving sustainability across different building types.
7. Policy and Regulatory Support: Stable regulatory frameworks and economic incentives are vital for promoting and sustaining green building practices. Case studies, including Allerton Bywater Millennium Community and Trinity Leeds, demonstrate how effective policies and incentives can positively impact sustainable development across various building types.

Implications for Future Projects

The case study analysis provides important insights for enhancing the long-term sustainability of green buildings, particularly for architects, urban planners, and policymakers.

1. Integrated Design Approach: A collaborative design process from the start is essential. By involving all stakeholders early, systems can be cohesively integrated, leading to better overall performance. This approach is crucial for diverse building types where varied requirements must be aligned to achieve sustainability goals.
2. Long-Term Maintenance Strategies: To maintain the performance of green buildings, it's critical to develop detailed maintenance plans and invest in skilled personnel. Regular monitoring and timely upgrades are necessary to ensure the long-term effectiveness of sustainable systems across residential, commercial, and mixed-use buildings.
3. Financial Planning and Incentives: Attracting investment in green buildings requires securing financial incentives and clearly demonstrating long-term savings. Policymakers should create a stable regulatory environment that supports green building initiatives, enhancing their economic viability and encouraging broader adoption.
4. Occupant Engagement: Engaging occupants through education and incentives is key to promoting sustainable practices and improving overall building performance. Designers and planners should focus on creating environments that encourage and facilitate sustainable behaviors, ensuring long-term success across different building contexts.
5. Advanced Data Management: The use of advanced data analytics and building management systems is vital for optimizing building performance and identifying areas for improvement. Effective data management is critical to achieving long-term sustainability, enabling informed decision-making and improved operational efficiency across various building types.

Framework Integration

The proposed framework for addressing the technical, financial, and operational challenges identified in the case studies provides a structured approach to enhancing the long-term sustainability of green buildings. This framework categorizes challenges and solutions into three main areas technical, financial, and operational allowing stakeholders to systematically address the issues relevant to different building types.

Incorporating this framework into the design, planning, and operational stages of green building projects is essential. It serves as a valuable guide for future projects, helping stakeholders navigate the complexities of sustainable building practices. The framework ensures that diverse building types can achieve and maintain long-term sustainability, ultimately contributing to global sustainability goals.

Policy Recommendations

To encourage the widespread adoption of green building practices across various building types, the following policy actions are recommended:

1. Implement Integrated Design Standards: Policymakers should establish and enforce standards that mandate integrated design strategies in green building projects, ensuring collaboration among architects, engineers, and other key stakeholders to achieve comprehensive sustainability goals.
2. Increase Financial Incentives: Expanding financial support, such as tax breaks, grants, and subsidies, for green building projects can help mitigate upfront costs and encourage broader adoption of sustainable practices.
3. Support Ongoing Training Programs: Develop continuous training initiatives for facilities managers and building operators, in partnership with industry and government, to equip them with the skills needed to manage advanced sustainable technologies effectively.
4. Promote Occupant Engagement: Foster programs that encourage sustainable behaviors among building occupants, including incentives for energy conservation and waste reduction, as well as educational campaigns to raise awareness about sustainability.
5. Advocate for Advanced Building Management Systems: Encourage the adoption of advanced building management systems in both new and existing green buildings to enable real-time monitoring and enhance long-term sustainability.

The analysis of diverse green building types highlights the need to address technical, financial, and operational challenges to ensure their sustainability. By adopting integrated design approaches, creating comprehensive maintenance plans, utilizing financial incentives, engaging occupants, and implementing advanced management systems, architects, urban planners, and policymakers can significantly improve the sustainability of green buildings. The insights and recommendations from this analysis can guide future projects and promote more sustainable urban environments.

6. CONCLUSION

This chapter has thoroughly analyzed the long-term sustainability of various green building types, focusing on energy efficiency, renewable energy integration, sustainable materials, economic viability, and operational challenges. Despite facing significant upfront costs and technical hurdles, green buildings offer considerable environmental and economic advantages over their lifecycle. The integration of energy-efficient technologies, such as high-performance glazing, insulation, and advanced HVAC systems, has consistently demonstrated substantial reductions in energy consumption and carbon emissions. For example, the Green Building in Holbeck Urban Village achieved up to 45% energy savings, highlighting the effectiveness of these technologies in reducing environmental impacts. Renewable energy sources, including solar panels, wind turbines, and ground source heat pumps, play a critical role in decreasing reliance on non-renewable energy. Buildings like Greenhouse, Citu, and York Environmental Centre have achieved significant energy savings, contributing to urban sustainability goals.

The use of materials with low embodied energy and high recyclability is essential for improving a building's environmental performance. Projects such as Gledhow Bank Eco-Houses and Allerton Bywater Millennium Community have demonstrated how these materials can reduce ecological footprints and enhance thermal comfort, promoting the long-term sustainability of buildings. Economic analyses from case studies, including Trinity Leeds and Innovate Green Office, show that despite higher initial construction costs, green buildings deliver substantial financial benefits through energy savings and increased property values, confirming their economic viability. Effective facilities management and regular maintenance are crucial for maintaining the performance of green buildings. The involvement of skilled personnel and ongoing training is key to ensuring the efficient operation of sustainable technologies, as seen at York Environmental Centre and White Willows in Sheffield. Occupant behavior also plays a vital role in the sustainability of green buildings. Behavioral changes, supported by awareness programs and user-friendly technologies, can significantly enhance energy efficiency, underscoring the importance of occupant engagement in achieving sustainability goals.

Finally, strong policy and regulatory support are essential to encourage the adoption and sustainability of green building practices. Stable regulations and economic incentives, as observed in Allerton Bywater Millennium Community and Trinity Leeds, are critical for widespread adoption and the achievement of sustainability objectives.

7. FUTURE RESEARCH DIRECTIONS

Despite the progress made, further research is needed to improve the long-term sustainability of diverse green buildings. Future studies should focus on:

1. Longitudinal Performance Studies: Conduct long-term studies to assess the durability and efficiency of sustainable materials and energy systems over time, providing insights into maintenance needs and sustained performance.
2. Comprehensive Economic Analyses: Develop economic models to evaluate total ownership costs, including construction, operation, maintenance, and potential property value increases, aiding informed investment decisions.
3. Emerging Technologies Integration: Explore the effectiveness of new sustainable technologies like advanced energy storage, smart management systems, and innovative materials to keep buildings at the forefront of sustainability.
4. Occupant Behavior and Engagement: Investigate strategies to influence occupant behavior to enhance energy efficiency and building performance, including education, incentives, and user-friendly technologies.
5. Policy and Regulatory Impact: Assess the impact of various policies on green building adoption and performance, offering recommendations to support regulatory frameworks and economic incentives.
6. Multi-Objective Optimization: Use advanced modeling and simulation tools to balance sustainability goals, such as energy efficiency, cost-effectiveness, and occupant comfort, for more holistic green building strategies.
7. Climate Adaptation and Resilience: Study the role of green buildings in climate adaptation, focusing on designing structures that withstand extreme weather and contribute to urban resilience.

Addressing these research areas will significantly improve green building development and operation, ensuring long-term sustainability and urban sustainability contributions.

CONFLICT OF INTEREST

The authors declare no conflict of interest. The research presented in this chapter was conducted independently, without any financial or personal relationships that could have influenced the work.

ACKNOWLEDGMENT

The authors would like to extend their gratitude to the scholars and practitioners whose work provided valuable insights into the secondary data analyzed in this chapter. The authors also acknowledge the support of the academic and professional communities, whose guidance and feedback were instrumental in shaping the research and analysis presented here.

REFERENCES

Abas, D. N., Zakaria, R., Aminuddin, E., Abdullah, R., Khan, J. S., & Sahamir, S. R. (2018). An outlook on strategic framework development needs of roles and involvement of joint property management in high rise green buildings in Malaysia. *Chemical Engineering Transactions*, *63*, 427–432. DOI: 10.3303/CET1863072

Abdalla, H., Fattah, K. P., Abdallah, M., & Al-Tamimi, A. K. (2021). Environmental footprint and economics of a full-scale 3d-printed house. *Sustainability (Basel)*, *13*(21), 11978. DOI: 10.3390/su132111978

Alam, F., & Jin, Y. (2023). The utilisation of small wind turbines in built-up areas: Prospects and challenges. *Wind*, *3*(4), 418–439. DOI: 10.3390/wind3040024

Altomonte, S., & Schiavon, S. (2013). Occupant satisfaction in leed and non-leed certified buildings. *Building and Environment*, *68*, 66–76. DOI: 10.1016/j.buildenv.2013.06.008

APEC. (2016). *Review of APEC Green Building Project: Final Report.* Nathan Associates Inc.

Asyera, E., Gamal, A., & Pandjaitan, T. H. (2020). Occupant behavioral change for energy efficiency in office buildings. *AIP Conference Proceedings, 2255(1*, 020451. DOI:DOI: 10.1063/5.0020451

Aung, T., Liana, S.R., Htet, A., & Bhaumik, A.. (2023). Implementing green facades: A step towardssustainable smart buildings. *Journal of Smart Cities and Society, 2*, 41-51. htttps://.DOI: 10.3233/SCS-230014

Bakar, N. N. A., Hassan, M. Y., Abdullah, H., Rahman, H. A., Abdullah, M. P., Hussin, F., & Bandi, M. (2015). Energy efficiency index as an indicator for measuring building energy performance: A review. *Renewable & Sustainable Energy Reviews*, *44*, 1–11. DOI: 10.1016/j.rser.2014.12.018

Basten, V., Latief, Y., Berawi, M.A., Budiman, R., & Riswanto. (2017). Evaluation of green building rating tools based on existing green building achievement in Indonesia using Life Cycle Assessment Method. *AIP Conferences Proceedings, 1818*, 020008, .DOI: 10.1063/1.4976872

Binsacca, R. (2010, September 02). *Interview With Green Construction Consultant*. From Builder: https://www.builderonline.com/building/building-science/interview-with-green-construction-consultant_o

Blackburne, L., Gharehbaghi, K., & Hosseinian-Far, A. (2021). The knock-on effects of green buildings: High-rise construction design implications. *International Journal of Structural Integrity*, *13*(1), 57–77. DOI: 10.1108/IJSI-06-2021-0062

Cannavale, A., Zampini, G., Carlucci, F., Pugliese, M., Martellotta, F., Ayr, U., Maiorano, V., Ortica, F., Fiorito, F., & Latterini, L. (2022). Energy and daylighting performance of building integrated spirooxazine photochromic films. *Solar Energy*, *242*(7), 424–434. DOI: 10.1016/j.solener.2021.10.058

Cao, M., Li, R., & Wang, Z. (2024). Research on integrated art design of building exterior finishing with photovoltaic application technology. *E3S Web of Conferences*, *512*, 01004. .DOI: 10.1051/e3sconf/202451201004

Dulguun, E. (2023). Implementation of cutting-edge technologies in facility management challenges and opportunities in the digital era. *Journal Dimensie Management and Public Sector*, *4*(4), 24–33. DOI: 10.48173/jdmps.v4i4.243

Eslamnoor, N., & Vural, S. (2021). User behavior and energy efficiency in north cyprus university dormitories: A case study. *Journal of Housing and the Built Environment*, *37*(3), 1159–1178. DOI: 10.1007/s10901-021-09878-1

Fan, Y., & Xia, X. (2018). Energy-efficiency building retrofit planning for green building compliance Author links open overlay panel. *Building and Environment*, *136*, 312–321. DOI: 10.1016/j.buildenv.2018.03.044

Farhadi, A., & Iqbal, M. T. (2024). Optimal sizing and techno-economic analysis of a hybrid power system for postville. *European Journal of Engineering and Technology Research, 9*(2), 5–12. DOI: 10.24018/ejeng.2024.9.2.3127

Fu, C., Lu, L., & Pirabi, M. (2023). Advancing green finance: A review of sustainable development. *Digital Economy and Sustainable Development, 1*(1), 44265. DOI: 10.1007/s44265-023-00020-3

Ghosh, B., & Karmakar, S. (2024). 3d printing technology and future of construction: A review. *IOP Conference Series. Earth and Environmental Science, 1326*(1), 012001. DOI: 10.1088/1755-1315/1326/1/012001

Hajibeigi, P., Pazhouhanfar, M., Grahn, P., & Nazif, H. (2023). Enhancing citizens' perceived restoration potential of green facades through specific architectural attributes. *Buildings (Basel, Switzerland), 13*(9), 2356. DOI: 10.3390/buildings13092356

Halicioglu, F. H., & Gurel, K. (2023). The perceived quality of green buildings: From a review of occupant satisfaction towards a conceptual framework. *International Journal of Quality &Amp. International Journal of Quality & Reliability Management, 41*(3), 757–777. DOI: 10.1108/IJQRM-10-2022-0307

Hassebo, A., & Tealab, M. (2023). Global models of smart cities and potential iot applications: A review. *IoT, 4*(3), 366–411. DOI: 10.3390/iot4030017

Holladay, M. (2023, May 26). *Interview with Ann Deminster.* From Green Building Advisor: https://www.greenbuildingadvisor.com/article/interview-with-ann-edminster

Htet, A., Liana, S. R., Aung, T., & Bahumik, A. (2023). The Intersection of Facade Engineering and Building Information Modeling: Opportunities and Challenges. *Journal of Technology Innovations and Energy, 2*(3), 94–110. DOI: 10.56556/jtie.v2i3.643

Htet, A., Liana, S. R., Aung, T., & Bhaumik, A. (2023). Economic and Environmental Impact Analysis of Green Building Materials in Facade Engineering. *International Journal of Membrane Science and Technology, 10*(2), 2805–2814. DOI: 10.15379/ijmst.v10i2.2973

ILFI. (2022). *Certified Case Studies.* From International Living Future Institute: https://living-future.org/program/lbc/

Jankovic, S. D., & Curovic, D. M. (2023). Strategic integration of artificial intelligence for sustainable businesses: Implications for data management and human user engagement in the digital era. *Sustainability (Basel), 15*(21), 15208. DOI: 10.3390/su152115208

Kamal, M. A., & Ahmed, E. (2023). Analyzing energy efficient design strategies in high-rise buildings with reference to hvac system. *Architecture Engineering and Science*, *4*(3), 183. DOI: 10.32629/aes.v4i3.1275

Kibert, C. (2016). *Kibert, C. J. (2016). Sustainable construction: green building design and delivery*. John Wiley & Sons.

Kim, Y. U., Yang, S., Yun, B. Y., & Kim, S. (2022). Evaluation of energy consumption in apartment buildings with biochar and phase-change material aggregate-applied artificial stone finishing materials. *International Journal of Energy Research*, *46*(9), 12772–12786. DOI: 10.1002/er.8043

Korkmaz, S., Horman, M., & Riley, D. (2009). Key attributes of a longitudinal study of green project delivery. *In Construction Research Congress 2009: Building a Sustainable Future* (pp. 558-567.). ASCE.DOI: 10.1061/41020(339)57

Li, S. (2024). Sustainable material selection in architectural design. *Advances in Social Science, Education and Humanities Research*, *4*(10), 38476. DOI: 10.2991/978-2-38476-222-4_2

Li, S., Zheng, X., & Zeng, Q. (2023). Can green finance drive the development of the green building industry?—Based on the evolutionary game theory. *Sustainability (Basel)*, *15*(17), 13134. DOI: 10.3390/su151713134

Li, W., Wang, F., Liu, T., Xue, Q., & Liu, N. (2023). Peer effects of digital innovation behavior: An external environment perspective. *Management Decision*, *61*(7), 2173–2200. DOI: 10.1108/MD-06-2022-0865

Liu, Z., & Guo, A. (2021). Application of green building materials and multi-objective energy-saving optimization design. *International Journal of Heat and Technology, 39(1). International Journal of Heat and Technology*, *39*(1), 299–308. DOI: 10.18280/ijht.390133

Lu, Y., Chang, R., Shabunko, V., & Lay Yee, A. T. (2019). he implementation of building-integrated photovoltaics in Singapore: Drivers versus barriers. *Energy*, *168*, 400–408. DOI: 10.1016/j.energy.2018.11.099

Mahdi, B. S., Sulaiman, N., Shehab, M. A., Shafie, S., Hizam, H., & Hassan, S. L. B. M. (2024). Optimization of operating cost and energy consumption in a smart grid. *IEEE Access : Practical Innovations, Open Solutions*, *12*, 18837–18850. DOI: 10.1109/ACCESS.2024.3354065

Mahmoud, A. S. (2022). Overview of Green Roof Technology as a Prospective Energy Preservation Technique in Arid Regions. *Engineering Technology and Applied Science Research*, *12*(4), 8982–8989. DOI: 10.48084/etasr.4919

Maida, M. J. (2015). Case Studies: An update on the SPD. Leeds City: Leeds City Council.

Maida, M. J. (2015). *Sustainable Construction in Practice: Performance of sustainable buildings in Leeds*. Leeds City.

Mallinson, D. J., Azevedo, L., Best, E., Robles, P., & Wang, J. (2023). The future of ai is in the states: The case of autonomous vehicle policies. *Business and Politics*, *26*(2), 180–199. DOI: 10.1017/bap.2023.19

Moghaddam, F. B., Mir, J. M. F., Yanguas, A. B., Delgado, I. N., & Dominguez, E. R. (2020). Building orientation in green facade performance and its positive effects on urban landscape: Case study of an urban block in Barcelona. *Sustainability (Basel)*, *12*(21), 9273. DOI: 10.3390/su12219273

Najed, F., Djunaedy, E., Utami, A. R. I., Sari, I. P., Rithwan, F., & Sofyan, S. (2023). The effect of baseline establishment on energy savings estimate in green building certification. *Journal of Physics: Conference Series*, *2596*(1), 012051. DOI: 10.1088/1742-6596/2596/1/012051

Najed, F., Djunaedy, E., Utami, A. R. I., Sari, I. P., Rithwan, F., & Sofyan, S. (2023). The effect of baseline establishment on energy savings estimate in green building certification. *Journal of Physics: Conference Series*, *2596*(1), 012051. DOI: 10.1088/1742-6596/2596/1/012051

Nguy n, T., & Waibel, M. (2023). Promoting urban health through the green building movement in vietnam: An intersectoral perspective. *Sustainability (Basel)*, *15*(13), 10296. DOI: 10.3390/su151310296

Ohaedeghasi, C. I., Ezeokoli, F. O., Agu, N. N., Okolie, K. C., & Ugochukwu, S. C. (n.d.). rospects for improving building maintenance management in nigerian public universities: A case study of nnamdi azikiwe university. *Journal of Engineering Research and Reports*, *21*(4), 53–59. DOI: 10.9734/jerr/2021/v21i417458

Paumgartten, P. (2023). The business case for high performance green buildings: Sustainability and its financial impact. *Journal of Facilities Management*, *2*(1), 26–34. DOI: 10.1108/14725960410808096

Prozuments, A., Borodinecs, A., Zaharovs, S., Banionis, K., Monstvilas, E., & Norvaišienė, R. (2023). Evaluating reduction in thermal energy consumption across renovated buildings in latvia and lithuania. *Buildings (Basel, Switzerland)*, *13*(8), 1916. DOI: 10.3390/buildings13081916

Qin, L. (2023). Impacts of green rating systems on the economy and society. *IOP Conference Series. Earth and Environmental Science, 1204*(1), 012006. DOI: 10.1088/1755-1315/1204/1/012006

Reddy, V. J., Hariram, N. P., Ghazali, M. F., & Kumarasamy, S. (2024). Pathway to sustainability: An overview of renewable energy integration in building systems. *Sustainability (Basel), 16*(2), 638. DOI: 10.3390/su16020638

Regona, M., Yigitcanlar, T., Hon, C. K. H., & Teo, M. (2023). Mapping two decades of ai in construction research: A scientometric analysis from the sustainability and construction phases lenses. *Buildings (Basel, Switzerland), 13*(9), 2346. DOI: 10.3390/buildings13092346

Sayed, E. T., Olabi, A. G., Alami, A. H., Radwan, A., Rezk, A., & Abdelkareem, M. A. (2023). Renewable energy and energy storage systems. *Energies, 16*(3), 1415. DOI: 10.3390/en16031415

Shailesh, K., Ramswaroop, R. Y., & Nidhi, C. (2024). Ai-driven energy management systems for smart buildings. *Power System Technology, 48*(1), 322–337. DOI: 10.52783/pst.280

Shufrin, I., Pasternak, E., & Dyskin, A. (2023). Environmentally friendly smart construction review of recent developments and opportunities. *Applied Sciences (Basel, Switzerland), 13*(23), 12891. DOI: 10.3390/app132312891

Song, J. (2023). The green building development and assessment method – energy and cost analysis in china. *Highlights in Science. Engineering and Technology, 29*, 149–154. DOI: 10.54097/hset.v29i.4534

Soyombov, D. (2024). The role of policy and regulation in promoting green buildings. *World Journal of Advanced Research and Reviews, 22*(1), 139–150. DOI: 10.30574/wjarr.2024.22.1.1047

Surya, R., Ali Berawi, M., & Sari, M. (2024). Development of a green building assessment system at the green building planning stage in indonesia. *E3S Web of Conferences, 517*, 05017. .DOI: 10.1051/e3sconf/202451705017

Tasan, M., Ozkan, Y., Ozgur, A. O., & Ozpinar, A. (2024). Data plateau: A unified analytics platform with intuitive interfaces for real-time and ml-driven insights. *Orclever Proceedings of Research and Development, 4*(1), 73–89. DOI: 10.56038/oprd.v4i1.457

Umoh, A. A., Nwasike, C. N., Tula, O. A., Adekoya, O. O., & Gidiagba, J. O. (2024). A review of smart green building technologies: Investigating the integration and impact of ai and iot in sustainable building designs. *Computer Science & IT Research Journal*, *5*(1), 141–165. DOI: 10.51594/csitrj.v5i1.715

USGBC. (2024). *Green Building*. From U.S. Green Building Council: https://www.usgbc.org/articles

Weisser, W. W., Hensel, M., Barath, S., Culshaw, V., Grobman, Y. J., Hauck, T., Joschinski, J., Ludwig, F., Mimet, A., Perini, K., Roccotiello, E., Schloter, M., Shwartz, A., Hensel, D. S., & Vogler, V. (2022). Creating ecologically sound buildings by integrating ecology, architecture and computational design. *People and Nature*, *5*(1), 4–20. DOI: 10.1002/pan3.10411

Wu, S. R., Fan, P., & Chen, J. (2015). Incorporating culture into sustainable development: A cultural sustainability index framework for green buildings. *Sustainable Development (Bradford)*, *24*(1), 64–76. DOI: 10.1002/sd.1608

Yang, Q., Pu, L., & Huang, S. (2024). Review of urbanization-associated farmland research in china: A sustainability perspective. *Land (Basel)*, *13*(4), 534. DOI: 10.3390/land13040534

Yin, S., Liu, J., & Zeng-lin, H. (2022). Relationship between urban morphology and land surface temperature a case study of nanjing city. *PLoS One*, *17*(2), e0260205. DOI: 10.1371/journal.pone.0260205 PMID: 35139084

You, W. & Zhong, L. (2023). Exploring hydrogen-based energy storage systems for canadian residential buildings: an energy evaluation methodology. *E3S Web of Conferences, 396*, 04001. .DOI: 10.1051/e3sconf/202339604001

Yudelson, J. (2009). *Green Building Trends: Europe* (1st ed.). Island Press.

Zhang, W. (2023). Smart building skin design with dynamic climate adaptability of smart cities based on artificial intelligence. *Advances in Multimedia, 2023*, 1–12. DOI: 10.1155/2023/2859195

KEY TERMS AND DEFINITIONS

BIPV (Building-Integrated Photovoltaics): Photovoltaic materials that replace conventional building elements, serving as both construction materials and power generators.

BREEAM: A leading sustainability assessment method for master planning, infrastructure, and buildings, measuring sustainable value across categories like energy and ecology.

Building Information Modeling (BIM): A digital representation of a facility's physical and functional characteristics, aiding in decision-making throughout its lifecycle.

Building Management Systems (BMS): Systems that control and monitor building equipment like HVAC and lighting, crucial for maintaining green building efficiency.

Carbon Footprint: The total GHG emissions by a person, organization, or product. Green buildings aim to reduce carbon footprints through energy efficiency and sustainable practices.

Embodied Energy: The total energy required for production, from extraction to installation. Sustainable construction favors materials with low embodied energy.

Energy Efficiency: Reducing energy use in buildings through technologies and practices that maintain or improve comfort and productivity.

Green Building: Structures designed to minimize environmental impact by efficiently using resources, protecting health, and reducing waste and pollution.

Lifecycle Assessment (LCA): Evaluating the environmental impacts of building materials and systems throughout their lifecycle, from creation to disposal.

Passive Design: Architectural strategies leveraging local climate to maintain comfortable temperatures, minimizing the need for active heating and cooling systems.

Renewable Energy: Energy from naturally replenished sources like solar and wind, integrated into buildings to reduce reliance on non-renewable resources.

Sustainability: Practices promoting long-term ecological balance, energy efficiency, and reduced resource consumption, ensuring future generations can meet their needs.

Sustainable Materials: Environmentally friendly building materials with low embodied energy, recyclability, and minimal environmental impact.

Urban Sustainability: Designing and managing cities to ensure long-term environmental, economic, and social health, with green buildings playing a key role.

Water Efficiency: Reducing water use through low-flow fixtures, rainwater harvesting, and greywater recycling in green buildings.

Chapter 8
Role of Civil Engineers in Sustainable Water Management

Sandipan Babasaheb Jige
SR College, India

ABSTRACT

Abstract- The access to clean drink water is basic water service needed for the civilized life. In the global around 60% population live in water stress area where available not sustainably meet demand for at least part of year. The conventional water sources provision by snowfall, groundwater and rainfall extraction is insufficient to complete growing demand, with need to find sustainable alternatives. The engineering solution offer human societies means to bypass ecosystem service and appropriate planetary resource directly to replicate desire service. The civil engineer provides facilities for day-to-day life and work, they design build and help. They also maintain and successfully operate structure like dams, tanks, pipelines, pumping stations, reservoirs and treatment plants for water and waste water. The rivers play important role in development of the civilization they provide natural water. It is significant source for ecosystem and human life for keeping health good and sustainable development.

INTRODUCTION

Today, the world is confronted with a global water quality crisis. This is due to the continued growth of industrialization, urbanization, population and environmental pollution. Water is an essential part of life. About 97.5% of the water is found in the ocean and reaming used as fresh water which is used. In some parts of the

DOI: 10.4018/979-8-3693-2448-6.ch008

world, rivers, lakes, dams and seas are polluted, so the fresh water does not get to the people. The human needs water for drinking. The brain and heart of humans comprise 73% of water. The lungs contain 83% and the skin contains 64% water. In the body, muscles and kidneys contain 79% and bones 31% water composed (Birdie J. S., Birdie G.S. 2014). In the day, adult males consume a minimum of three liters of water and females up to 2.2 liters of water. In the world, ten in one do not have fresh and clean drinking water. Globally, about 700 million people will be displaced by intense water safety by 2030. In the future, about two-thirds of the population will live in water-stressed areas and about 1.8 billion will live in areas which are plagued by water safety by 2025. According to the World Health Organization (WHO) and United Nations Children's Fund (UNICEF), more than 884 million people have no access to improved drinking water. The lack of clean water is responsible for many diseases and deaths in the world. Today, around the globe, one out of every six people does not have adequate access to water and more than double that number lacks basic sanitation for which water is needed. In many countries, safe drinking water is not accessible, so, due to diarrhea, many children lose their lives. Water for drinking and personal use is a small part of society. Household water usually accounts for less than five percent of total water use. Water is an essential part of civilization, and it is a global issue. Without proper design of water distribution and wastewater collection, one cannot attain sustainability with respect to water. Civil engineers are currently implementing smart farm technology which prevents communities from depleting clean water resources (Daniel Mother 2018). They monitor soil level using data it collects to determine exactly how much water is allotted for field and to prevent waste water. The new technologies produce clean drinkable water by treating saline water by a process known as desalination. Water preservation is important in times of drought. The Geographic Information System (GIS) and Artificial Intelligence (AI) are leveraged to enhance access to water, sanitation and hygiene services and advance global development targets.

The sustainable water resources system is designed and managed to best serve people living and in the future. Sustainability means different things to different people. It includes consideration of the welfare of those living in the future (Allen 2015). The environment and ecological sustainability have resulted from growing concern about the long-term health of our planet. The inclusion of sustainability criteria along with more common economic, environmental, ecological and social used to evaluate alternate water resources development management strategies to identify the need to change how we commonly develop and use our water resources (David Chin 2020). The use of sustainable water systems is those designed and operated in ways that make them more adaptive, robust and resilient to an uncertain and changing future. The changing of social and institutional components of the water resources system is often the most challenging because they involve in changing the way individuals think and act. Sustainable water management is a process which meets current water needs without compromising future water needs (Benny J. 2017). It ensures that everyone has access to clean water today, but not at the cost of future generations to come. Sustainable water management is a complex issue with social, economic and environmental consequences. Everyone on the planet has access to adequate water resources for drinking, sanitation purposes and vital uses for sustaining life. Fresh water is essential for the life of the earth for each person to access enough water to live. We deplete our water supply, of which there are no alternative water resources. So we ensure that they meet the water use needs of every individual by more effectively managing the earth's water resources (Suryanarayanan S. and et.al 2024). For sustainable management, achievement requires a multidisciplinary and holistic approach in which environmental, technical, societal, cultural issues, economic and landscape esthetics are addressed. Civil engineers play an important role in it. They plan, design, construct and help to maintain and optimize the operation of structures and facilities like treatment plants, dams, canals, pipelines, pump stations and bridges (Biradkar A. H. 2020). They also create, improve and protect the environment in which we live. Civil engineers need to focus on sustainable engineering practice to make sure the natural environment's integrity is maintained while meeting the needs of the present generation without compromising the needs of future generations.

Objectives

- To study the role of civil engineers in sustainable water management
- To focus on sustainable water management for better future
- To study the future demand of sustainable water management
- To aware young generation about water issues and need of sustainable management

Analysis and Results

All animals, plants and living things need water to survive life. There can be no life on earth without water, because water is an essential part of our life. The human body shows a 60% weight by containing water. Water is a unique and non- substitutable resource for life, society and the economy. It has many value and benefits. It has different things for different people, like in households, schools, and workplace, for health, hygiene, dignity and productivity (Murty V.N. 2018). The water is found on earth from polar ice caps to steamy geysers. Water is an inorganic, transparent, testless, odorless and colorless chemical substance, having the chemical formula H2O. The chemical formula of water was discovered by Henry Cavendish. Water is the main constituent of the earth's hydrosphere, and it acts as a solvent. On the earth, water covers 70.9% of the surface of the earth. It occurs mostly in the sea and oceans. In the form of groundwater, its volume is 1.7%. In the ice caps and glaciers, its volume is 1.7%; in the form of vapor, clouds and precipitation, its volume is 0.001%. The water cycle in the environment continuously occurs and water moves in it by evaporation, transpiration, condensation, precipitation and runoff and reaches into the sea (Deka Mirdul 2018). The water word is derived from the old English word water and the photo Germanic word.

Global Water Status and Issues

The adult male consumes three liters per day of water per day, while the adult female needs 2.2 liters per day. In the world today, 771 million people who are ten in one, do not have fresh and poor water and 1.7 billion people and one in four lack access to toilet facilities. The earth has covered about 71% of the water in the ocean; 96.5% of the water is held in the oceans. It also exists in the air as water vapors in rivers and lakes, and in the icecaps and glaciers in the ground as soil moisture (Dr. Sumalatha J. 2023). In the world, about 700 million people in the world are displaced by intense water scarcity by 2030. Today, nearly half of the population lives in potentially scarce water for at least 30 days a year and it will reach up to 5.7 billion in 2050. In the world, the agricultural sector needs one trillion cubic feet of water per year. It is equal to the annual flow of twenty Nile Rivers. In Asia, the region has shown a 73% potential water affects peoples' regions. In the future, two-thirds of the world's population will live in water-stressed areas and about 1.8 billion people will live in areas plagued by water scarcity by 2025. In the future also in ten people, one person gets clean water. About 785 million people do not get pure water for drinking. On the surface, water, about 144 million people, is dependent on it; globally, about two billion people use contaminated water for drinking purposes (Robinson R. J., et al. 2007).

The water has produced different types of diseases, like typhoid, diarrhea, cholera, dysentery and polio etc. The disease diarrhea produced by contaminated water causes about 4, 85,000 deaths each year globally. Diarrhea also causes 2, 97,000 deaths of children fewer than five years of age. In 2017, schistosomiasis affected about 220 million people. It is caused by the parasitic worms contracted by exposure to infected water. It is a chronic disease of humans. In the extremely high-water vulnerable areas today, 1.42 billion people and 450 million children live. In the world, less than three percent of water resources are fresh water. It is growing scarce because of misuse, poor management and over-extraction of groundwater (Shimon C. 2024). The population is rapidly growing throughout the world, so the demand for water also increases day by day, but 72% of water withdrawals are for agriculture, 16% for municipal household services and 12% for industrial areas.

Sustainable water management is a process for meeting current water needs without compromising future water needs. Everyone ensures access to clean water without the expense of future generations to come. Sustainable water management is a complex process related to economic, social and environmental consequences. On earth, everyone accesses water for drinking and sanitation purposes and other uses for sustaining life (Larry W. Mays 2006). Society and technology evolve and the population grows. The water demands continuous rise. Sustainable water management is an important way of implementing new government water polices with better planning. For the achievement, sustainable water management requires a holistic multidisciplinary approach in which technical, economic, landscape aesthetic, environmental, economic, societal and cultural issues are addressed (Glenn O. & et al. 198). The global scale for sustainable water provides each person with affordable access to a minimum of 20 to 50 liters daily of water needed for sustainable life. The United Nations General Assembly gives the right to safe water, cleaning, drinking water and sanitation as human rights that are essential for full enjoyment of life and all human rights (Pankaj Kumar, and et.al 2022). So, in the future on the planet, everyone needs water to enjoy life.

Historical Water Conservation Methods

The archaeological evidence of India shows the water conversation and management system is deeply rooted in ancient India. An Indus valley civilization city shows the excellent water conservation, harvesting and drainage system. The King of Bhopal Bhoja built the largest artificial lake in India. In Arthur, Chanakya mentions the irrigation use of water harvesting systems. The Chola king Karikala built kallanai across the river Cauvery to divert water for the irrigation of land. In India there were many systems of water conservation in the ancient period. Some historical water conservation methods are as follows:

- **Rooftop harvesting:** The water that falls on the ground does not deposit its runoff quickly in spring and in small streams when there is no rain. The roof harvesting systems are used for the storage of drinking water in all states. The houses built with sloping roofs with iron sheets are conducive to rain water harvesting. In the harvesting system, rain gutters are used on the side of the roof and are made from corrugated iron sheets. The water pours into a pipe which has connected to tanks; the tanks are cementing concrete located under the house. The collected water is used in dry seasons for many purposes (Slupheny M. M. 2018).

- **Talab (Lake):** It occurs in every part of India. It is contracted for the reservoirs for water that sore water for household consumption and drinking purposes. The pokhariyan ponds are natural ponds, but some are built by humans. The reservoir with areas having different types the area having less than five big has it is known as Talai and the size is medium, called Bandhi, and when the size is very big, the lakes are called Sagar.
- **Apatani:** This water conservation system is used for many purposes, like land, water and farming systems. It also protects soil erosion and conserves water for irrigation and also for paddy and fish culture. The system is a rice cultivation and fish farming system occurring in an elevated region of about 1600 meters high and in gentle sloping valleys. In this system, both ground and surface water are used for the irrigation. The Apatani tribes of Zero in Arunachal Pradesh used this system; the valleys were terraced into plots separated by a 0.6-meter-high dam supported by bamboo farms. The separate plots have shown inlets and outlets for the water. The deeper channel connects from the inlet point to the outlet point (Gurmel Singh 1990). The plots drained with water with the help of opening and blocking the inlet and outlet points. The stream water is also trapped by well-constructed 2–4 meters high and one meter thick near forest hill slopes. This structure of system has been highly successful and continuously used for many years.
- **Kund:** It has a saucer-shaped catchment area of water. It has a small slope towards the central circular underground well. The water is used for harvesting water and also for drinking it in sandier tracts in Rajasthan and Gujarat. These well pits traditionally consist of lime and ash covers, but in the modern period they are constructed of cement.
- **Taanka:** It is a traditional rainwater harvesting technique that mainly occurs in the indigenous in the Thar Desert region of Rajasthan state. It is a cylindrical paved occur in an underground pit into which rainwater from rooftops, courtyards or artificially prepared catchment flows occur (Mohammad H. D. 2023). It has completely filled once. The water stored in a taanka can last throughout the dry season and is sufficient for a family of 5–6 members. It is also an important element of water security in these arid regions. These tanks have saved families from the everyday drudgery of fetching water from distant sources.
- **Jhalara:** It is a step well, having a rectangular shape. The wall shows the steps on the three or four sides. The well collects the subterranean seepage of the upstream reservoir. The Jhalara has been built to ensure an easy and regular supply of water for royal ceremonies and for community use. In the Rajasthan state, Jodhpur city has these types eight Jhalaras and in which the oldest one is occurs in 1660 AD.

- **Johads:** It is the oldest system used for water conservation and the recharge of groundwater; it has a beneficial system for groundwater levels increasing. The small earthen check dams that capture and store rainwater show the construction in an area that has naturally occurring high elevation on three sides and, on the fourth side, excavated soil is used to create a wall, in which storage pits are also made from the excavating area.
- **Kuhls:** It is mainly found in the Himachal Pradesh region of India. It has surface channels occurring in mountain areas. A good example is the Kanga valley water system. It has shown that about seven hundred fifteen major kuhls and two thousand five-hundred minor khuls have irrigated more than 30,000 hectors land in the valley. The kuhls are built by public funds it has donated by the public and also by royal rulers. It has also been recognized as an important cultural tradition of the region (Pietro Laureano 2013). Sometimes it shows a single outlet opening into a stream or river in which many johads are interconnected through deep channels. This type of structure of water conservation, also known as madakas, prevents structural damage to water pits. This type of construction of johads occurs in the Pemghara area of Odessa Karnataka.
- **Baoli:** It is a secular structure of construction so that everyone is able to draw water; it is built by nobility civilians, in a strategic region. It also shows rooms on its sides. It is made with nice step wells having beautiful arches and curved motifs. In villages, it is used for utilitarian purposes and also for social gathering. The step will mostly be used for the agricultural drainage system by the channel into the field.
- **Ahar Pynes:** It mostly occurs in Bihar state. It has a traditional flood water harvesting system. It is built at the end of a diversion channel like a Pye; it has a reservoir having embankments on the three sides. In dry months, water is collected from rivers by artificial rivulets known as pynes. This system is mostly used in paddy cultivation areas with low rainfall, so irrigation mostly depends on ahar pynes.
- **Panam Keni:** The native tribe of Wayanad is Kuruma. It used a special feature well known as Panam Keni. It is used for storing water. The stem of toddy palms is used in this system. The wooden cylinder stem of plants is made by soaking the stems in water for a long time so that the core rots away, and only the outer hard layer remains for use. It is shown to be four feet in diameter and is good in depth and has been immersed in groundwater springs occurring in the forest and in the field. It is very useful in the hottest summer seasons for people's drinking water.
- **Kere:** It occurred in Karnataka. The tanks are called in the Kannada language as kere. It has the predominant traditional irrigation system. It occurs in the

Karnataka plateau. It is fed by channels branching off from check dams built across streams or streams in a valley. The outer let flow of one tank connected to the next and the course of the stream, the tanks are built in series located in an area of a few kilometers (Khan Naseer 2021). Water wastage by overflow is not allowed in it and also seepage from the tank higher up in series; it has been collected in the lower side.

- **Zabo:** in the Zabo type of system, water conservation occurs in the forestry region, for agriculture and animal care. This system, also called the Ruza system of water conservation, has been plasticized in the Nagaland area. The pond-like structure created near terraced hill sides, the rainwater falls on forested hill tops has been collected by channels, and it has been deposited by runoff water in this structure. The water channel passes from the cattle yard so that it carries the dung and urine of animals and goes into the paddy field which is located at the foot of the hills. The ponds created near the paddy field are also used for fish and foster the growth of medicinal plants and also different ornamental plants.
- **Eri:** It is one of the oldest water management systems in India. It mostly occurs in Tamil Nadu state. The eri means tank. It also acts as a flood control system. It also controls wastage of water at the time of heavy rainfall. It also helps fuel ground recharge of water and increases the level of water. It also prevents soil erosion. This system is fed by channels, and it diverts the river water. Another eri is a non-system that is fed slowly by rainwater (Barah B.C. 2015). In the system of eri, tanks are interconnected in order to enable access to the farthest village. It balances the water when their supplies become excessive. The irrigation of all complete water use has enabled the eri water system. It also supports paddy cultivation. Without the eri system, paddy cultivation in Tamil Nadu state would have been impossible.
- **Zing:** It occurs in the Jammu and Kashmir region of India. It has small tanks that collect the melting glacier water. A guiding channel network brings water from glaciers into the tanks. The trickle in the morning melts and the water of the glacier turns into a flowing stream in the afternoon (Dr. Panda S. C. 2018). In dry regions, water is equitably distributed. It relies on the melting of glacial water to meet its farming needs.
- **Jack Well:** In the Great Nicobar Islands, the tribal people like Shipmen lived in a region of rugged topography. They used these types of wells for harvesting water. In this water conservation system, low-lying regions of the island are converted into a jack well in which pits are encircled by bunds made of logs of hard wood. A long-length of bamboo is also cut longitudinally and placed under trees to collect runoff water from leaves (Yadav B., & et al.

2022). The big jack wells are interconnected with more bamboo so that overflow from one jack to the other ultimately leads towards the big jack well.

- **Nadi:** It occurs in Rajasthan state. It is the type of village pond it stores rainwater and is collected from. The location of Nadi was selected after careful deliberation of its catchment area and runoff characteristics. The Nadi has a strong bearing on storage capacity. It has received water supply from rainfall, erratic and large amounts of sandy sediments deposited regularly in the water supply. The nadis promote afforestation of their drainage basin, preventing siltation by adding spillways and silt traps into the old nadis.

Role of Civil Engineers in Water Sustainability

Civil engineering creates, improves and protects the environment in which we live. It also designs, builds and helps to maintain and successfully operate large facilities. The facilities are dams, reservoirs, tanks, pumping stations and pipelines with water and wastewater treatment plants. The engineering consider different factors, like construction cost, expected lifetime of project, government and environmental regulations for potential natural hazards like hurricanes and earthquakes. The engineering also works on problems like soil nature, groundwater contamination, energy development and conservation (Juvenile Fiction 2009). They also focus on sustainable practices for natural environmental integrity without compromising the needs of future generations. The structural engineers specialize in selection and designing materials for buildings, dams and other facilities. The environmental engineers train in water quality, treatment, wastewater management, air pollution control and solid waste management and related fields.

Methods of Water Sustainability

The Geographic Information System (GIS) and Artificial Intelligence (AI) innovations are being leveraged to enhance access to water for sanitation, hygienic services and global development targets. Water is a global issue. It is necessary for civilization. Water-scared nations are net imports of water for the needs of their growing population. Sustainable water management is only part of sustainable development. Proper distribution and wastewater collection are sustainability not attained with respect to water. The loss of water causes and the use of energy unnecessarily occurs in improper water distribution. The collection of waste water leads to leakage and water pollution. The wastewater is a combination of one or more domestic effluents consisting of black water which contains urine, excreta and fecal sludge. The gray water includes kitchen and bathing water from commercial establishments and institutions like education and hospitals, industrial effluent, storm water and urban runoff, agricultural and horticultural run-off (Mohammad H. & et al. 2023). The wastewater contains a wide range of chemical, biological and physical contaminates, like inorganic solids, heavy materials, biodegradable organics, pathogens, nanoparticles, endocrine disruptors, refractory organics and nutrients, macro solid emulsion and pharmaceuticals. The wastewater discharge of partially treated or untreated in environment is commonly in low-income countries. It affects human health, the environment and shows economic impact.

Engineering produces many new technologies and processes for clean and drinkable water. The saline water is treated by the desalination process. These processes do not purify salt water sources from the ocean; they also treat groundwater, which has an unhealthy concentration of sediment or salts. Solar desalination produces vapor compression distillation facilities. The desalination plants are of different complexities and are used all over the world. The desalination process often mirrors the natural process of producing rain; human-orchestrated desalination has been used to remove salt for thousands of years. The simple desalination method distillates water by converting it into steam and collecting vapors. The most feasible method and large-scale desalination comprised a multiple-stage flash evaporation method in 1960. The irrigation system for farms, gardens and lawns saves a significant amount of water. The indoor fixtures and efficient outdoor irrigation technologies like smart controllers help to save water (Wolfram Mauser 2009). The out spray sprinkles in drip irrigation deliver water directly to plants buried under the lawns. The rainwater harvesting system is based on the attachment of a water butt to the rainwater down pipe, a completely designed large which collects rainwater from a large area. It is used for domestic properties it collects from rainwater. It is one of the sustainable water-saving methods. In agriculture, more than half of the annual water is used. Preserved water is important in times of drought for agriculture. The water balance between communities' needs and agricultural needs is important, so civil engineers implement smart farm technology, which prevents communities from depleting their clean water resources. It monitors soil moisture level data used for how water to water is allocated to fields to prevent waste of water (Murthy Satya N., Challa 2020). The cutting-edge irrigation system provides an accurate amount of water uniformly to field, more portable water is available for the border public and thus water shortages are avoided due to the use of smart sustainable technologies in agriculture. Agriculture is the thirstiest consumer of water globally, with about 70% of water withdrawals worldwide. Raining water is the predominant agricultural production system around the world. By 2050, world agriculture needs to produce more than 60% of food globally and 100 percent more in developing countries. The industry and energy need about 20% of water demand. The more developed countries use a larger proportion of freshwater for industry than less developed countries. Municipal water use accounts for 10% of total water use. The world needs more water for drinking and improving sanitation. It is the most important challenge for sustainable development.

Modern Methods for Water Sustainability

In society, more wastewater is produced by human actions. It treats and eliminates harmful bacteria, chemicals and virulent toxins found in it. In wastewater treatment, modern engineering technologies have progressed and converted it into clean water that is suitable for discharge back into the environment. Civil engineers developed progressive techniques and converted waste water into clean water by basically three methods.

i) **Physical wastewater treatment-** These processes rely on natural physical forces like gravity and electrical attraction to separate harmful substances from water. The physical wastewater treatment includes adsorption, flotation and sedimentation.

Chemical wastewater treatment- It requires the use of a chemical solution to treat the wastewater. The chemical neutralization, precipitation and disinfection-like processes using ozone, chlorine and ultraviolet light. The clean usable water comes from the chemical wastewater treatment method.

Biological wastewater treatment-In this process, bioreactors like bacteria, algae and fungi are used to transform organic matter within wastewater into energy. This process of bioconversion relies on the biological oxidation and biosynthesis processes. It also leaves some end products which must be discharged, like minerals.

ii) **Pure water purifying sachet-** In developing countries, water-borne diseases are common due to facilities. They do not build facilities for water purification of contaminated water. Many social organizations work in such areas and develop low-cost solutions to this problem. The world's largest company, Procter and Gamble, develops one water purification product, which is Pure Water Purifying Sachet. It employs a powder mixture formulated to remove suspended matter and pathogenic microorganisms from contaminated water. One pocket of the powder potentially turns about ten liters of dirty water into clean drinking water and saves lives who are exposed to waterborne illness. It is an amazing water resource engineering example of sustainable water production.

iii) **Water purification tablets-**They help to clean water by disabling harmful viruses, bacteria and other microorganisms with their active ingredients like iodine, chlorine and chlorine dioxide. In this treatment process, the chlorine is simply allowed to evaporate from the water, being removed by the lid of the water container, and iodine-neutralizing tablets are also used to mask unpleasant tests of iodine. The tablet serves as an intermediary solution to clean water shortages in areas that were recently struck by natural disasters and struggle to maintain a clean water source.

iv) **Small decentralized distillation units**- In the central water supply system, the pipe connected from the home to the central water source is the most readily available way of servicing a large community's need for clean drinking water. In the developing areas of the world, such infrastructure is not available due to many logistical features which are built, connected and maintained (Zaman A., Md. Hedayetullah 2022). The decentralized water distillation offers the potential to create clean and reliable drinking water for areas which are not connected to a centralized water supply due to physical or economic restraints. The decentralized unit utilizes one of many methods or systems to distill or filter contaminated surfaces and groundwater. This system is classified based on the quality of clean water they supply. The point of use supply system creates 25 liters of water per household. The point of entry supply system treats all water that enters the household. The small-scale system provides water to entire communities with quantities ranging from one thousand to ten thousand liters per day. It uses heat and radiation to destroy pathogens, chemical treatment to purify contaminated water and physical processes to separate contamination from water. The development of an advanced decentralized distillation unit which is cost-effective and has a broad longer range, civil engineers provide clean water to communities at a sustainable cost. The environmental and civil engineering fields provide water resources to millions of people around the globe with a sustained source of clean water.

v) **Gray water recycling system-**The water used in building structures is thought to be clean water coming from sewage or black water going out. Gray water is something between that. In domestic settings, gray water systems collect water from hand basins, baths, and showers (Garg S. K. 2018). The collected water is reused for washing machines, toilet flushing and other external uses.

vi) **Water meter-** It is a common term used by people when discussing water conservation techniques. The water meter measures water use in each residential apartment. It is installed in each home of an apartment, and it records the amount of water being used for billing purposes and tracks consumption of water.

vii) **Pressure-reducing valves-**The high water-pressure waste a lot of water, so installing a water-pressure-reducing valve turns out to be a great solution. The valve is used in commercial, residential and institutional applications to lessen incoming water pressure to lower the predetermined level (Pal J. G. 2023). This method protects the downstream plumbing system components by reducing water consumption.

viii) **Water-efficient taps-**These caps work in two ways: either reduce the water flow rate by tap or support users to avoid wastage of water by automatically running it off. Today at airports, shopping mall sensor taps that automatically sense the users' activity based on supply or turning off water using infrared sensors.

ix) **Insulated pipes-**The used piping and storage tank insulation is important for any hot water system. In most buildings, hot water return pipes are un-insulated or not properly insulated. It shows the need to wait for faucets or showers for hot water to flow when it is needed. Its results show significant wastage of water (Meenakshi P. 2019). The pipe, when insulated, then ensures hot water is immediately available and after the tap closed, it is supplied back to the plant. It helps to reduce the energy demand of heating units. The insulated with pipe likely waste by waiting for it with the tap open.

x) **Water-saving showers-**In the home, hotels personal bathing use a lot of water. It is about 80 liters. The use of showers for five minutes uses only 50 liters of water. The water-saving shower heads control the water flow rate and spray in a pattern. The showers have different sizes, shapes, colors and designs which have a direct effect on water consumption during baths.

xi) **Waterless urinals-**The conventional toilets and urinals contain water that harbors disease, causing bacteria to leak into the air when flushed on the surface. In waterless, urinals have no flushing action and are designed to dry out between uses so they do not release contamination (Neil S. Grigg 2023). All urinals are the same principle. The gravity carries urine from the basin by a seal in a cartridge into the plumbing drain system; the seal prevents odors and sewer gases from rising up the pipe and into the restroom. The seal is formed by a layer of liquid sealant within the disposable cartridge within an integrated trap.

xii) **Efficient toilets-**The National Energy Policy Act 1995 mandates new toilets to have low flow use no more than 1.6 gallons per flush. That is half the old used toilets. The water-efficient toilet's design and operation shifted from removing waste through flush water volume to increasing flush water velocity (Slupheny M. M. 2018). The limited supply and continuous waste resources is crucial to waking up and using wisely, so it protects and preserves our environment, minimizes the effects of drought, serves energy and water bills. The upgrade of low-flow toilets plays a crucial role in reducing the water footprint.

xiii) **Closed-loop water recycling in offsite construction factories-**The offsite construction factories are at the forefront of adopting advanced water conservation techniques. It implemented a closed-loop water recycling system within the manufacturing process. It captures and treats water used during different construction stages of component fabrication, allowing it for continuous reuse within the factory (Ralph A., & et al., 2015). The loop closed on water consumption by the factories significantly reduces demand for fresh water and currently minimizes wastewater generation. The approach showcases the commitment of offsite construction to sustainable practice both outside and within production. It contributes to a more circular and resource-efficient construction ecosystem.

xiv) **Low flow fixtures-** The installation of low flow faucets, shower heads and toilets in home and commercial buildings significantly reduces water consumption. These fixtures use less water while maintaining adequate performance, saving both water and energy which is used for water heating.

xv) **Xeriscaping-** It is a landscaping technique that focuses on drought resistance and native plants minimize the need for irrigation. This practice does not conserve water and also reduces the use of fertilizers and pesticides (Suresh R. 2005). Leak detection technology- It is modern technology for the detection of leaks in plumbing systems. The smart meters and leak sensors send alerts to homeowners or businesses when unusual water usage patterns detect and prevent water wastage.

Water Sustainability in India

Drought and flooding are two crucial adverse effects of the water imbalance on the earth. The drought affects human activity throughout the world. The historical record of drought shows they come from every part of the world. The drought is the chief cause of most famines throughout the history of humanity. Different civilizations perished because of abnormalities long persistent deficiency of rainfall. In the Indian historical record, the Vāyu Purana describes the earliest drought; the Ramayana describes the drought during the period of King Dasaratha. In Mahabharata, the drought period is described in the region Mandhata and also in the period of King Shantanu, ruler of Hastinapur. The king Chandragupta Maurya reigned due to serious famine (Banerjee G. and Ganguly S. 2018). In the period of pre-independence, drought and famine caught the attention of British rulers in the ninetieth centaury. The Indian Finance Commission describes severe famine and drought conditions in Punjab and Great Bengal.

India is one of the most water-rich countries in the world; it covers 04% of the world's water resources. Indian Cultures Rivers have become the heart and soul of their development. The major rivers cover about 253 million hectors and the medium rivers have about 24.6 million hectors and many river systems are perennial and seasonal (Sonarkar V. K. 2021). In India, the Ganga, Brahmaputra and Meghana are the largest river systems in India. Other major systems are the Indus, Sabarmati, Narmada and Mahi rivers and Tapi, Brahmani, Godavari, Krishna, Pennar and Kaveri. In the medium river system, the Subarnarekha River has 1.9 million hectors as a catchment area. In history, many floods and droughts regularly occurred in ancient times, but every part of the country has its own traditional water conservation ideas and systems. That reflects the geographical peculiarities and cultural uniqueness of that region or part. These techniques have evolved in specific conditions due to the peculiar climate and geology; the projects are not large and located in a decentralized manner and community-owned at times, they serve multiple purposes as well. The traditional system is an ecologically safe, viable and cost-effective alternative (Mridual Deka 2018). The small projects adequately intercept the large quantum of rainfall available for exploitation in a sustained manner. In the storage behind large dams in any basin, ensure that there is cushion of reserve as well as in build carry-over storage. In India, the average annual surface water flow is estimated as 1869 cubic km. The Central Water Commission indicates the water resources utilizable by the construction of a structure of 690 cubic km. The monsoon rain is the main source of water, with 76% rainfall occurring from June to September. The average annual precipitation in volumetric terms is 4000 cubic km. The demand for water requires engineering intervention to develop water resources to meet the required demands of humans. In India, the national water policy was prepared and adopted in 1987 it revised in 2002. The contamination of groundwater due to over-exploitation has also emerged as a serious problem in India. The water policy plan irrigation takes account of the irritability of land, and cost-effective irrigation is possible for all available sources of water (Singh T. K., Kuma R. 2022). The adaption of irrigation techniques optimizes water use efficiency; its benefits maximize the production of crops. The water allocation in the irrigation system has been done with regard to equality and social justice. The country has 14 major river basins and 44 medium river basins. In addition, there are 2.4 million water bodies occurring in different parts of India (Jose Maria and et.al 2024). In the Waste Bengal state there are more pounds and reservoirs occur. In Andhra Pradesh, many tanks occur, Tamil Nadu shows most of the lakes and Maharashtra state leads to water conservation schemes. The civil engineers plan, design, construct and help to maintain and optimize the operation of structures and facilities like pipelines, roads, canals, pump stations, bridges and dams. So civil engineers play an important role in water sustainability in India. The data of renewable internal fresh water resources refers to quantity of internal freshwater from in flowing river basins and recharge the ground water aquifers shown by following table.

Table 1.

Sr. No	Region	Water availability
01	Latin America and Caribbean	13,868km3
02	East Asia and Pacific	10,106km3
03	Europe and central Asia	7,071km3
04	North America	5,668km3
05	Sub-Saharan Africa	3,884km3
06	South Asia	1,982km3
07	Middle East and north Africa	230km3
08	World	42,809 km3

Source- Food and agriculture organization United nation (2025)

Water Sustainability Challenges and Future of World

The climate change impacts alter the precipitation pattern, it projected drastically reduce renewable surface water and groundwater availability and intensify competition for these resources. The increased extreme events like droughts, floods and other water related disasters are expected become more frequent and serve posing risk in water security and ecosystem (Dr. Panda S. C. 2018). The high temperature also leads to increase evaporation rate and water and water security in many places. The water demand grows due to rapid urbanization and population growth. The agriculture is major consumer of the water the demand of food increase for feed larger population so that water demand also increases for the agriculture. The industry and households also grows the water demand and putts pressure on water resources (Challa Satya Murthy 2002). The effective governance is crucial for sustainable water management. The unclear leadership and responsibilities, lack of funding is common governance issues that hinder implementation of sustainable water management practices. The water security is pressing issues in the world it exacerbated by uneven spatial and temporal distribution of freshwater resources (Colin A. Booth 2016). The water security coupled with poor water quality it threats human health and ecosystem. The economic disparities and social factors also play critical role in water resource management. In the many region of world the economic development and urbanization pressure on water resources by living standard (Veera G. and et.al 2020). Today four billion peoples of world's population serve water scarcity for at least one month in the year. Inadequate water supply occur about two billion people live in countries. About 700 million people could displace by intense water scarcity by 2030.

CONCLUSION

The engineering solution offers human societies the means to bypass ecosystem services and appropriate planetary resources directly to replicate desired services. The civil engineer provides facilities for day-to-day life and work. They design, build and help. They also maintain and successfully operate structures like dams, tanks, pipelines, pumping stations, reservoirs and treatment plants for water and waste water.

The limits of water resource recognition are almost old as realization so the planet spheres leading concerns about sustainable water resources management. The water resources grow attention given its uneven distribution in many part of the world. The civil engineering solution to address water management challenges. It plays important role in area that access clean water and sanitation providing water for irrigation, offer protection against flood, allow power generation. The major source of environmental degradation and health risk in world is lack of sewer system. The combined sewer system build in some places, the waste water release untreated and lead serve water quality.

The urbanization and infrastructure development is continuing expand, the adaptation of sustainable is important for minimize environmental impacts and ensure healthier planet for next generation. The civil engineers design for sustainable infrastructure, water management is solution of minimize environment impacts while meeting the needs of society. The civil engineers involved in planning and development of waste management system, with prioritize recycling. They also help for reduce energy consumption and water management system. The water scarcity is growing worldwide so need of its conservation design sustainable irrigation system, implement rainwater harvesting techniques and develop water treatment plants. Water is an essential part of civilization, and it is a global issue. Without proper design of water distribution and wastewater collection, one cannot attain sustainability with respect to water. The sustainable water resources system is designed and managed to best serve people living and in future.

REFERENCES

Allen, D., & Shonnard, D. R. (2015). *Sustainable engineering: Concepts, design and case studies*. Prentice Hall.

Banerjee, G. and Ganguly, S. (2018) Water resources and environmental sustainability, IBP publication.

Barah, B.C. (2015) 'Traditional water harvesting systems: An ecological economics Survey' New age international publisher.

Bhuiyan C., Wolfgang-Albert, F. and Jain S. K. (2019) Water security and sustainability: Proceeding of Down to Earth-2019, Springer publication.

Biradkar, A. H. (2020). *Basic civil and environmental engineering*. Nirali Parkashan.

Birdie, J. S., and Birdie, G.S. (2014) Water supply & Sanitary engineering-including environmental engineering and pollution control, Dhanpat Rai books publication.

Chadwick, A., Morfett, J., & Borthwick, M. (2021). *Hydraulics in civil and environmental engineering*. Crc Press. DOI: 10.1201/9781003026839

Chadwick, A., Morfett, J., & Borthwick, M. (2021). *Hydraulics in civil and environmental engineering*. Crc Press.

Charlesworth, S. M., & Booth, C. A. (Eds.). (2016). *Sustainable surface water management: a handbook for SUDS*. John Wiley & Sons. DOI: 10.1002/9781118897690

Chin, D. (2020). *Water resource engineering*. Pearson Publication.

Dehghani, M. H., Karri, R. R., Tyagi, I., & Scholz, M. (Eds.). (2023). *Water, the Environment, and the Sustainable Development Goals*. Elsevier.

Deka, M. (2018) Water Sustainability, Notion press publication.

Garg, S. K. (2018) Environmental engineering swage waste disposal and air pollution engineering, Khanna publication.

Glenn, O., Schwab, R.K., and Telcott, E. (1981) 'Soil and water Conservation engineering', Wiley publication.

Grigg, N. (2023). *Water resource management: Principle methods and tools*. Wiley Publication.

Gude, V. G., Gadhamshetty, V., & Kandiah, R. (Eds.). (2020). *Sustainable Water: Resources, Management and Challenges*. Nova Science Publishers.

Joseph, B. (2017) Environmental science and engineering, Mc Grew Hill publication.

Kumar, P., Nigam, G. K., Sinha, M. K., & Singh, A. (Eds.). (2022). *Water Resources Management and Sustainability*. Springer Singapore. DOI: 10.1007/978-981-16-6573-8

Laureano, P. (2013) 'Water conservation techniques in traditional settlements' Copal publishing group publication.

Mauser, W. (2009). *Water resources: efficient, sustainable and equitable use*. Haus Publishing. DOI: 10.2307/jj.10984417

Mays, L. (2006) Water resources sustainability, McGraw Hill publication.

Mays, L. (2019). *Water resource engineering*. Wiley Publication.

Meenakshi, P. (2019) Elements of environmental and engineering, PHI learning publication.

Mirdul, D. (2018) Water sustainability, Notion press publication.

Mohammad, H. D., Karri, R. R., Tyagi, I. and Scholz, M. (2023) Water, the environment and the sustainable development goals, Elsevier publication.

Mother, D. (2018) Civil engineering basics: Water, wastewater and storm water conveyance, Create space independent publishing platform.

Murthy, C. S. (2002). *Water resources engineering: Principles and practice*. New Age International.

Murthy, S. (2020) Water resources engineering: Principles and practice, New age international publication.

Murty, V.N. (2018) 'Land and water management engineering' Jain Brothers publication.

Pal, J. G. (2023) An introduction to water management techniques for professional engineers, independently published.

Panda, S. C. (2018). *Principles and practices of water management, M/S Agrobios*. Publication.

Ponce-Ortega, J. M., Nápoles-Rivera, F., Lira-Barragán, L. F., & Ramírez-Márquez, C. (2024). Sustainable Water Systems.

Robinson, R. J., Jones, A. A., and Woo, M.K. (2007) Managing water resources in physical and social environment, IGU home of geography publication.

Roy, P.K., Roy, M.B. and Pal, S. (2021) Advances in water resources management for sustainable use, Springer publication.

Sharma, R. (2024) water management in India: A legal approach, Universal academic books publishers and distributors.

Shimon, C. (2024) Water management: Prioritizing justice and sustainability, Island press publication.

Singh, G. (1990) 'Manual of soil and water conservation practices' Oxford and IBH publication.

Singh, P., Raj, A., & Yadav, B. (2022). Impacts of agriculture-based contaminants on groundwater quality. In *Sustainability of Water Resources: Impacts and Management* (pp. 249–261). Springer International Publishing. DOI: 10.1007/978-3-031-13467-8_16

Slupheny, M. M. (2018) Introduction to environment management, PHI learning publication.

Sonarkar, V. K. (2021). *Water resource engineering*. Nirali Parkashan.

Sumalatha, J. (2023) Environmental geotechnical engineering, Notion press publication.

Suresh, R. (2005) 'Soil and water conservation engineering' Standard book publisher and distributor.

Suryanarayanan, S., Shivaraju, H. P., & Jenkins, D. (2024). *Water management in developing countries and sustainable development*. Springer Publication. DOI: 10.1007/978-981-99-8639-2

Tiwari, K., and Kuma, R. (2022) Advances in civil engineering for sustainable development- Geotechnical, transportation & water resources, Walnut publication.

Wurbs, R.A. and James, W.P. (2015) Water resources engineering, Pearson publication.

Zaman, A., & Hedayetullah, M. (2022). *Sustainable Water Resource Development and Management*. Apple Academic Press. DOI: 10.1201/9781003180494

KEY TERMS AND DEFINITIONS

Ecosystem: It is community of the organisms in specific area, the term 'eco' means part of the world and 'system' refers foe coordinating units. In ecosystem community of organism and their physical environment interact together.

Urbanization: It is process in which large numbers of people become permanently concentrated in relatively small areas and form cities. It is also natural increase in the population and rural to urban migration.

Civilization: It is advanced state of human society contains high developed forms of government, culture, industry and common social norms. The human society developed with well social organization, culture in the time.

Sewage: It is type of the wastewater which has produced from community of people. It is characterized by rate of flow, physical condition, chemical and toxic constituents with its bacteriologic status.

Sustainability: It is social goal of the peoples to co-exist on earth over long time. It has three dimensions environment, economic and social. The word 'sustain' means to last over time so sustainability is the ability of something to last.

Chapter 9
Seaport Pier Constructions, Diagnostics, and Evaluation Using Monitoring Data

Sérgio António António Neves Neves Lousada
https://orcid.org/0000-0002-8429-2164
Universidade da Madeira, Portugal

Dainora Jankauskiene
Klaipėda State University of Applied Sciences, Lithuania

Viktoras Dorosevas
Klaipėda State University of Applied Sciences, Lithuania

Birute Ruzgiene
Klaipėda State University of Applied Sciences, Lithuania

Indrius Kuklys
Klaipėda State University of Applied Sciences, Lithuania

Lina Kukliene
Klaipėda State University of Applied Sciences, Lithuania

ABSTRACT

The engineering structure known as breakwaters (piers) of seaports serves an essential purpose in protecting the coasts from the damaging effects of storms and waves. Even minor harm can lead to instability, impacting the structure's elements based on incidence and frequency. At Klaipėda State Seaport, monitoring points were placed on piers for behavior analysis and stability assessment. Continuous geodetic and geotechnical measurements were conducted, including crack widths and displacements in concrete pier constructions. The analysis of monitoring data

DOI: 10.4018/979-8-3693-2448-6.ch009

set has been done regarding determined positional changes of monitoring points and crack-gauges correlation with environmental temperature and wind speed (below 10 m/s; over 10 m/s). Due to the methodology developed, and the results obtained, no significant variations were identified in the displacements of the analysis points, relative to the breakwaters (piers)under study. It is recommended to apply this methodology as a form of monitoring to be carried out in the near future in relation to the case study.

INTRODUCTION

The seaports are a special character structure, with specific construction requirements, remained a multidimensional system, combined with an economic function, infrastructure system, geographical area, trade, transport technologies, and specific port management (García, 2013; Glavinovic et al., 2023; Lousada & Castanho, 2022, 2023; Tijan et al., 2021; Wu et al., 2021). In general, according to seaport's use can be conditionally distinguished two groups: currently exploited and designed. Worldwide, breakwaters (piers) are frequently utilized, mostly to give protection from wave action (Flor et al., 2023; Tran et al., 2023). Breakwaters are sometimes employed to save priceless habitats that are in danger from the sea's destructive effects or to prevent beaches from eroding. The protection is primarily intended for vessels in seaports and for seaport facilities. Protection against currents is crucial, even if wave motion typically poses the threat. Structure makes access possible, lessens overtopping, etc. Breakwaters can also lessen or stop the silting of navigation channels, however, sometimes they increase, as confirmed throughout history in the North Seaports (Freire & Taveira-Pinto, 2011; Zhang et al., 2022). Breakwaters occasionally provide facilities for loading cargo or passengers. The basic components of a typical breakwater and their purposes are as follows: foundation improvement (if necessary) provides embankment stability; scour protection apron prevents erosion; core provides the bulk of the structure and minimizes wave transmission; underlayer contains the core and provides foundation stretch; concrete crest structure provides access and minimizes overtopping; etc. (*BRITISH STANDARD 6349-7*, undated; Koreta, undate).

Numerous scholars have dedicated a significant amount of effort to studying the impact of ocean waves on the level of stability of breakwater. Their research commonly resorted to both numerical analyses and physical model testing. As a result, significant advancements in numerical analysis techniques have been achieved in the last couple of decades. Moreover, researchers have also been focusing on analyzing the different levels of stability of each type of breakwater as well as on their seabed foundation's dynamic response. This has mostly been achieved through the use of elastoplastic finite element models, either self-developed or sourced from commercial companies (Guler et al., 2015; Hanzawa et al., 2012; He & Ye, 2021; Hur et al., 2008; Jeng et al., 2001; Mizutani & Mostafa, 1998; Sawada et al., 2015; Tsai et al., 2006).

The piers in seaports are engineering constructions of many varieties, and they serve a crucial purpose in protecting the coast from the damaging effects of storms and waves (Teng et al., 2023). Since stone is typically used in the main construction of piers, even minor damage to such a construction can cause considerable damage. Main task is to determine the value of piers construction elements motion and frequency. Such movements (displacements) may be caused by the settlement of pier foundations (most of the time), the deterioration of the pier's constituent parts, or the influence of storms. The destruction of the same poses the greatest risk to bringing about these adverse effects. Recent advances in monitoring technology, which allow both geotechnical and geodetic monitoring, now make it possible to track and anticipate the condition of objects in real time (Choudhury & Rizos, 2010). As a result, constant pier state monitoring is essential to prevent any kind of pier damage (Beshr, 2015; Maida et al., 2021).

The observation of the seaport breakwater's (piers) structures state can involve the several monitoring methods (Clemente et al., 2021; Santos et al., 2004):

1) Visual inspections – periodic visual inspections of the breakwater structure can help identify any visible damage or deterioration. Obviously, trained personnel on foot or using aerial or underwater cameras can perform these inspections;
2) Structural measurements – breakwater structures can be monitored using various types of structural measurements, such as deflection, strain, and displacement (Antunes et al., 2012). These measurements can be obtained using sensors (Chen et al., 2008), such as strain gauges or accelerometers, which are attached to the structure and used to measure changes over time (Matsuya et al., 2010);
3) Geotechnical measurements – the condition of the soil or rock (*The Rock Manual*, undated) underlying the breakwater structure can be monitored using geotechnical measurements, such as soil or rock samples, or using geotechnical instruments, such as piezometers or inclinometers;
4) Hydrographic surveys – hydrographic surveys can be used to monitor the changes in the shape and elevation of the breakwater structure over time (Guo et al., 2012). These surveys can be conducted using a variety of techniques, such as laser scanning, sonar, or GPS (*GPS measurements of ground deformation induced by water level variations into a granitic aquifer (French Brittany) - Moreau - 2006 - Terra Nova - Wiley Online Library*, undated; Schneider, 2006);

5) Numerical modelling – numerical modelling can be used to simulate the behaviour of the breakwater structure under different wave and loading conditions (Khan et al., 2022; Sukumar et al., 2003). This can help identify potential weak points or areas of vulnerability and guide the design of repairs or upgrades (Asadpoure et al., 2006).

Overall, a combination of these methods can be used for monitor the state of seaport breakwater structures and ensure that they are functioning properly and safely (Ilia & O'Donnell, 2018).

It's crucial to choose building materials carefully while developing a new port, as demonstrated by the effectiveness of ultra-high-performance concrete (UHPC) in enhancing the seismic performance of RC bridge piers (Clemente et al., 2021; Maida et al., 2021; Santos et al., 2004). According to study findings (Antunes et al., 2012), the adoption of a UHPC jacket aids in improving the mechanical performance indices and is appropriate for usage in seismic and coastal zones. If lightweight reinforced concrete (LRC) walls with slits are used in the newly built port constructions, it is suggested, based on the findings of the study (*BRITISH STANDARD 6349-7*, undated), to put tiny steel dampers in the slits to create a regulated, semi-rigid connection between the LRC wall piers and RC frames. those who put in steel slit dampers (SSDs) at the openings in LRC walls to give the RC frame ductile lateral strength.

Numerous studies have examined the precision and detail of topographic datasets produced from Terrestrial Laser Scanning (TLS) and Structure from Motion (SfM) for a variety of objects (Chen et al., 2008; Guo et al., 2012; Matsuya et al., 2010; *The Rock Manual*, undated). The findings of the study (*GPS measurements of ground deformation induced by water level variations into a granitic aquifer (French Brittany) - Moreau - 2006 - Terra Nova - Wiley Online Library*, undated) demonstrated that SfM-photogrammetry and TLS are useful tools to track morphological changes in object settings. SfM-photogrammetry is advised due to its low cost and flexibility of the acquisition, especially in dynamic side channels showing significant and quick changes using 3-D models. By producing and comparing with the interconnection of the terrestrial point clouds through the bathymetric data, this method provides a trustworthy way to identify and assess accurately morphological changes at various spatial scales (Asadpoure et al., 2006; «Civil Engineering '17 : 6th International Scientific Conference "Research for Environment and Civil Engineering Development 17" : Proceedings», undated; Ilia & O'Donnell, 2018; Khan et al., 2022; Schneider, 2006; Sukumar et al., 2003).

The purpose of the study is to provide a unique approach for estimating the crack widths and displacements, cracks identified in concrete surfaces caused by hydrotechnical construction pier damage. For this, we will use a statistical analysis of data collected from monitoring stations at an experimental Seaport object.

MATERIALS AND METHODS

Study Location

The Southern and Northern piers of Klaipėda State seaport were chosen as object (Figure 1) for the experimental study and monitoring measurements. There are existing breakwaters (piers) for protection the port from waves, ice pack or drift, and to direct the current of the mouth of the Dane river to the Baltic Sea and protect the swimming channel (fairway) from pollution. On the other hand, it is a favorite place for residents of the Klaipėda city and tourists. The North pier is a highly regarded area for seaside walks, sunsets and ships arriving and departing among the residents of the port city and visitors to the city.

Figure 1. Study location: Southern and Northern piers at Klaipėda State seaport. (Source: Authors).

General characteristics of the seaport: two piers (Northern and Southern) are parallel, the structures are of inclined dams, built on piles; the surface part is reinforced by stones, concrete slabs and tetrapod. The length of the Southern berth is 1227 m, and the length of the Northern berth is 1158 m. After deepening the seaport, the piers were reinforced with stones and large concrete blocks. The Southern pier has been under construction from 1847 to 1861 and the Northern – from 1834 until 1884, finally reconstructed in 2002 along with the added new part of the North pier («Civil Engineering '17 : 6th International Scientific Conference "Research for Environment and Civil Engineering Development 17" : Proceedings», undated).

In the area where the Klaipėda State seaport is located, the tidal ranges are generally small, mostly between about 2 and 5 cm. In the western sea areas, tidal ranges of up to 10–30 cm are observed (Weisse et al., 2021).

The height of waves depends on the force and direction of the winds, in the Klaipėda observation sites are fully open to the dominant wind directions (SW and NW) but are mostly sheltered from waves excited by eastern (offshore) winds. For example, in the period 1999–2007, the height of waves reached from 0.25–0.5 m to 3.0 m. Under strong Eastern, Southeaster and Southern winds, the highest waves reached 0.75–1.25 m. Maximum waves up to 3.0 m could be observed under Northern and Northeast winds (Kasiulis et al., 2015; Lebedeva et al., 2012).

Thus, the overall structure of piers elements is heterogeneous and complicated. Therefore, monitoring of structures with such elements would require special diagnostic procedures. The conducted multi-year investigations formed the prerequisites for creating and proposing a monitoring system based on positional geodetic observations and hydro-technical construction elements cracks estimation of the pier.

Methodology of Special Monitoring Procedures

Generalized structurally healthy monitoring (SHM) strategies, which are composed of modern monitoring systems with the potential for geotechnical and geodetic monitoring, can be roughly divided into two groups: global and local. These SHM strategies allow for a preliminary assessment in order to identify and predict the state of objects in real time.

For measuring the stability of distinctive hydro-technical structures, such as the piers at the research site, two approaches are accepted: geodetic and geotechnical monitoring, including global and local processes (Lebedeva et al., 2012). Figure 2 shows how the algorithm for execution monitoring procedures was created; a brief description of the algorithm will be processed in section – “Analysis of measurements results”.

Figure 2. The algorithm of monitoring procedures. (Source: Authors).

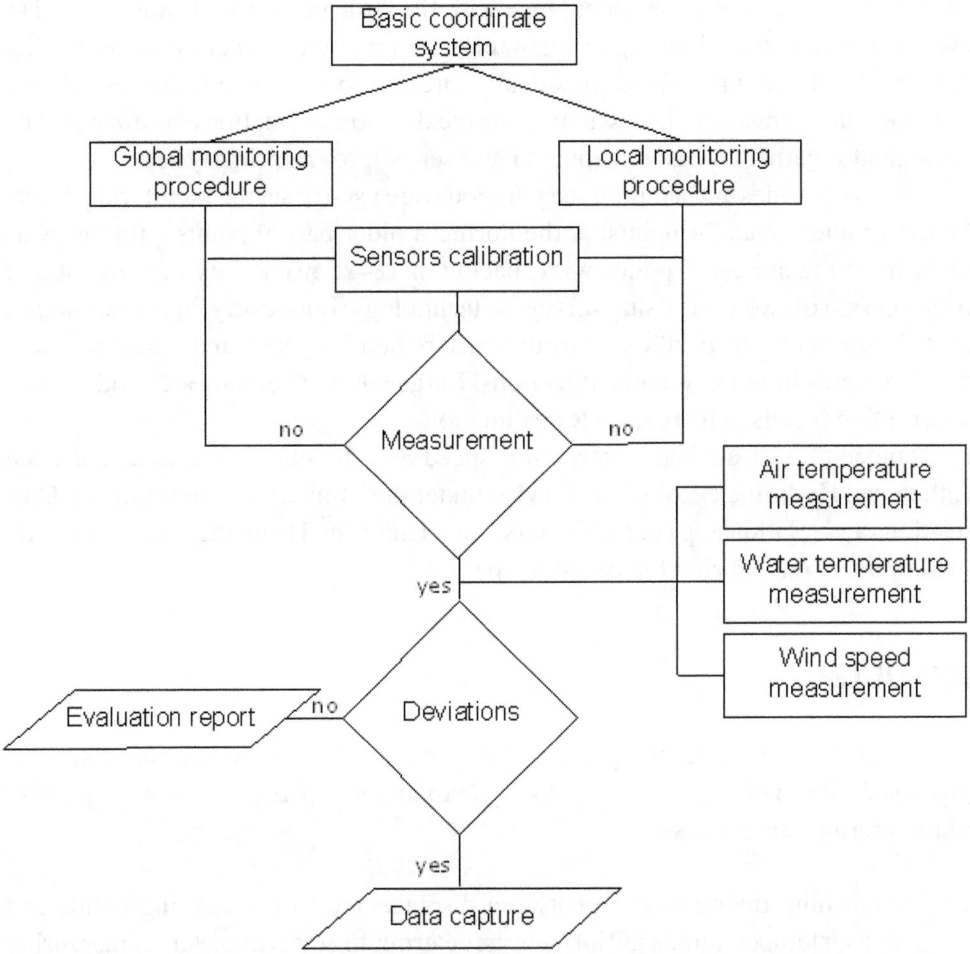

Geodetic (global) monitoring is related with determination of point's position using electronic geodetic instruments (total stations) (Kaloop et al., 2008), which provides measurements results of the high accuracy (Izvoltova et al., 2014; Kovačič et al., 2016).

Geodetic observations of pier's points on the study site in order to determine positional deviations were done with video-assisted robotic total station *TRIMBLE S8* employing Trimble Vision technology (Joyce et al., undated). The *TRIMBLE S8*'s a-priori precision is defined by the manufacturer with the following values: angle setup accuracy of 1" and distance setup accuracy of 1 mm. Total stations were mounted inside a special cabin (box) with apertures to avoid environmental effects and were fixed to a support. The instrument is shielded by the cabin from the elements—snow, rain, and vandalism. Automatic measurements are performed using an electronic total station in accordance with the indicated cycle. It was advised to take measurements at night to prevent the effects of refraction when operating during the day. Total stations were automatically targeted to reference locations prior to each measurement cycle, necessitating the recalculation of station coordinates. The accumulator gathers measurement data and sends it to the primary server.

Prisms on stones and a reinforced tetrapod were used to set the monitoring points (at the southern end, 34 points; at the northern older part, 9 points with intervals of 50 m; at the newest, 8 points with spacing of 18–28 m). Points were monitored over a period of two years using surveying technology: once every 24 hours at night, identifying 3D point positions during measurement's cycles, and detecting daily displacements in the x, y, and z directions. Large amounts of data are produced by observation results and are used for estimations.

Temperature of air and water, wind speed and direction were obtained from Lithuanian Hydrometeorological Service under the Ministry of Environment. Correlations and relationships of parameters were determined from the analysis of daily measurement data during the two years period.

RESULTS

The analysis of monitoring points data set regarding to evaluation of seaport pier's stability involves two main aspects (explanation of Figure 2 - the algorithm of monitoring procedures):

1. Establishing the correlation between displacements of measuring points and water, air temperature, wind influences, determining the parameters of measuring values – Global monitoring data.
2. Calculation of the parameters for evaluation of crack-gauges – Local monitoring data.

For the analysis of global monitoring data, the distinctive points at Northern and Southern piers (N1, N2, N3, S1, S2, S3) and reference stations (T1, T2) have been selected (Figure 3) (*Monitoring and Analysis of Seaport Piers Structural Elements Displacements | SpringerLink*, undated).

Figure 3. Distribution of monitoring points at the piers. (Source: Authors).

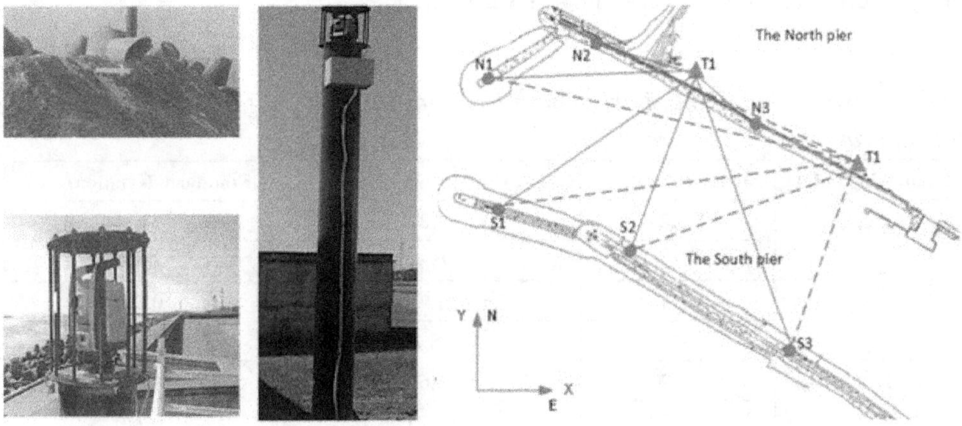

Analyzing selected data set from cycle measurements of monitoring points, have been detected how air and water temperature correlates with the values of displacements, equation (1):

$$\Delta s_k = a_k + b_k t$$

Equation 1

where: Δ_{Sk} – displacements; a_k – constant of the measuring points k (mm); b_k – temperature coefficient (mm/ C); t – water temperature (C).

The regression analysis was applied for determination of the correlation of the air and water temperature with the displacement of measured selected piers points (Lai et al., 1979). The results of the regression analysis are presented in Table 1.

Table 1. The points displacement's correlation with air and water temperature.

Monitoring points	ak	bk	Standard error, mm
S1, S2, S3	-8,527	1,722	4,730
N1 (new pier)	-6,276	1,217	3,039
N2, N3 (old pier)	3,382	0,191	0,777

Analyzing impact of air temperature to measured values, two temperature intervals were chosen: >12 C and <12 C; the zero hypothesis of the normal distribution of the points displacements was checked. The displacements parameters (mean and standard deviation) of selected monitoring points depending on distinguished temperature intervals are presented in Table 2.

Table 2. Evaluation of monitoring points displacements regarding to the influence of temperature.

Monitoring point	Temperature, C	Mean of displacements, mm	Standard deviation, mm
S1, S2, S3	<12	2,87	1,354
	>12	25,94	5,937
N1 (new pier)	<12	1,73	0,683
	>12	18,13	3,544
N2, N3 (old pier)	<12	4,64	0,733
	>12	7,20	0,770

The wind speeds up to 10 m/s is accepted, analyzing of piers point's measurements results, regarding to the displacement's relation with water and air temperatures.

As a local monitoring procedure (Tofallis, 2008), the data set of a structural damage (a crack of the concrete surface) has been taken for statistical analysis (Bedon & Fasan, 2019; Zhao, 2019). At the study object on the pier's (breakwaters) construction was found cracks and observation of 6 locations on Southern pier (breakwater) and 4 on Northern pier (breakwater), this was detected during the two years period. Measurements of pier's (breakwaters) concrete constructions cracks have been made with the cracks-gauges *Geosense VWCM4000* with accumulator *Wi-SOS400 VWNode*. Cracks values were determined with accuracy of ±0.1% of measured value and cracks changes have been defined. As example, at the selected locations on Northern pier (crack-gauges A and B, Figure 4) the crack's width is about 7 mm at point A and at B – 10 mm (Do et al., 2020; Jankauskiene et al., 2018).

Figure 4. Crack-gauges locations and concrete surface's cracks images. (Source: Authors).

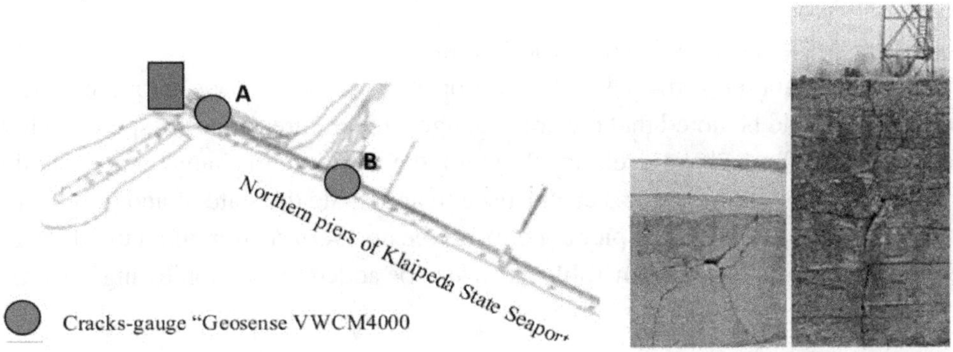

It is feasible to assess the accuracy and reliability of the readings by looking into the relationship between crack width and temperatures. When using the least squares method to solve a problem, a line that depicts the potential link between an independent variable (temperature) and a dependent variable (crack's width change ratio with initial, %) is considered to be the best fit.

Taking into account the linear dependence between crack's width and temperatures, cracks changes are expressed by equation (2):

$$\Delta_T = aT + b$$

Equation 2

where: Δ_T – crack's width change, %; T – temperature, °C; a, b – constant for certain crack.

The proposal is to take into account crack of concrete structure change calculating by equation (2), and in geotechnical monitoring systems data analysis to determine real width of crack, by equation (3):

$$\Delta_R = \Delta_0 \pm \Delta_0 \frac{\Delta_T}{100}$$

Equation 3

where: Δ_R – real crack's width, mm; Δ_0 – initial crack's width, mm.

So, a prediction rule of crack condition can be obtained by equation (4):

$$-\Delta_{all} < \Delta_R < +\Delta_{all}$$

Equation 4

where Δ_{all} – allowable crack's width, mm.

Reported data set from cracks observation at the seaport was used for local monitoring. It should be noted that the crack gauge temperature and wind speed (below 10 m/s; over 10 m/s) have a relationship with the crack width change ratio with the original width (%). Eqns. (2-4) can be used to anticipate the state of and how structural damages will affect the pier concrete in the northern section of Klaipėda State Seaport. The parameters from Table 3 can also be added to the monitoring systems.

Table 3. The parameters for evaluation of crack-gauges.

Crack-gauge location	Constants		Initial crack's width, mm
	a	b	
Wind speed under 10 m/s			
A	–1,63	9,11	7
B	–1,64	9,26	10
Wind speed above 10 m/s			
A	–1,69	11,18	7
B	–1,74	13,92	10

Both global and local monitoring strategies provide different types of information and support flexibility of data analysis.

CONCLUSIONS

The developed methodology and previously presented for modern geodetic and geotechnical monitoring of piers involves the required procedures and realizes basic technical requirements for seaport structures measurement. During period of two years the daily cyclical measurements at the experimental study object a huge data set was constructed, that requires special analysis. Therefore, taking into account determination of data quality, the objective was to evaluate the correlation of monitoring points displacements and crack-gauges with air temperature and wind speed (Tables 1 and 2). The statistical evaluation of data showed that application of the normal distribution and least square techniques for displacement's data analysis allows to evaluate the measurements results, in coherence with the proposed environment temperature. The proposed analysis technique permits to estimate the influence of pier's concrete structural damages, establishing the correlation between displacements of measuring points and water, air temperature, wind influences ((below 10 m/s; over 10 m/s)), determining the parameters of the measurement.

When defined displacements are typically in the range of 3–4 cm to 10–11 cm, the analysis of the test data reveals that significant variations of pier displacements of the selected places were not identified. All measured elements are stable (no deformations or slumps), and no changes in pier cracks at the designated locations were found. However, it is advised that the monitoring of piers be done continuously while using the suggested observations methodology and unique approach to data analysis regarding the impact of environmental temperature and wind speed characteristics. It establishes predictive foundations for more complex forecasts of pier structural degradation combined with the mitigation of effects.

REFERENCES

Antunes, P., Travanca, R., Rodrigues, H., Melo, J., Jara, J., Varum, H., & André, P. (2012). Dynamic Structural Health Monitoring of Slender Structures Using Optical Sensors. *Sensors (Basel)*, *12*(5), 5. Advance online publication. DOI: 10.3390/s120506629 PMID: 22778661

Asadpoure, A., Mohammadi, S., & Vafai, A. (2006). Modeling crack in orthotropic media using a coupled finite element and partition of unity methods. *Finite Elements in Analysis and Design*, *42*(13), 1165–1175. DOI: 10.1016/j.finel.2006.05.001

Bedon, C., & Fasan, M. (2019). Reliability of Field Experiments, Analytical Methods and Pedestrian's Perception Scales for the Vibration Serviceability Assessment of an In-Service Glass Walkway. *Applied Sciences (Basel, Switzerland)*, *9*(9), 9. Advance online publication. DOI: 10.3390/app9091936

Beshr, A. A. E.-W. (2015). Structural Deformation Monitoring and Analysis of Highway Bridge Using Accurate Geodetic Techniques. *Engineering*, *7*(8), 8. Advance online publication. DOI: 10.4236/eng.2015.78045

Chen, J., Zhao, W., Luo, H., Li, P., & Wen, Y. (2008). A real-time system for monitoring distant dynamic displacement of structures. *Geo-Spatial Information Science*, *11*(1), 49–55. DOI: 10.1007/s11806-007-0158-1

Choudhury, M., & Rizos, C. (2010). Slow structural deformation monitoring using Locata—A trial at Tumut Pond Dam. *Journal of Applied Geodesy*, *4*(4), 177–187. DOI: 10.1515/jag.2010.018

Clemente, D., Calheiros-Cabral, T., Rosa-Santos, P., & Taveira-Pinto, F. (2021). Hydraulic and Structural Assessment of a Rubble-Mound Breakwater with a Hybrid Wave Energy Converter. *Journal of Marine Science and Engineering*, *9*(9), 9. Advance online publication. DOI: 10.3390/jmse9090922

Do, T. A., Hoang, T. T., Bui-Tien, T., Hoang, H. V., Do, T. D., & Nguyen, P. A. (2020). Evaluation of heat of hydration, temperature evolution and thermal cracking risk in high-strength concrete at early ages. *Case Studies in Thermal Engineering*, *21*, 100658. DOI: 10.1016/j.csite.2020.100658

Flor, G., Sánchez-Fernández, M., Flor-Blanco, G., & de Sanjosé Blasco, J. J. (2023). Application of Geomatic Techniques for the Assessment of Anthropogenic Changes in the Urban Beaches of "La Magdalena" (Santander, Spain). *Remote Sensing (Basel)*, *15*(1), 1. Advance online publication. DOI: 10.3390/rs15010256

Freire, R., & Taveira-Pinto, F. (2011). *ANÁLISE DO COMPORTAMENTO HIDRO-MORFOLÓGICO DO QUEBRAMAR DA AGUDA. 5.as Jornadas de Hidráulica.* Recursos Hídricos e Ambiente.

García, B. (2013). Ports: Definition and study of types, sizes and business models. *Journal of Industrial Engineering and Management*, *6*(4). Advance online publication. DOI: 10.3926/jiem.770

Glavinovic, R., Vukic, L., & Peronja, I. (2023). Analysis of Management Models of Regional Ports in the Republic of Croatia. *Journal of Marine Science and Engineering*, *11*(2), 332. DOI: 10.3390/jmse11020332

Guler, H. G., Arikawa, T., Oei, T., & Yalciner, A. C. (2015). Performance of rubble mound breakwaters under tsunami attack, a case study: Haydarpasa Port, Istanbul, Turkey. *Coastal Engineering*, *104*, 43–53. DOI: 10.1016/j.coastaleng.2015.07.007

Guo, J., Zhou, M., Wang, C., & Mei, L. (2012). The application of the model of coordinate S-transformation for stability analysis of datum points in high-precision GPS deformation monitoring networks. *Journal of Applied Geodesy*, *6*(3–4), 143–148. DOI: 10.1515/jag-2012-0028

Hanzawa, M., Matsumoto, A., & Tanaka, H. (2012). STABILITY OF WAVE-DISSIPATING CONCRETE BLOCKS OF DETACHED BREAKWATERS AGAINST TSUNAMI. *Coastal Engineering Proceedings*, *33*(33), 33. Advance online publication. DOI: 10.9753/icce.v33.structures.24

He, K., & Ye, J. (2021). Physical Modeling of the Stability of a Revetment Breakwater Built on Reclaimed Coral Calcareous Sand Foundation in the South China Sea—Regular Wave. *Applied Sciences (Basel, Switzerland)*, *11*(5), 5. Advance online publication. DOI: 10.3390/app11052325

Hur, D.-S., Kim, C.-H., Kim, D.-S., & Yoon, J.-S. (2008). Simulation of the nonlinear dynamic interactions between waves, a submerged breakwater and the seabed. *Ocean Engineering*, *35*(5), 511–522. DOI: 10.1016/j.oceaneng.2007.12.002

Ilia, A., & O'Donnell, J. (2018). An Assessment of Two Models of Wave Propagation in an Estuary Protected by Breakwaters. *Journal of Marine Science and Engineering*, *6*(4), 4. Advance online publication. DOI: 10.3390/jmse6040145

Izvoltova, J., Villim, A., & Kozak, P. (2014). Determination of Geometrical Track Position by Robotic Total Station. *Procedia Engineering*, *91*, 322–327. DOI: 10.1016/j.proeng.2014.12.068

Jankauskiene, D., Dorosevas, V., & Pukite, V. (2021). Monitoring and Analysis of Seaport Piers Structural Elements Displacements. MAPAN, 1-6.

Jankauskiene, D., Dorosevas, V., & Volkovas, V. (2018). The crack analysis in monitoring procedures of Seaport pier concrete constructions. *Journal of Measurements in Engineering*, *6*(1), 1. Advance online publication. DOI: 10.21595/jme.2018.19763

Jeng, D. S., Cha, D. H., Lin, Y. S., & Hu, P. S. (2001). Wave-induced pore pressure around a composite breakwater. *Ocean Engineering*, *28*(10), 1413–1435. DOI: 10.1016/S0029-8018(00)00059-7

Joyce, K., Koehler, M., & Vogel, M. (undated). *Trimble Engineering and Construction Group Westminster, Colorado, USA*.

Kaloop, M., Beshr, A. A. A., & Elshiekh, M. Y. (2008). *Using Total Station for monitoring the deformation of high strength concrete beams*. https://www.semanticscholar.org/paper/Using-Total-Station-for-monitoring-the-deformation-Kaloop-Beshr/f8b73755dfb4fad206da019c8cfda66753e557a3

Kasiulis, E., Punys, P., & Kofoed, J. P. (2015). Assessment of theoretical near-shore wave power potential along the Lithuanian coast of the Baltic Sea. *Renewable & Sustainable Energy Reviews*, *41*, 134–142. DOI: 10.1016/j.rser.2014.08.044

Khan, M., Salahuddin, T., Khan, Q., Al Alwan, B., & Almesfer, M. (2022). Numerical study of binary mixture and thermal analysis near a solar radiative heated surface. *Solar Energy*, *231*, 262–269. DOI: 10.1016/j.solener.2021.11.003

Koreta, O. (n.d.). *Maritime structures — Part 7: Guide to the design and construction of breakwaters*. Available at https://www.academia.edu/37123655/Maritime_structures_Part_7_Guide_to_the_design_and_construction_of_breakwaters

Kovačič, B., Kamnik, R., Pustovgar, A., & Vatin, N. (2016). Analysis of Precision of Geodetic Instruments for Investigating Vertical Displacement of Structures. *Procedia Engineering*, *165*, 906–917. DOI: 10.1016/j.proeng.2016.11.800

Lai, T. L., Robbins, H., & Wei, C. Z. (1979). Strong consistency of least squares estimates in multiple regression II. *Journal of Multivariate Analysis*, *9*(3), 343–361. DOI: 10.1016/0047-259X(79)90093-9

Lebedeva, R., Skripkiūnas, G., & Vasiljeva, L. (2012). The effects of seawater on the durability of hydrotechnical (concrete) structures in the port of Klaipeda. *Engineering Structures and Technologies*, *4*(4), 111–118. DOI: 10.3846/2029882X.2012.753745

Lousada, S., & Castanho, R. A. (2022). The Role of Ports in Tourism: Porto Santo Harbour. *Water (Basel)*, *14*(19), 19. Advance online publication. DOI: 10.3390/w14193176

Lousada, S., & Castanho, R. A. (2023). Port Structures, Maritime Transport, and Tourism. *Water, 15*(22). *Water (Basel), 22*(22), 3898. Advance online publication. DOI: 10.3390/w15223898

Maida, Y., Sakata, H., Qu, Z., Maegawa, T., & Suzuki, H. (2021). Cyclic loading test of lightly reinforced concrete wall piers with slit dampers in RC frames. *Engineering Structures, 236*, 112099. DOI: 10.1016/j.engstruct.2021.112099

Matsuya, I., Katamura, R., Sato, M., Iba, M., Kondo, H., Kanekawa, K., Takahashi, M., Hatada, T., Nitta, Y., Tanii, T., Shoji, S., Nishitani, A., & Ohdomari, I. (2010). Measuring relative-story displacement and local inclination angle using multiple position-sensitive detectors. *Sensors (Basel), 10*(11), 9687–9697. DOI: 10.3390/s101109687 PMID: 22163434

Mizutani, N., & Mostafa, A. M. (1998). Nonlinear Wave-Induced Seabed Instability Around Coastal Structures. *Coastal Engineering Journal, 40*(2), 131–160. DOI: 10.1142/S0578563498000091

Moreau, F., Dauteuil, O., Bour, O., & Gavrilenko, P. (2006). GPS measurements of ground deformation induced by water level variations into a granitic aquifer (French Brittany). *Terra Nova, 18*(1), 50–54.

Santos, J., Neves, M., & Silva, L. (2004). *Rubble-Mound Breakwater Inspection in Portugal.*, DOI: 10.1061/40733(147)21

Sawada, Y., Miyake, M., Miyamoto, J., & Kawabata, T. (2015). Numerical Analysis on Stability of Caisson-Type Breakwaters under Tsunami-Induced Seepage. *Transportation Infrastructure Geotechnology, 2*(3), 120–138. DOI: 10.1007/s40515-015-0022-y

Schneider, D. (2006). *TERRESTRIAL LASER SCANNING FOR AREA BASED DEFORMATION ANALYSIS OF TOWERS AND WATER DAMNS.*

Sukumar, N., Chopp, D. L., & Moran, B. (2003). Extended finite element method and fast marching method for three-dimensional fatigue crack propagation. *Engineering Fracture Mechanics, 70*(1), 29–48. DOI: 10.1016/S0013-7944(02)00032-2

Teng, F., Zhang, Y., Yan, W., Wang, X., & Zhang, K. (2023). Restoring Force Model of Precast Segmental Reinforced Concrete Piers after Seawater Freeze–Thaw Cycles. *Coatings, 13*(1), 1. Advance online publication. DOI: 10.3390/coatings13010016

Tijan, E., Jović, M., Panjako, A., & Žgaljić, D. (2021). The Role of Port Authority in Port Governance and Port Community System Implementation. *Sustainability (Basel), 13*(5), 5. Advance online publication. DOI: 10.3390/su13052795

Tofallis, C. (2008). Least Squares Percentage Regression. *Journal of Modern Applied Statistical Methods; JMASM*, *7*(2), 526–534. DOI: 10.22237/jmasm/1225513020

Tran, C.-T., Lin, C., & Tsai, C.-C. (2023). Bragg Resonance of Water Waves by Multiple Permeable Thin Barriers over Periodic Breakwaters. *Water (Basel)*, *15*(3), 3. Advance online publication. DOI: 10.3390/w15030495

Tsai, C.-P., Chen, H.-B., & Lee, F.-C. (2006). Wave transformation over submerged permeable breakwater on porous bottom. *Ocean Engineering*, *33*(11), 1623–1643. DOI: 10.1016/j.oceaneng.2005.09.006

Weisse, R., Dailidienė, I., Hünicke, B., Kahma, K., Madsen, K., Omstedt, A., Parnell, K., Schöne, T., Soomere, T., Zhang, W., & Zorita, E. (2021). Sea level dynamics and coastal erosion in the Baltic Sea region. *Earth System Dynamics: ESD*, *12*(3), 871–898. DOI: 10.5194/esd-12-871-2021

Wu, Z., Mao, Z., & Shen, W. (2021). Integrating Multiple Datasets and Machine Learning Algorithms for Satellite-Based Bathymetry in Seaports. *Remote Sensing (Basel)*, *13*(21), 21. Advance online publication. DOI: 10.3390/rs13214328

Zhang, K., Li, Q., Zhang, J., Shi, H., Yu, J., Guo, X., & Du, Y. (2022). Simulation and Analysis of Back Siltation in a Navigation Channel Using MIKE 21. *Journal of Ocean University of China*, *21*(4), 893–902. DOI: 10.1007/s11802-022-5052-9

Zhao, L. (2019). Fast Decentralized Data Analytics in IoT Wireless Networks. *IEEE Access*, *PP*, 1–1. DOI: 10.1109/ACCESS.2019.2960494

KEY TERMS AND DEFINITIONS

Concrete Surface Cracks: Are fissures or fractures that develop on the exterior layer of concrete structures. These cracks can result from various factors such as shrinkage, thermal expansion and contraction, structural overloading, or environmental conditions. They may compromise the structural integrity and aesthetics of the concrete. Monitoring and repairing concrete surface cracks are crucial for maintaining durability and preventing further damage. Techniques like sealing, epoxy injection, or structural reinforcement are often employed to address these cracks, ensuring the longevity and functionality of the concrete surface. Regular inspection and maintenance help mitigate the impact of concrete surface cracks in construction and infrastructure.

Data Correlation: Involves analyzing and identifying relationships or patterns between different sets of data. This process aims to uncover connections, dependencies, or trends that may exist between variables. Statistical methods, such as correlation coefficients, are often employed to quantify the strength and direction of these relationships. Data correlation is valuable in fields like data science, finance, and scientific research, providing insights into cause-and-effect relationships or uncovering hidden connections within diverse datasets. This analytical approach enhances decision-making processes by revealing dependencies and helping predict future outcomes based on observed patterns in the data.

Displacements: In a general sense, refer to the act of moving or shifting from a previous position. In various contexts, the term is used to describe changes in location, often measured in terms of distance and direction. In physics and engineering, displacements signify the alteration in the position of an object, typically from its equilibrium point. In the context of geology or seismology, displacements may indicate the movement of tectonic plates, or the shift caused by seismic activity. The term is versatile, finding application across disciplines to denote alterations, movements, or changes in spatial arrangements.

Geodetic Observations: Involve the systematic measurement and analysis of Earth's shape, gravitational field, and other related phenomena. Geodesy, the science behind these observations, employs precise techniques such as satellite positioning, gravity measurements, and laser ranging to understand the Earth's size, shape, and dynamic processes. Geodetic observations contribute to the development of accurate maps, monitoring tectonic plate movements, and supporting applications like navigation systems. By capturing and analyzing data over time, geodetic observations play a crucial role in scientific research, infrastructure development, and addressing challenges related to global positioning and geophysical changes.

Monitoring Points: Are specific locations or designated positions where data, observations, or measurements are systematically collected and assessed. These points serve as critical nodes in various monitoring systems, enabling the continuous tracking of conditions, performance, or changes within a given environment or process. In environmental science, monitoring points may include stations for measuring air or water quality. In technology, they could be nodes in a network where data traffic is observed. The term is versatile, finding application in contexts like security, healthcare, and manufacturing, emphasizing the importance of strategically positioned points for effective and comprehensive monitoring.

Seaport's Piers: Refer to the docking facilities or platforms constructed along the waterfront of a seaport to facilitate the loading and unloading of ships. These piers serve as vital points of connection between land and sea, enabling the efficient transfer of cargo and passengers. Equipped with berths, cranes, and storage areas, seaport piers play a crucial role in maritime trade and transportation, serving as hubs for shipping activities. They are essential components of a port's infrastructure, contributing to the seamless flow of goods and fostering economic activities associated with maritime commerce.

Statistical Analysis: Involves the systematic interpretation and examination of data through mathematical and computational techniques. It encompasses the collection, organization, and interpretation of quantitative information to identify patterns, trends, and meaningful insights. Utilizing statistical methods, such as hypothesis testing, regression analysis, and probability distributions, analysts draw conclusions about the relationships within the data. This process aids decision-making, supports scientific research, and informs predictions. Statistical analysis is widely applied across various fields, including economics, biology, sociology, and business, providing a rigorous framework for extracting meaningful information from complex datasets.

Chapter 10
Sustainable Mining Waste Management for Building and Construction

Manas R. Das
KIIT University, India

Suchismita Satapathy
https://orcid.org/0000-0002-4805-1793
KIIT University, India

Lalit K. Pothal
KIIT University, India

ABSTRACT

The mining sector produces a large amount of waste and is a major contributor to environmental deterioration. However, as a sustainable strategy, new efforts have been undertaken to investigate novel techniques to repurpose mining waste in the manufacturing of bricks. The manufacturing of bricks for construction industries mostly dependent on natural resources like minerals(and clay. Huge amount of iron ore tailing waste and fly ash may be utilized for construction and building industry, which will resolve the pollution issues in sustainable way and help both mining and construction industries. Many economic and environmental advantages can be realized by integrating mining waste into the processes used in the manufacture of bricks and building materials. Hence in this chapter an effort is taken and bricks of mining wastes are made and tested to show the utility of mining waste in sustainable way.

DOI: 10.4018/979-8-3693-2448-6.ch010

INTRODUCTION

The quantity of mineral resources that are now extracted worldwide is thought to be in the billions of tons. Environmental regulations are frequently neglected in favour of a continuous expansion in the quantity of mineral extrication and manufacture. The mining sector has intense adverse effect on the environment. When mining waste is not properly handled, they cause serious threats to both the environment and human being. Hazards caused because of the waste's flammability, reactivity, corrosivity, acute, chronic, or extrinsic toxicity, among other factors. Changes in soil use, ecosystem fluctuations, pollution, water scarcity and disruption of groundwater flows, adjustments to infrastructure networks, imbalanced industrial development, forced relocation, and adjustments to the local population and economic structure are just a few of the numerous changes that occur in the region and society due to mining activities. Mining operations produce around 250 million tonnes of solid waste annually. Moreover, rising population means higher degree of consumption, which puts unprecedented burden on natural resources.

Iron is one of the most widely used metals in the world. It is the main raw material for making steel, which is used in everything from buildings and bridges to cars and household items. India, like many other countries, has a large iron and steel industry that plays a big role in its economy. But iron mining and processing create a huge volume of waste. When iron ore is mined, the useful metal is separated from the rock. The left-over material is called iron ore tailings can make up more than 60% of the original material. This means that for every ton of iron ore processed, more than half a ton of waste is produced. In addition to tailings, the iron industry also produces slag and other residues from steel production and energy use. As a consequence, the iron mining sector annually generates bulk amount of tailing wastes, amounting approximately to 7 billion tons, making it one of the major generators of trash globally. Researchers are assuming that by the year 2025, 19 billion of solid tailings will accumulate worldwide; of these, 20% cannot be recycled at all because of their structural complexity (chemical and physical). Conventionally, this waste is dumped in ponds, landfills, or heaps. Over time, these waste sites grow very large and can become dangerous. For example, if a tailings dam breaks, it can release toxic materials into rivers and farmland, harming people and the environment. These waste materials can also take up valuable land, pollute groundwater, and increase dust and air pollution. Because of these problems, there is growing interest in finding out better methods to manage iron industry waste. Instead of just dumping it, researchers and industries are now exploring how to reuse this waste to make useful products, especially in building and construction. This is a part of sustainable mining waste management, which aims to protect the environment and use resources more prudently. In the long run it will facilitate to achieve the sustainable development goals pledged in COP 26 of UNFCCC.

One of the most promising ways to reuse mining waste is by turning it into construction materials. Iron ore tailings, for example, can be used to make bricks, concrete, road base, and even tiles. This is possible because the tailings still contain minerals like silica and alumina, which are useful in building materials. Some researchers have developed geopolymer bricks, which are made by mixing iron tailings with other industrial wastes like fly ash and blast furnace slag. These bricks are strong, water-resistant, and more environmentally friendly than regular clay bricks or concrete blocks. They also reduce the need for natural materials like sand and gravel, which are becoming scarce in many parts of the world. In India, several studies have shown the success of using iron ore tailings for making bricks and road materials. For example, researchers in Karnataka created bricks using a mix of tailings and cement, and the bricks passed all the strength and durability tests. By using IOT the optimum strength of bricks was achieved with 5% perlite, with the ideal mix being IOT-alccofine-perlite at 30-10-2.5% (Rao and Gayana 2021). In Odisha, where iron mining is common, industries are testing the use of tailings in construction purpose. Using IOT as a partial replacement for fine aggregates, saving up to 20%. This reduces production costs and environmental pollution, promoting sustainable construction. (Sabat et al.2015) These efforts are helping reduce the burden on natural resources while also solving the problem of utilization of the huge volume of mining waste. Moreover, this approach is best suited for economic growth. It generates new business opportunities, lowers the cost of building materials and provides job opportunities in areas near the mine. This kind of waste-to-resource solution is a key part of the circular economy, where materials are reused instead of being thrown away.

While the idea of reusing mining waste in construction is exciting, it also comes with some challenges. First, not all types of waste are the same. The quality and safety of the waste can vary depending on the mine and the processing methods used. Some tailings may contain heavy metals or chemicals that could be harmful if not properly treated. As because of this, it is very important to test and process the waste carefully before using it in construction. Second, there are legal and public concerns. Some people worry that using mining waste in homes or roads might affect their health or the environment. Worldwide, several countries have adopted innovative methods to mining waste reuse in construction. In Australia, iron ore tailings are used to create road base and geopolymer concrete, reducing the need for virgin materials (Kuranchie et al.). Chile, a leading copper producer, has implemented policies encouraging the reuse of tailings in building infrastructure, supported by government-funded research. South Africa uses gold mine waste to produce eco-friendly bricks. (Machiels et al.)

China has large-scale applications of steel slag in cement and concrete production. (Yi et al.). These international examples demonstrate how mining waste, when managed effectively, can support sustainable construction while reducing environmental impact and enhancing circular economy practices. To solve this, we need clear government policies, strict quality standards and public awareness programs. People must be informed that these materials, when used properly are safe and even better for the environment than traditional options.

Another matter is that the construction industry is adoption of new materials unless they are workable and accessible. This is why pilot projects and case studies are so important. When people see that waste-based bricks or concrete work just as well as regular ones they are more likely to accept and use them. In recent years, many successful examples from India, Australia and South America have shown that it is possible to build schools, roads, and homes using iron mining waste. Governments, industries and researchers must continue to work together to promote this kind of innovation. With the right support and investment, sustainable mining waste management can become a normal part of construction and help create a greener, cleaner future.

The use of mining waste in construction is gaining momentum, but its widespread adoption largely depends on supportive policy and regulatory frameworks. In many countries, including India, current regulations mainly focus on waste disposal and environmental protection rather than reuse and valorisation. For example, mining operations are required to manage tailings and waste rock under strict environmental laws, but there are limited incentives or guidelines encouraging the reuse of these materials in construction. To promote sustainable practices, regulatory bodies need to update existing policies to include standards for the safe processing, testing, and certification of mining waste-derived construction products. Government departments like the Ministry of Environment, Forest and Climate Change (MoEFCC) and the Bureau of Indian Standards (BIS) can play a critical role in formulating these guidelines. Similarly, state-level mining and construction authorities must be involved in monitoring compliance and promoting local implementation. Many stakeholders including mining companies, construction firms, environmental groups and researchers must collaborate to advocate for clear policies, public awareness and industry incentives. Policy recommendations include offering tax benefits for industries reusing waste, funding for pilot projects and mandatory inclusion of recycled materials in public infrastructure. These actions will create a regulatory ecosystem that supports innovation, safety and environmental sustainability. Figure.1 and 2 shows life cycle of mining waste and waste generation during iron ore extraction.

Figure 1. Life cycle ofmining waste and its sustainable reuse in construction materials.

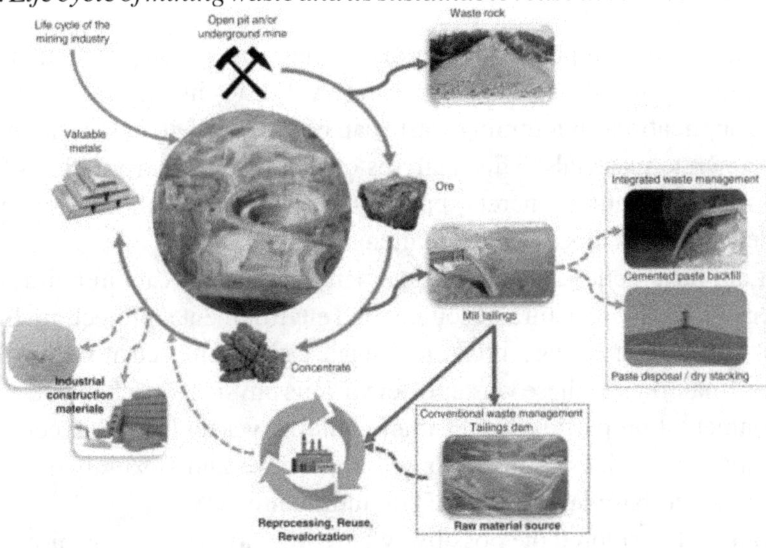

Figure 2. Mining waste generated during iron ore extraction and processing.

LITERATURE REVIEW

Prabhu et al., have emphasized on transforming waste materials into beneficial resources, especially in the construction industry. Wastes like fly ash are prominent in concrete applications, but mining waste can be used as better alternative for conventional concrete materials. This analyses challenges in incorporation of mining waste into self-compacting concrete applications, emphasizing the interrelationships and interdependencies among these challenges.

Ally et al., have noticed waste management is a major issue in buildings, as it provides an alternative to natural resources and environmental protection. Industrial wastes, such as mining tailings, rocks, and slags, can be replaced as raw materials in construction materials as these wastes offer various physical, mechanical, chemical, and environmental properties. For example, mining waste, including cementitious materials and fired bricks, offers strong compressive and flexural strengths, can also serve as a sub-base layer and backfill until nontoxic.

Cobîrzan et al., explored the possibility of reutilization of mining bi-products in the production of masonry units, creating new employments and supporting socio economic development.

Shreekant et al. found out exploitation of mineral resources can cause economic and societal development, but it also generates huge quantity of waste, which can adversely pollute the environment. To protect the environment and enhance sustainable resource development, comprehensive waste utilization is needed. Non-fired bricks can be produced from mine waste, reducing CO2 emissions and promoting natural resource conservation as an eco-friendly alternative.

Drechsler et al, aimed at Zero Waste for sustainable development by recycling and reutilization of all waste. Innovative technologies, like geopolymers, can create commercially valuable products from huge quantity of waste materials tailing, quarry waste, clays, bauxite residue, fly ash and blast furnace slags. These technologies can economically and environmentally use waste materials in construction, building, mining, and waste management sectors.

Segui et al., indicated that the mining industry generates significant environmental waste, prompting research into using these bi-products as alternative aggregates for pavement materials and concrete production. They focused on recycling of mining wastes as road construction material, focusing on infrastructure with poor geotechnical properties. This method decreases waste volume and offers sustainable raw materials.

Simão et al., evaluated the potential use of Sulphide mining waste in facing bricks. After characterisation, a company-specific blend was changed, partially by replacement of primary raw materials with waste materials like Plombières tailings and Neves Corvo waste rocks found that pre-treated Neves Corvo waste rock materials were not suitable due to high sulphur and metal content.

Helser et al., focused on the challenges that the mining industry faces in managing and storing mine bi-product, including waste rock and tailings. traditional mine waste can leach contaminants into the environment even after closure. To placate this risks and economic benefits, valorisation of waste can be achieved by recovering beneficial metals and removing hazardous wastes. The leftover materials can be used to create environmental-friendly construction materials like geopolymers, ceramics, or cement.

Li et al., mentioned that "green" and "sustainable" construction material has gained popularity in recent years, particularly in the mining sector. Acid mine drainage occurs due to mineral processing industry's structure, causing heavy metal ions to leach into groundwater. This contamination affects rivers, farmland, and drinking water. Key technologies for green mine construction include backfill mining and the reuse of mining waste for road and bridge construction.

Almeida et al., reviewed on sustainable strategies for incorporating mining residues into construction materials production, focusing gaps and barriers in achieving a safe and sustainable inclusion of mine residues and the potential for recovering these residues as secondary supplies.

Klang et al., described the building sector, has environmental impact. To mitigate this, a sustainable construction approach is being implemented, focusing on social, economic, and environmental indicators. Building Information Modelling (BIM) is being utilised to manage construction and demolition waste (CDW) in the design phase. BIM provides a platform for determining construction waste amount, disposal, and costs, resulting in enhancing environmental performance.

Chavarry-Vallejos et al., aimed to create ecological bricks by utilising mining tailings, wastes of mining activity, for sustainable housing development. Focused on particle size distribution, workability, plasticization, mineralogical analysis, and rheology. Results show that incorporating alkaline binder, lime fibres, and clay cork improves brick development in severe climates. Tailings with lead-zinc sulphide mitigate carbonization and contraction coefficient, fly ash, while shale and cement achieve compressive strength.

Kuranchie et al., A major mining area, generates large volumes of waste from its mineral deposits. This waste, including mine tailings, waste rocks, fly ashes, and slags, creates economic, environmental, and land problems. Public awareness encourages alternative handling methods, with reuse being the most sustainable solution. This paper explores the potential of mine waste as cost-effective and sustainable construction materials.

Soni et al., reviewed on exploration and the environmental efficacy of solid waste management and its relationship with regulatory structure, economic structure, science, and time and explored the sustainability, social and environmental impact of green building materials and its long-term scope for utilization solid waste in construction.

Ibrahim et al., focused on sustainable building, emphasizing on resource and waste management in construction sectors. Building construction is vital consumer of non-renewable resources and natural materials, causing significant waste generation.

Velásquez et al., assessed the environmental and health complications of recycling mine tailings in artisanal and small-scale mining communities. It focused on health risks and toxicity leaching protocols, with negligible consideration of field application and social concerns. It proposed comprehensive approach to recycling tailings framework to help communities and researchers better understanding of the environmental and health implications.

Kakkos et al., used life cycle assessment to compare conventional and circular construction practices in Switzerland. It emphasized on the increase in residential building stock between 2012 and 2016. The study reveals minimum cumulative impacts at the building level and resulting benefits at the country level over five years. Utilizing existing or recycled components instead of pure materials can enhance the benefits.

Verhagen et al., revealed that recycling demolition waste is crucial to mitigate primary material demand in the construction sector, which consists of for 50% of global consumption. The Dutch government's 50% reduction in primary materials consumption by 2030 is not sufficient, needing more improvements in demolition waste processing to ensure material quality and sustainable building use.

Rose et al., noticed the construction industry produces significant waste and resources, making sustainable development crucial. Reutilization of building components could help meet this demand. However, reclaiming components from demolition is challenging due to cost and labour costs. It reviewed waste interventions and examines systemic mechanisms leading to component disposal. A triage process is proposed to collect information about discarded components, focusing on reusability and recyclability ensuring public awareness.

Pappu et al., pointed out India generates 960 million tonnes of solid waste annually, the total waste produced is 350 million tonnes of organic waste, 290 million tonnes of inorganic waste, and 4.5 million tonnes of hazardous waste. Development in waste management have led to alternative construction materials and efforts to recycle and use non-hazardous and hazardous wastes for value-added applications.

Akadiri et al., proposed a sustainable framework for the building industry, based on the triple bottom line principle. It focused strategies for resource conservation, cost efficiency, and human adaptation, aiming to balance economic, social, and environmental issues, thereby promoting sustainability in the industry.

Chung et al., mentioned four criteria for assessing the sustainability of solid waste management in Hong Kong. The criteria include economic optimization, social acceptability, environmental desirability, equity and administrative diligence. It revealed that Hong Kong's management of construction, demolition, clinical, and chemical waste fails in most sustainability criteria. Preventive waste management programs are inadequate, and economic incentives for waste producers are lacking. The government's efforts to promote environmentally responsible behaviour are often insufficient, leading to the loss of future generations' welfare for community interests.

Coelho et al., used a top-down Life Cycle Analysis (LCA) approach to quantify environmental hazards of buildings adjusted to various waste/material management options. Results show shallow, superficial demolition may not decrease environmental impacts, but core material separation in demolition operations and recycling/reuse can bring environmental benefits, with a reduction of around 77% in climate change impact, 57% in acidification potential, and 81% in summer smog impact.

Tang et al., indicated increasing generation of solid waste has caused environmental concerns worldwide. To promote sustainability in the construction industry, efforts have been made to recycle waste for sustainable construction materials. The study explores recycling municipal and construction solid waste for geopolymer composites, revealing successful incorporation of waste as precursor, aggregate, additive, reinforcement fibre, or filling material.

Martins et al., pointed out severe environmental hazard can be caused if high-volume wastes such as Sulfidic tailings are not handled properly. They can weather and produce acid mine drainage, making them unsuitable for construction materials like cement, concrete, and ceramics. The production of calcium sufflaminate cement may be a suitable method for valuing these tailings.

Yellishetty et al., explored the use of mining solid waste in construction, addressing pollution concerns in mining industry. The wastes, which contain acid-producing minerals and very high intensity of heavy metals, are categorized into coarse-grained rock, sand, silt and clay.

Rao and Gayana, explored the use of iron ore tailings in construction, focusing on the production of bricks and concrete. The study found that initially, bricks were denser, but increased perlite content led to a decrease in density. The optimum strength was achieved with 5% perlite, with the ideal mix being IOT-alccofine-perlite at 30-10-2.5%. This approach promotes a greener environment and reduces dump maintenance issues.

Sabat et al., explained that urbanization and industrialization in the 21st century have led to environmental problems. Mining generates waste, including iron ore tailings (IOT). Proper waste management and disposal are crucial. A study recommends using IOT as a partial replacement for fine aggregates, saving up to 20%.

This reduces production costs and environmental pollution, promoting sustainable construction.

Kuranchie et al., presented a cost-effective and environmentally friendly method for making bricks using tailings from iron ore mines in Western Australia. The geopolymerisation process conserves energy and reduces greenhouse gas emissions by avoiding high temperature kiln firing. The study investigates the effects of initial setting time, curing temperature, and activator content on brick properties. The geopolymer bricks meet American Society of Testing and Materials and Australian Standards requirements and are lower in cost than commercial clay bricks.

Machiels et al., explained on NEMO project as a new business in the Closing workshop of (Re) Mining extractive waste on 17-18 May 2022 at Lamot Conference centre in Mechelen, Belgium. The event, a joint-event with the H2020 ETN SULTAN project, aimed to present the latest developments in the field, novel concepts, and business cases, including case studies from historical and active mining sites.

Yi et al., described that China's steel slag utilization rate is only 22%, behind developed countries. This leads to environmental pollution and land occupation. Research on slag utilization in steel enterprises, road construction, cement, waste water treatment, construction, and agriculture production is ongoing. Key routes and critical problems for large-scale utilization are proposed.

From the above literature reviews, it was found that utilization of mining waste for construction industries is the sustainable way of utilization of mining waste. To check the mining waste utilization, iron or tailing wastes are collected and utilized in construction industry as bricks and tiles and properties are tested.

RESEARCH METHODOLOGY

The bricks of fly ash and iron ore tailing waste are made and various tests were conducted to check the suitability of mining waste bricks with natural bricks. For making fly ash bricks fly ash, lime, gypsum and sand are taken in percentage (55+15+5+10) and dimensions of 9cm × 14cm × 29cm. The tailing waste bricks are made with tailing waste of iron ore, fly ash, gypsum and sand in the proportion (40+15+15+5+10) and same dimension bricks are made. A pan mixer is used for blending process of raw materials and kept for 4-5 minutes and then compressed. The Curing process takes 14 days and then further air dried for 7 days. Two cylindrical pieces of fly ash and iron ore tailing 50 mm in diameter are made for sound absorption test. Fig.3 and Fig.4 shows bricks and Fig.5 and Fig.6 show cylindrical pieces for noise testing. Then compression test, sound absorption test etc were conducted.

Figure 3. Iron ore tailing bricks

Figure 4. Fly ash bricks

Figure 5. Cylindrical piece of fly ash

Figure 6. IOT cylindrical piece

RESULTS AND DISCUSSION

The fabrication of bricks is done at constant temperature. for 7 days,14 days and 28 days respectively. Then the tests were conducted to check specific gravity, compressive strength, bulk density etc.

Table.1, Table.2 and Table.3 shows the specific gravity, compressive strength, bulk density etc. result of 3 types of bricks (I.e., Iron ore tailing bricks, Fly ash bricks and Normal bricks). It is found that iron ore tailing bricks show maximum compressive strength and porosity, water absorption compared to other bricks in all the cases. Good porosity shows good sound absorption capacity. Hence the sound absorption test is conducted in Figure.7.

Table 1. Bricks after 7 days

Bricks	Temperature (degree centigrade)	Specific gravity	Compressive strength in Mpa	Bulk density (gm/cm3)	Porosity percentage	Water absorption percentage
Iron tailing	950	1.79	26.31	1.91	26.06	10.19
Fly ash	950	1.53	11.04	1.21	14.98	5.03
Normal Bricks	950	1.28	8.41	0.72	10.06	3.86

Table 2. Bricks after 14 days

Bricks	Temperature (degree centigrade)	Specific gravity	Compressive strength in Mpa	Bulk density (gm/cm3)	Porosity percentage	Water absorption percentage
Iron tailing	950	1.26	27.61	1.79	32.07	18.01
Fly ash	950	1.22	18.99	1.64	25.32	8.08
Normal Bricks	950	1.09	14.53	0.86	15.52	6.41

Table 3. Bricks after 28 days

Bricks	Temperature (degree centigrade)	Specific gravity	Compressive strength in Mpa	Bulk density (gm/cm3)	Porosity percentage	Water absorption percentage
Iron tailing	950	1.63	24.63	1.34	36.74	13.36
Fly ash	950	1.41	20.41	1.09	18.42	8.61
Normal Bricks used	950	1.03	1.05	0.74	15.39	9.05

Figure 7. Impendence tube

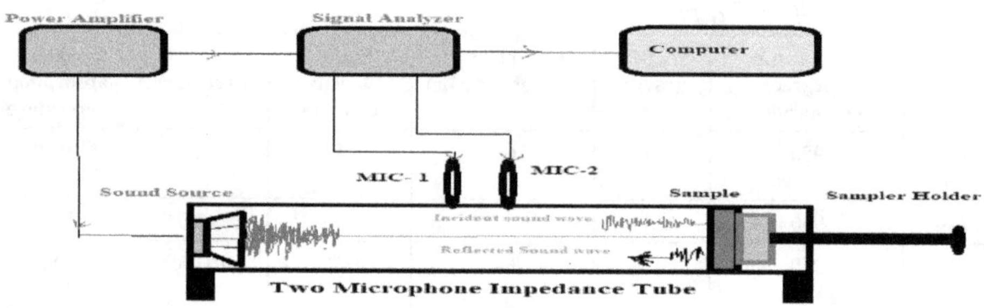

The Sound absorption test is conducted by impedance tube method as per Fig.7. As per Table.4 it is seen that Iron ore tailing bricks shows noise absorption capacity more than other bricks and it is more than 0.65 which is acceptable worldwide.

Table 4. Sound absorption coefficient

Bricks	125hz	250hz	500hz	1000hz	2000hz	4000hz
Iron tailing	0.48	0.49	0.53	0.59	0.63	0.71
Fly ash	0.18	0.24	0.34	0.41	0.43	0.59
Normal bricks	0.19	0.22	0.24	0.30	0.36	0.40

Hence, it is found that the mining wastes can be utilized as bricks as construction materials. Particularly iron ore tailing and fly ash waste bricks are showing good strength compared to normal bricks. It will solve the waste minimization and utilization problem in a sustainable way.

CONCLUSION

The iron industry produces huge amounts of waste like tailing waste, it not only creates pollution, but also decreases the fertility of lands. Hence, proper utilization of tailing waste of iron mines waste can help society to reduce pollution, environmental damage, save natural resources and lower construction costs. Many experiments are going on utilization of mining wastes. In this chapter, the iron ore tailing brick shows better results compared to the normal brick and fly ash produced brick used. It will help construction industries in supplying bricks, tiles etc. and also, due to consumption of large quantities of waste material, the earth will be pollution-free and safe. But to make it an extensive practice, we must overcome technical and social challenges through research, policy and education. Sustainable mining waste management is not just about waste it's about changing the way we think about materials, construction and the environment.

REFERENCES

Akadiri, P. O., Chinyio, E. A., & Olomolaiye, P. O. (2012). Design of a sustainable building: A conceptual framework for implementing sustainability in the building sector. *Buildings (Basel, Switzerland)*, *2*(2), 126–152. DOI: 10.3390/buildings2020126

Ally, A. N., Blanche, M. M., Nana, U. J. P., Grâce, M. M., François, N., & Pettang, C. (2021). Recovery of mining wastes in building materials: A review. *Open Journal of Civil Engineering*, *11*(4), 379–397. DOI: 10.4236/ojce.2021.114022

Almeida, J., Ribeiro, A. B., Silva, A. S., & Faria, P. (2020). Overview of mining residues incorporation in construction materials and barriers for full-scale application. *Journal of Building Engineering*, *29*, 101215. DOI: 10.1016/j.jobe.2020.101215

Chavarry-Vallejos, C. M., Gutiérrez, A. A. V., Bancayán, C., Oré, L. J. C. R., Rodríguez, J. S. T., Andrés, Í., & Serrano, J. E. (2023). Use Of Mining Tailings in The Manufacture of Ecological Bricks to Encourage the Construction of Sustainable Housing. *International Journal (Toronto, Ont.)*, *10*(3), 2555–2567.

Chung, S. S., & Lo, C. W. (2003). Evaluating sustainability in waste management: The case of construction and demolition, chemical and clinical wastes in Hong Kong. *Resources, Conservation and Recycling*, *37*(2), 119–145. DOI: 10.1016/S0921-3449(02)00075-7

Cobîrzan, N., Muntean, R., Thalmaier, G., & Felseghi, R. A. (2022). Recycling of mining waste in the production of masonry units. *Materials (Basel)*, *15*(2), 594. DOI: 10.3390/ma15020594 PMID: 35057312

Coelho, A., & De Brito, J. (2012). Influence of construction and demolition waste management on the environmental impact of buildings. *Waste Management (New York, N.Y.)*, *32*(3), 532–541. DOI: 10.1016/j.wasman.2011.11.011 PMID: 22182407

Drechsler, M., & Graham, A. (2005, October). Geopolymers-an innovative materials technology bringing resource sustainability to construction and mining industries. In *Proceedings of the IQA Annual Conference* (pp. 12-15).

Helser, J., Perumal, P., & Cappuyns, V. (2022). Valorizing (cleaned) sulfidic mine waste as a resource for construction materials. *Journal of Environmental Management*, *319*, 115742. DOI: 10.1016/j.jenvman.2022.115742 PMID: 35849929

Ibrahim, M. I. M. (2016). Estimating the sustainability returns of recycling construction waste from building projects. *Sustainable Cities and Society*, *23*, 78–93. DOI: 10.1016/j.scs.2016.03.005

Kakkos, E., Heisel, F., Hebel, D. E., & Hischier, R. (2020). Towards urban mining—Estimating the potential environmental benefits by applying an alternative construction practice. A case study from Switzerland. *Sustainability (Basel), 12*(12), 5041. DOI: 10.3390/su12125041

Klang, A., Vikman, P. Å., & Brattebø, H. (2003). Sustainable management of demolition waste—An integrated model for the evaluation of environmental, economic and social aspects. *Resources, Conservation and Recycling*, *38*(4), 317–334. DOI: 10.1016/S0921-3449(02)00167-2

Kuranchie, F. A., Shukla, S. K., & Habibi, D. (2013). Mine wastes in Western Australia and their suitability for embankment construction. In Geo-Congress 2013: Stability and Performance of Slopes and Embankments III (pp. 1443-1452). DOI: 10.1061/9780784412787.145

Kuranchie, F. A., Shukla, S. K., & Habibi, D. (2016). Utilisation of iron ore mine tailings for the production of geopolymer bricks. *International Journal of Mining, Reclamation and Environment*, *30*(2), 92–114. DOI: 10.1080/17480930.2014.993834

Li, S., Yu, L., Jiang, W., Yu, H., & Wang, X. (2022). The recent progress China has made in green mine construction, Part I: Mining groundwater pollution and sustainable mining. *International Journal of Environmental Research and Public Health*, *19*(9), 5673. DOI: 10.3390/ijerph19095673 PMID: 35565067

Machiels, L., Paajanen, M., Jones, P. T., & Binnemans, K. (2022). Near-zero-waste recycling of low-grade sulphidic mining waste for critical-metal, mineral and construction raw-material production in a circular economy.

Martins, N. P., Srivastava, S., Simão, F. V., Niu, H., Perumal, P., Snellings, R., Illikainen, M., Chambart, H., & Habert, G. (2021). Exploring the potential for utilization of medium and highly sulfidic mine tailings in construction materials: A review. *Sustainability (Basel)*, *13*(21), 12150. DOI: 10.3390/su132112150

Pappu, A., Saxena, M., & Asolekar, S. R. (2007). Solid wastes generation in India and their recycling potential in building materials. *Building and Environment*, *42*(6), 2311–2320. DOI: 10.1016/j.buildenv.2006.04.015

Prabhu, R. S., Anuradha, R., & Jude, A. B. (2022). Understanding the mining waste as resources in self-compacting concrete: A numerical study on sustainable construction. *Resources Policy*, *78*, 102806. DOI: 10.1016/j.resourpol.2022.102806

Rao, S., & Gayana, B. C. (2021). Experimental Study on Iron Ore Tailings as Aggregate in Development of Bricks and Concrete. *Research Transcripts in Materials*, *1*, 51–68.

Rose, C. M., & Stegemann, J. A. (2018). From waste management to component management in the construction industry. *Sustainability (Basel), 10*(1), 229. DOI: 10.3390/su10010229

Sabat, V., Shaikh, M., Kanap, M., Chaudhari, M., Suryawanshi, S., & Knadgouda, K. (2015). Use of iron ore tailings as a construction material. *International Journal of Conceptions on Mechanical and Civil Engineering*, *3*(2), 1–6.

Segui, P., Safhi, A. E. M., Amrani, M., & Benzaazoua, M. (2023). Mining wastes as road construction material: A review. *Minerals (Basel), 13*(1), 90. DOI: 10.3390/min13010090

Shreekant, R. L., Aruna, M., & Vardhan, H. (2016). Utilisation of mine waste in the construction industry–a critical review. International journal of Earth sciences and engineering, 9(01), 182-195.

Simão, F. V., Chambart, H., Vandemeulebroeke, L., Nielsen, P., Adrianto, L. R., Pfister, S., & Cappuyns, V. (2022). Mine waste as a sustainable resource for facing bricks. *Journal of Cleaner Production*, *368*, 133118. DOI: 10.1016/j.jclepro.2022.133118

Soni, A., Das, P. K., Hashmi, A. W., Yusuf, M., Kamyab, H., & Chelliapan, S. (2022). Challenges and opportunities of utilizing municipal solid waste as alternative building materials for sustainable development goals: A review. *Sustainable Chemistry and Pharmacy*, *27*, 100706. DOI: 10.1016/j.scp.2022.100706

Tang, Z., Li, W., Tam, V. W., & Xue, C. (2020). Advanced progress in recycling municipal and construction solid wastes for manufacturing sustainable construction materials. *Resources, Conservation and Recycling. X*, *6*, 100036. DOI: 10.1016/j.rcrx.2020.100036

Velásquez, J. R., Schwartz, M., Phipps, L. M., Restrepo-Baena, O. J., Lucena, J., & Smits, K. M. (2022). A review of the environmental and health implications of recycling mine tailings for construction purposes in artisanal and small-scale mining communities. *The Extractive Industries and Society*, *9*, 101019. DOI: 10.1016/j.exis.2021.101019

Verhagen, T. J., Sauer, M. L., van der Voet, E., & Sprecher, B. (2021). Matching demolition and construction material flows, an urban mining case study. *Sustainability (Basel), 13*(2), 653. DOI: 10.3390/su13020653

Yellishetty, M., Karpe, V., Reddy, E. H., Subhash, K. N., & Ranjith, P. G. (2008). Reuse of iron ore mineral wastes in civil engineering constructions: A case study. *Resources, Conservation and Recycling*, *52*(11), 1283–1289. DOI: 10.1016/j.resconrec.2008.07.007

Yi, H., Xu, G., Cheng, H., Wang, J., Wan, Y., & Chen, H. (2012). An overview of utilization of steel slag. *Procedia Environmental Sciences*, *16*, 791–801. DOI: 10.1016/j.proenv.2012.10.108

Chapter 11

The Requirement of Sustainable Water Management and Its Progress in India

Aditya Singh
https://orcid.org/0000-0001-9347-5627
Amrita Vishwa Vidyapeetham, India

ABSTRACT

The book chapter explains the concept and importance of Sustainable Water Management in general and how it can help India to solve most of its current problems. Additionally, it discusses a number of problems can arise suddenly in the future related to water in the country, which can be solved with a proper long-term planning of sustainable water management practices. Then, the current gaps in the research in regard to Sustainable Water Management are covered in the chapter. It also talks about some major projects and policies of India to walk on the path of Sustainable Water Management. Also, some challenges and benefits of these projects and policies are briefly mentioned in the book chapter. Further, data from different sources are collected to perform graphical analyses to support the study.

INTRODUCTION

Before moving to the Sustainable Water Management, it is important to understand what Sustainable Water means. When Sustainable Water is mentioned, then it signifies that any country to be water self-sufficient which means there should be plenty of water to satisfy various needs, which includes industrial, municipal,

DOI: 10.4018/979-8-3693-2448-6.ch011

agricultural and so on. This further means, that there should be a consistency in the supply of water regardless of the impact of climate change, which includes flood resiliency or excessive rainfall, as well as the case of drought or lack of sufficient rainfall in a given place. Moreover, the stack up of economics is also essential in meeting the demand as well as supply along with the delivery process of water to be effective without any limit. Then, it could also mean to include energy neutrality through the pairing of renewable energies with the conventional technologies of water treatment. Then on a large scale such as the worldwide scale, if each person is able to have at least twenty to fifty litres of water per day at a reasonable price to sustain their respective lives, then it can be considered as having sustainable water. General Assembly of the United Nations recognise that it is the human right to have water for sanitation in addition to clean as well as safe drinking water which is important to completely enjoy all other human rights along with life. Whereas sustainable water was defined by the regulator of United Kingdom as a cycle of water which is sustainable in nature in a way that the requirements of sewerage services as well as water could be met at the same time allowing the future generations to enjoy those services as well. Whereas the term Sustainable Water Management corresponds to water utilization in ways that not only the present social, ecological as well as economic requirements are met, but also those requirements could be met in the future as well. This means that the people and the concerned authority are supposed to not only see the immediate water supply operations as well as the boundaries of the judiciary, but beyond them, in a way to go for resilient solutions for each region while collaboratively manage water which could decrease the possible risks. However, a holistic as well as multidisciplinary method is needed to obtain sustainable water management in which economic, technical, cultural, environmental, societal in addition to landscape aesthetic must be discussed. This leads to holistic in addition to effective management of the water resources, nonetheless the several increasing water resources demand at present, seeks for holistic, integrated as well as sustainable management of water. This further means that proper planning of wastewater as well as water systems is required to not only mange their infrastructure but also operations, along with making sure that they serve the concerned community's sustainability in the case of water utilities, as per EPA in USA. The advantages of taking sustainability into consideration while planning the infrastructure, as per the agency, comprises of understanding regularity across the range of choices which have the potential to take care of the goal of communities but also utilities as well. Then optimization of social, economic, as well as environmental advantages through projects selection, as well as defining goals, via an inclusive in addition to transparent process in the presence of the concerned communities. Also, improving the long term managerial, financial as well as technical capacity of the given utilities, as part of the advantages. When water utilities are discussed, then normally they have a long-term commitment regarding maintenance as well as infrastructure operation, along with similar planning horizon. This is why the possible gains of the decisions on the investment as well as costs, are understood over the course of time in the long run. Hence, the Sustainability Policy of EPA, seeks for wastewater as well as water systems to take on complete and vigorous planning to make sure that the investments on water infrastructure are not only resource efficient, economical over their lifecycle but also matches with the other important community goals. However, it is essential to know that the between utilities, the water sustainability's levels will also vary, as some communities as well as utilities had been implementing the considerations regarding sustainability into their respective planning process, along with methods to not only polish but also enhance their present efforts. On the other hand, others might pay attention to the ways the earlier mentioned considerations could assist to economically meet the current service needs or current regulatory. Then by the year 2012, the Sustainable Water Management Profile a tool was made by the Water Foundation with its trustworthy partners Sonoma Ecology Center, Environmental Science Associates in addition to The Bay Institute, in a way to assist the water agencies in not only assessing their progress in the direction of sustainable water management, but also conditions of water stress to be measured, by helping them in examining the responses of their own management to such conditions. The mentioned group further focus on their tool development which could help business as well as residential customers, public utilities, federal as well as state agencies, financial analysts and agencies responsible for land use planning. Moreover, as per International Water Association, the system of water in regard to sustainable development might not plainly comprise of water utilization, nonetheless also comprise of system where water utilization has conventionally been needed. For instance, to secure the sustainable supply of water as well as ease the water stress, waterless car washes in addition to waterless toilets could be part of it. As per the Association, there are some solutions to use various sources of water in a sustainable way:

Groundwater: IWA talks about that it is more than half of the freshwater worldwide, but it is essential to make sure that the quantity of water leaving, entering, as well as stored could be sustained, but for now there had been decline in its quantity as well as quality in many places across the world.

Figure 1. Water cycle and depiction of groundwater

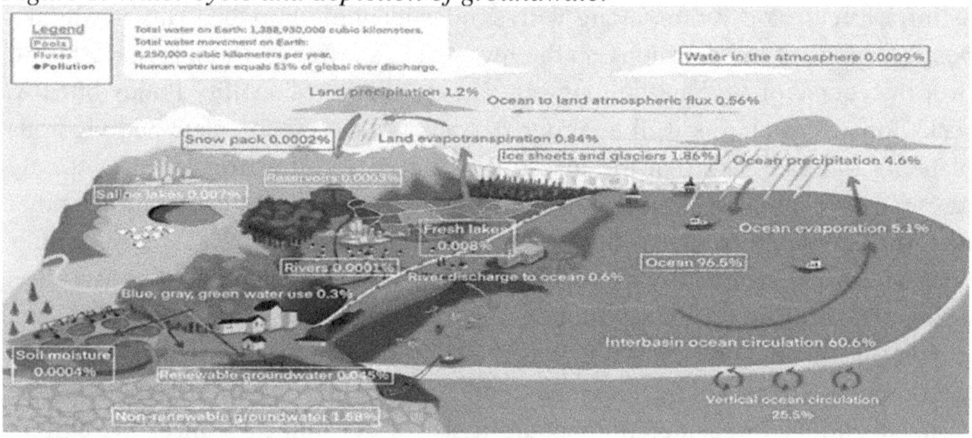

Sources: LangeLeslie and Anna Wright, Wikimedia

Surface Water: IWA says that if followed proper practices, then dams have the ability to give the supply of water in a sustainable manner, as it helps in irrigation, diversion of water, generation of power, navigation through waterbody, as well as prevents flood.

Figure 2. Surface water in UK

Source: Perry Dark, Wikimedia

Reclaimed Water: Also known as reused water, which is a way to reutilize water again after treatment in easing the stress generated on primary resources of water, but the capacity of each country and region varies in this regard, but still the capacity of reclaimed water could be improved further.

Figure 3. Reclaimed water in Florida, USA

Source: P,TO 19104, Wikimedia

Desalination: As per IWA, desalination could be made sustainable by using renewable energy as the source in the process, like solar power could be used as renewable energy in the process in a decentralized manner in the case of the six thousand islands of Greece to have sustainable supply of water.

Figure 4. Victorian Desalination Plant in Victoria, Australia

Source: JoelCarterRamadge, Wikimedia

Figure 5. Satellite map of Water Foundation in California, USA

Source: Google Earth

Figure 6. Satellite map of Sonoma Ecology Center in California, USA

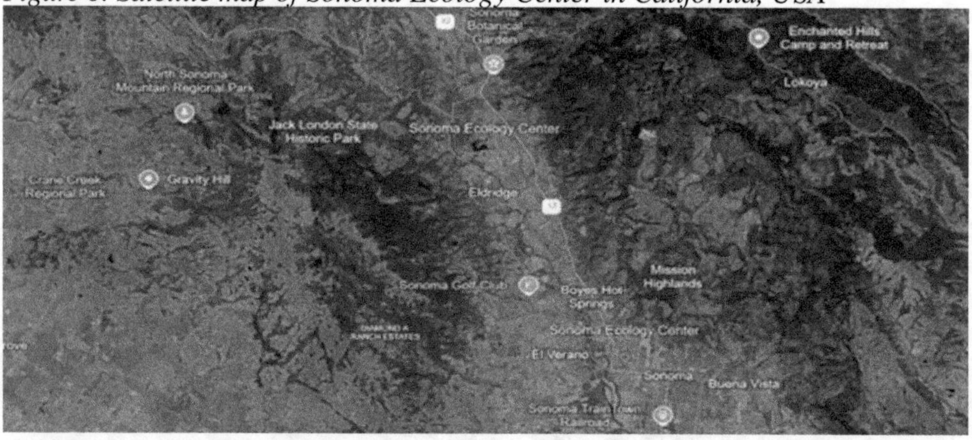

Source: Google Earth

Figure 7. Satellite map of Environmental Science Associates in California, USA

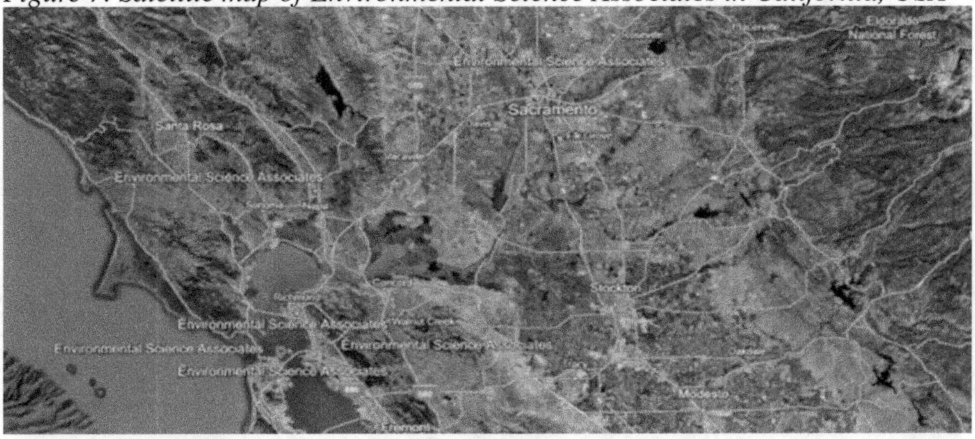

Source: Google Earth

Figure 8. Satellite map of the Bay Institute in California, USA

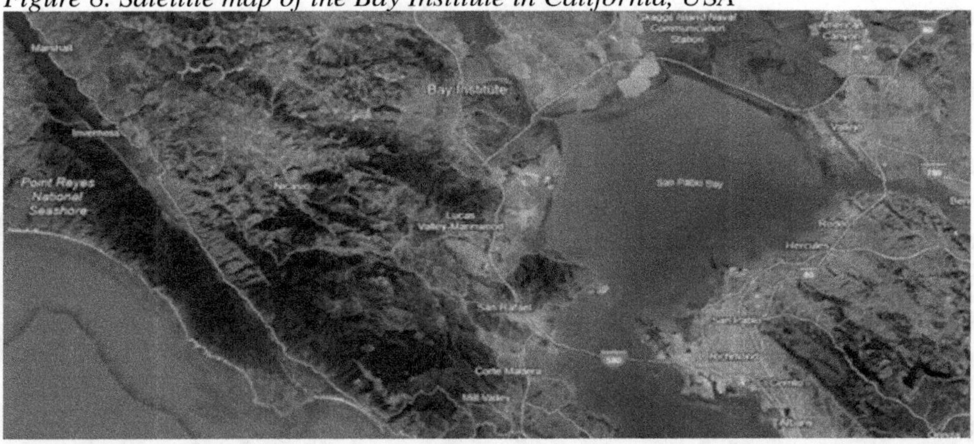

Source: Google Earth

Figure 9. Satellite map of International Water Association's Headquarters in London, UK

Source: Google Earth

Then United Nations had marked the several areas where management of water resources is essential in a sustainable manner, which supports "Securing Sustainable Water for All" as its primary goal. This means, that such areas include agriculture requiring seven tenths of the water globally, industry as well as energy requiring two tenths of the water globally, and then domestic sector, where the first two sectors must need to enhance their productivity based on their high consumption of water, and of course wastage of water resources must be avoided in the three sectors and must be managed and utilised in an efficient manner. Further cities require to upgrade their management of water resource in a sustainable manner, as the population is expected to grow further and migration from villages to cities will take place more in the future, especially in developing countries at a significant rate. Ecologically, the current progress and success of sustainable water management in the world is far from the required level, which raises the curtain of uncertainty for the future as the situation becomes more critical with the passing time. Moreover, to devise a successful strategy for sustainable water management, it is important to know that the national as well as regional governments will plan it differently, and further other varied factors will be responsible to create a major difference among them. These factors include climate change risks as well as impact in an area, water infrastructure's maturity level, access to finance, regulations in the concerned area, ambition level of the government, and so on. Then its conceptual framework takes water as a renewable but finite resource having regional as well as international restrictions into consideration. That is why incorporating economic, ecological in addition to social aspects must be taken into account, with the assistance of regulatory or legal as well as institutional authorities in order to move in the direction of managing water resources in a sustainable manner. Of course, there are many case studies available in the world, where governments have developed strategies or some sustainable water management projects were developed in several countries in the world, for example:

Beijing in China: Beijing's Huaifang Underground Water Reclamation Plant is a good example of sustainable water project, as it also won Gold in the twelfth IWA (International Water Association) project innovation awards for the city's sustainable solution to reutilize water while considering the ecological aspects. This wastewater recycling plant is of thirty-one hectares to enhance the Liangshui Reiver's quality of water in addition to comfort the southern part of the city from the pressure of sewage treatment.

Figure 10. Satellite map of Huaifang in China, where the famous underground water reclamation project is located

Source: Google Earth

Figure 11. Satellite map of Liangshui River in Beijing, China, which was improved with the underground reclamation project

Source: Google Earth

Abu Dhabi in UAE: The country United Arab Emirates is also in the race of desalination plants with renewable energy, for example Mascara Renewable Water which is a French company to build the solar photovoltaic desalination plant with reverse osmosis in Abu Dhabi. The plant entered in service in the year 2016 and the cost of project was three hundred euros.

Figure 12. Satellite map of various desalination plants in UAE

Source: Google Earth

Rotterdam in Netherlands: The list of Sustainable Cities Water Index of Arcadis showed that Rotterdam topped it. It had been practical as well as groundbreaking in its method to manage water, which comprised of investing generous amount of money its system of reservoir catchment. In terms of water reserves, the city is counted among the top ones in the world, and it is also a good example of sustainable water supply.

Figure 13. Satellite map of Rotterdam, Netherlands

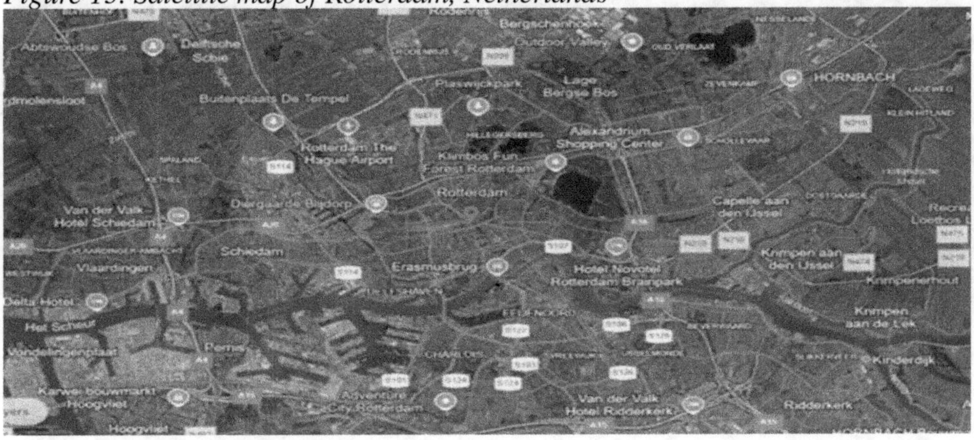

Source: Google Earth

Figure 14. Rotterdam city, Netherlands

Source: Leuvehaven, Wikimedia

Omdurman in Sudan: The city's water supply as well as optimization plan is one of the most important steps in the direction of millennium development goals in the region of Africa's sub-Saharan area in the latter half of the first decade of the 21st century. It was having the main objective to ease the shortage of drinking water in Khartoum as well as nearby areas, which is the country's capital.

Figure 15. Night view of Khartoum in Sudan

Source: Aamirco, Wikimedia

Figure 16. Satellite map of Khartoum, Omdurman and Nile River in Sudan

Source: Google Earth

California in USA: The American state has been working on fast tracking recharge of groundwater as well as to acquire the flood waters in order to store it underground, a way to recharge the basin. The target is to increase the recharge of groundwater by five hundred-thousand-acre feet at the minimum. The Merced Project receiving half a year permit by the State Water Board and at least approval of twenty permits, are some of the noticeable progresses by the state in the direction of the improving its management of water sustainability in the past years.

Figure 17. Satellite map of Merced where mentioned Merced Project is taking place in California, USA

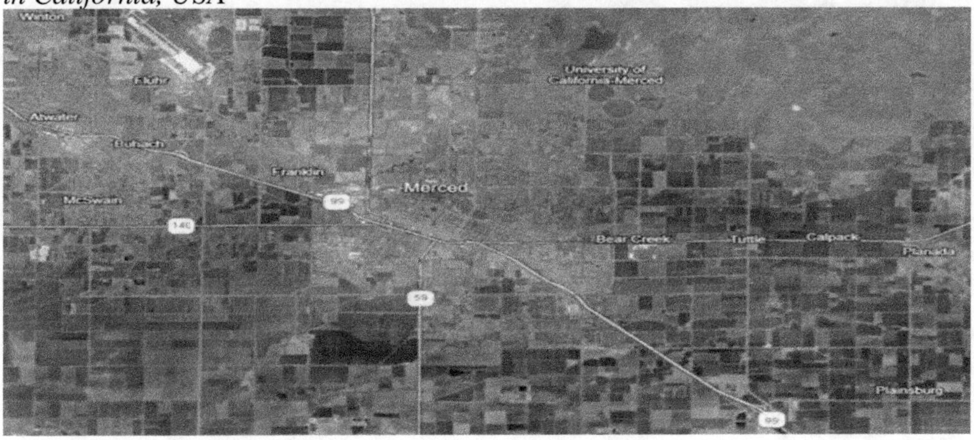

Source: Google Earth

Figure 18. Satellite map of California, USA

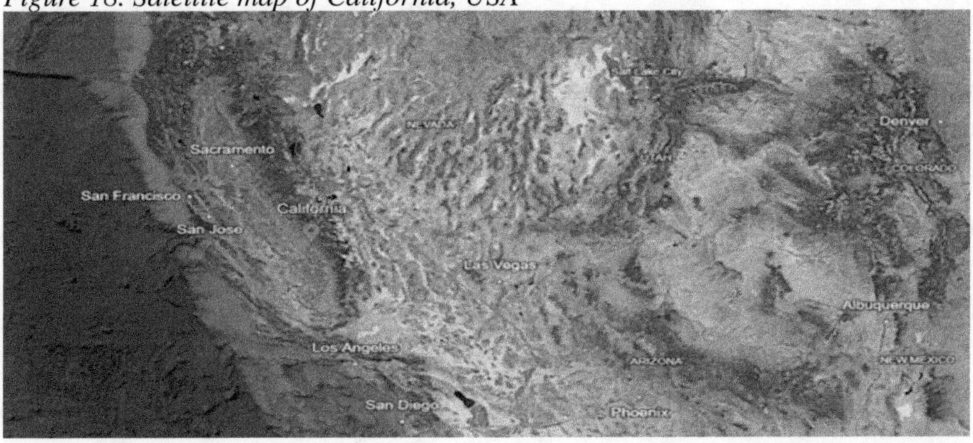

Source: Google Earth

Saudi Arabia: The company Metito signed a contract with the King Abdullah Economic City after facing a competition from ten international and regional companies to build a seawater desalination plant which could utilize the solar energy to operate in the year 2018. The plant was valued at over $58 million USD and the initial capacity was thirty thousand metre cube per day and expandable to twice. However, it is worth noting that Saudi Arabia has several similar projects and plants with the vision 2030 in mind.

Figure 19. Satellite map of King Abdullah Economic City, Saudi Arabia

Source: Google Earth

Northern Ireland: The focus was to manage water sustainably in lakes, rivers, aquifers, as well as loughs which could be utilized for industrial, agricultural as well as domestic purposes. Also, a way to manage the surplus rainwater in a sustainable manner to avoid flooding as much as possible, while taking care of the affected communities to reduce their problem of incoming floods. Further, to inspect the way amount of energy could be decreased while treating or transporting wastewater as well as drinking water.

Figure 20. Satellite map of Northern Ireland, UK

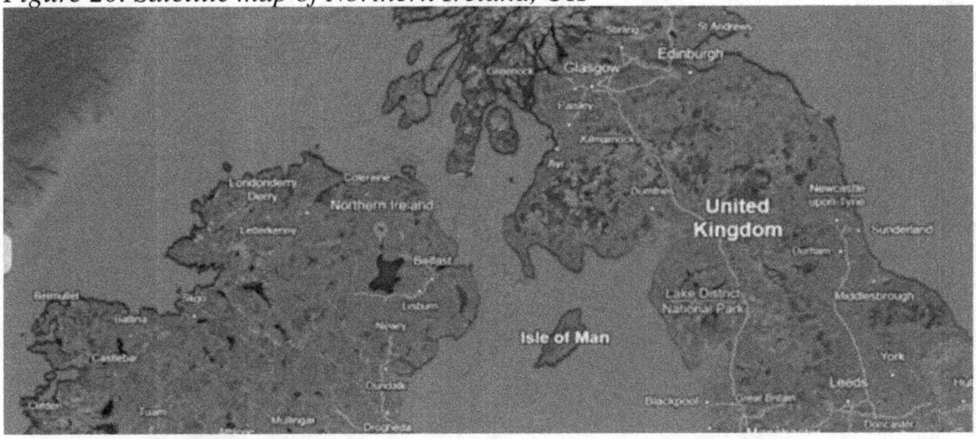

Source: Google Earth

Singapore: The project NEWater run by the Singaporean government makes it one of the leading countries in the reclamation of water. The project incorporates multiple process which sends the water already used to purify it in 3 steps, which includes reverse osmosis, microfiltration in addition to ultraviolet disinfection, then making majority of the water to be utilized for industrial purposes, but being drinkable, in the times of the scarcity of water, it could be added in the reservoirs.

Figure 21. UV disinfection unit's cutaway model in NEWater water treatment plant in Singapore

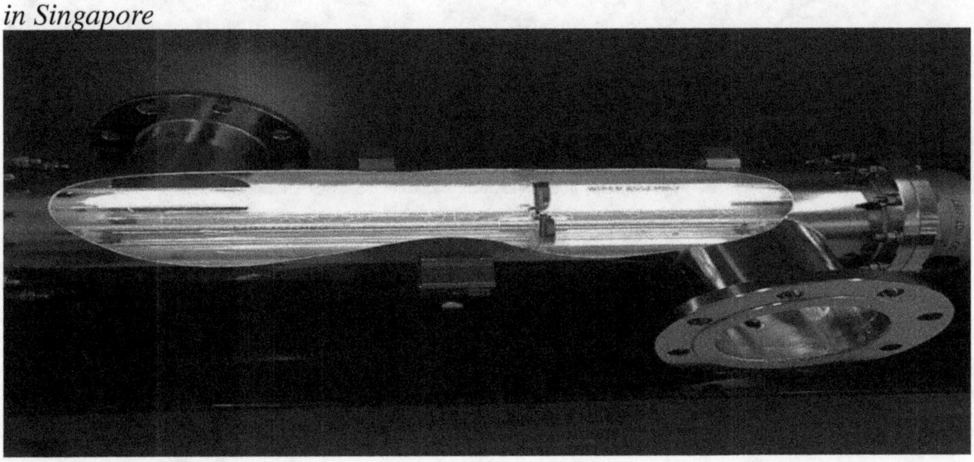

Source: Z22, Wikimedia

Figure 22. NEWater Bedock Plant in Singapore

Source: Z22, Wikimedia

Figure 23. Satellite map of NEWater Bedock Plant in Singapore

Source: Google Earth

Figure 24. Satellite map of Singapore, which is famous for its reclamation of water in the world

Source: Google Earth

Australia: In the Australian state Victoria, 4 regional strategies were developed to manage water sustainably, which were followed The Water Act of 1989. In at least every ten years, the above-mentioned strategies must be reviewed, for possible updates if required. In 2016, the strategy on sustainable water of the Central Region was review commenced and later on completed. Similarly, 5 yr examination of the strategies of Gippsland Region in addition to Western Region commenced in 2017 and later on completed, and then it was expected in the latter half of 2019 for Northern Region's strategies to be reviewed. However, the country in different parts also utilizes managed aquifer recovery in order to use reclaimed water as well as stormwater, to recharge the aquifers. This way, it becomes useful when water storage facilities or surface reservoirs aren't sufficient.

Figure 25. Satellite map of Victoria, Australia

Source: Google Earth

Andean Region: The region around the countries Columbia, Bolivia, in addition to Peru, are having the project called AICCA, with the target on dealing with climate change, as well as clean water. This is a biodiversity project to make sure the Andean people can enjoy the sustainable water practices.

Figure 26. Satellite map of Andean Region, showing Peru, Columbia and Bolivia

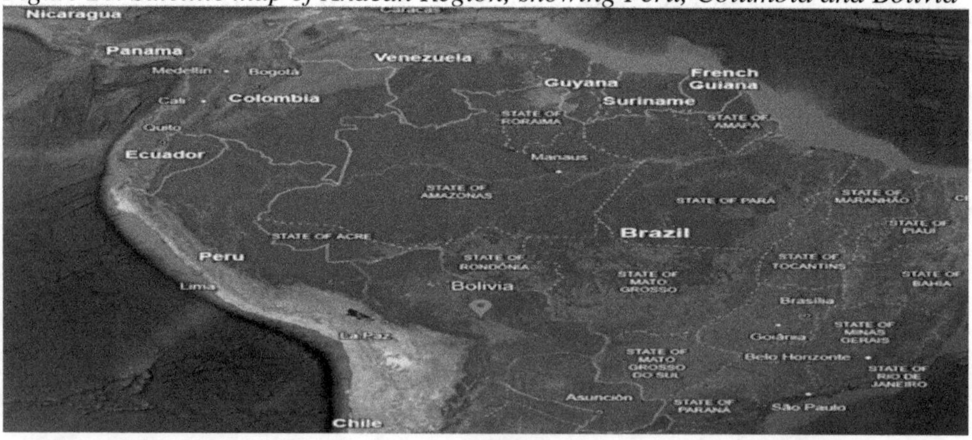

Source: Google Earth

Then, the International Desalination Association has set a target of one fifth of the novel desalination plants to be powered by renewable energy like solar energy in the span of 5 years from 2020 till 2025, in order to decrease the carbon footprint generated by desalination plants in a short period of time. Small scale trials across countries like Spain, countries in Middle East, as well as in India has been going on, but the challenge to how to increase the size of the plants as well as to reduce the dependence on grid power during night, is yet to address.

Indian Scenario

For the sustenance of ecosystems as well as for the survival of humans, water is essential. However, with rapid urbanization, climate change in addition to industrialization, direct as well as indirect problems are created affecting the water resources, which is why efficient conservation methods of water has become supreme priority in countries like India, where scarcity of water is a critical problem. It is evident that three fourths of the planet Earth is covered by water, but only less than 1% is fresh and drinkable water found in rivers, lakes, groundwater, as well as in ponds. However, in India, its population is around nineteen percent of the world's population but only has four percent of the freshwater resources of the world. This is why the per capita water availability is scarily around the international water scarcity threshold as per 2019 space inputs of the Central Water Commission water availability for the year is 1486-meter cube and the international threshold is 1000 metre cube annually, which shows India going near to that threshold in the future with growing population. Hence, the insufficient freshwater resources are put into tremendous pressure with not only population growth, but rapid economic development as well as industrialization. By the next year, 2025, it is expected that the water demand might go up by seventy percent in India. India being mostly agricultural country, as a vast majority of its population are involved in agriculture, but agriculture is also the biggest consumer of water resources, so the limited water resources are put into more stress, as agricultural output must be increased to meet the needs of the growing population of the country not only now but also in the future. Then, pollution as well as insufficient storage as well as infrastructure for the current needs of the country to manage water efficiently, are also major issues to be taken care. However, the country had been focussing on a complete method centring around sustainable development, also including the ways to make sure the precious water resources are secure as well as not exploited, and the way to deal with environmental worries at the same time. As a response, the policymakers took several steps towards sustainable policies as well as equitable distribution of water resources in the country. Till 1970s, the country focused on improving the productivity of its agricultural activities by strengthening its irrigation capacity, in order to make the country food sufficient, which include 1970's Green Revolution to transform the countries agriculture sector noticeably. Then from the decade of 1970s till the last decade of the 20th century, the country experienced rapid industrialization as well as urbanization, which led to the overexploitation as well as depletion of water sources in many areas making the country plan about a national level policy, i.e. National Water Policy in the latter half of the second last decade of the 20th century to manage the water resources properly, which was also the start of the water sector's sustainable development. From, the last decade of the 20th century, more reforms started taking place, along with the participation of the private sector as well as methods which were driven by the market to manage water resources. This led to another important initiative by the latter half of the first decade of the 21st century, i.e. National Water Mission, whose framework was laid down later on by 2011. Then by the 2nd decade of the 21st century onwards, many other policies and initiatives were taken by the country to develop in the direction of sustainable water management. The country implemented afforestation as well as conservation of soil, rainwater harvesting, efficient techniques for irrigation, watershed management, awareness campaigns, in addition to reuse as well as recycling of water in different parts of India, over the years.

Figure 27. Union Minister of Water Resources Prof. Saifuddin in an international conference on sustainable water management and world aqua congress in New Delhi, India

Source: Ministry of Water Resources (GODL-India), Wikimedia

Why is Sustainable Water Management Essential and How Does it Affect India?

Since, the human body on an average is around 60% made of water, of course, different parts of the body have different water percent and age, sex, level might again make variation in an individual. Hower, it is safe to say that the human body is mostly contains water, so fresh water supply to survive is important for humans.

Plus, water is important for all types of plants, animals at varying levels, that also means that water is not only needed for drinking purposes, but also for food production including agricultural activities, as well as domestic chores. Then, for development, industries are also needed, but they also rely on water, which makes that not only the survival of humans but also the development of humans rely on water. However, the water resources which can be used is limited in the world and that also getting depleted with time, due to mismanagement, overutilization, illegal activities, faults and leakages. Then, the increase in population at a fast rate and development of human societies, villages, towns, cities, states, countries, demand for more water resources as time passes. This problem is severe in India, because the country is most populated in the world and has only 4% of global water resources, where the per capita water availability in the country shows that India is in the list of water stressed countries in the world. Then, climate change is also affecting water resources, where vast majority of the events that are considered disasters in the world due to climate change are water related ones including droughts, floods, change in rainwater pattern, rise in the sea level, and so on. Of course, this includes India as well, that is why sustainable water management is important in the world and particularly in India in this context. Effective management of water resources in a sustainable manner will reduce not only the current water problems but also the future occurrence of major water related problems as well as disasters.

Objectives

There are several objectives of this chapter which are as follows:

- To understand the concept of sustainable water as well as sustainable water management.
- To understand the current developments in the world in the direction of sustainable water management.
- To understand the current stand of India and its government in regard to sustainable water management.
- To see the development of India in the context of sustainable water management and the supporting government policies as well as initiatives to address sustainable water management.
- To understand the future trend of sustainable water management with the assistance of graphical analyses.

How Do Smart and Advanced Technologies Affect Sustainable Water Management and What is India's Stand?

Smart and advanced technologies are being used with time to improve the efficiency of humans as well as to make life comfortable for people. This also includes reducing wastages and decreasing the carbon footprint in the process. Singh (2023) discussed how AI is improving sustainability in the construction sector and ANN utility in transforming civil engineering area covered by Singh (2023). However, it is important to understand that these advanced technologies could also be used to build infrastructure for water sector to promote sustainable practices. Then, these smart technologies could be used to encourage smart water management, which ultimately supports the sustainable water management, wastage, leakages or faults could be detected and timely repaired to reduce mismanagement of water resources which are already limited. This will further help in adapting to different water demands, water pump operations in a precise manner while reducing the wastage of energy, automation of water pumps to improve the efficiency of energy, and so on leading to sustainable development and management of water resources. Singh (2023) discussed that India working on infrastructure projects to attain sustainable goals and development in the country, and Singh (2023) mentioned that the country sustainable planning as well as development to create the built environment. This clearly shows that India is focusing on developing in a sustainable manner and that also includes the water sector and related infrastructure in the country as mentioned by Singh (2023) about construction of dams in the country along with their need to improve the future of human societies.

MOTIVATION

The author found that a number of water-related problems have been going on and with time these problems are becoming worse. The author covered the status of several countries in the world to understand how other countries are trying to lessen their water problems. Then the author thought to focus on how the most populated country in the world is trying to deal with this water related problems. The author found that the Central Government had taken a number of initiatives and policies to mitigate the water related problems as well as promote sustainable water management in the country. This led the author to focus on such policies as well as initiatives which will be helpful in making his home country walk on the path of sustainable water management.

LITERATURE REVIEW

This section is divided into three subsections which are smart cities and smart water management; smart agriculture and irrigation; as well as sustainable management:

Smart Cities and Smart Water Management

Sushma et al. (2024) studied the Indian smart cities, and they worked on water resource as well as energy management by performing an experimentally examining the wireless integrated smart system for their studied area. They found that the existing standard metering in such cities consumes time and expensive, which is why an integrated smart system for metering would be the solution. They stated that their proposed metering system would effectively measure the usage of water as well as energy.

Fong et al. (2024) studied the environment of smart cities, and they worked on making supply of water sustainable with the assistance of pipeline management technologies in their studied area. They stated that the water pipeline infrastructure's health maintenance is a problem, which could be solved with the use of technologies like artificial intelligence as well as wireless sensing to help water pipelines in acquiring preventive maintenance facilities in order to decrease the contamination risk as well as leakage of water throughout their infrastructure.

Oberascher et al. (2022) studied smart water cities, and they worked on a thorough review of data requirements, applications as well as technologies for communication which could be used for management integration to help their studied area. They stated to covert the gap between the capabilities of Internet of Things and water infrastructure in urban areas through their study.

Sundaresan et al. (2021) studied smart cities, and they worked on the system of intelligent water management with the assistance of blockchain architecture to help with their studied area. They then proposed a system, where design of reservoir as well as smart water tank were considered with 2 modes operation, which were running mode as well as filling mode to accomplish their study.

Kumar et al. (2024) studied smart cities, and they worked on the worldwide scenario of the water footprints in them. They stated that the degraded quality as well as scarcity of water are the major problem in the urbanized areas internationally, and they found a significant tool which is water footprint analysis for the efficient water resource management which could lead to green and sustainable future. They presented various methods to calculate the consumption of water in cities, as well as covered the water smart city concept in their study.

Kavya et al. (2023) studied smart water management, and they worked on modelling a water demand prediction model for short term purpose to assist their studied area. They stated that to predict the demand of water, ML as well as DL models were utilized. They then considered not only multivariate but also univariate time series prediction models, where Long-Short Term Memory model outperformed other models to provide the desired outcomes.

Wu et al. (2023) studied the operation management of smart water grid, and they worked not only on its localization but also on anomaly detection on the basis of high-fidelity digital twin. They stated that for smart water grids, a framework of generic digital twin was developed by them, and then to construct a high-fidelity smart water grid digital twin, they incorporated hydraulic model calibration as well as data analytics to accomplish their research.

DiCarlo et al. (2023) studied community water system, and they worked on the way it adopts smart technology as well as on the management of customer complaints. They stated that complaint management was considered by of five hundred four community water systems based on their survey. They found that the records of the complaints are stored digitally normally by the large-scale community water systems, whereas small scale ones give priority to other necessities as they don't receive many complaints, but to improve the infrastructure as well as detect the trend complaint data is necessary in such cases.

Hu et al. (2022) studied smart metering, and they worked on not only on water demand management in urban areas on it, but also on a trading framework allowed through water. They proposed such framework for domestic – industrial water and stated that trading benefits require water price as well as tradable water permit.

Singh and Ahmed (2021) studied systems which were about smart water management on the basis of Internet of Things, and they worked on a systematic review paper to accomplish their study. They covered different components as well as architecture of the above stated systems in depth and they performed a detailed survey of similar systems which were already present.

Figueiredo et al. (2021) studied entities of the water management as well as about smart cities, and they worked on a digital water solution for their studied area. A Water Wise System was proposed by them, which was the consequence of a research and development project, aided by not only the Grant of the Portuguese Government, but also by the European Union.

Zhang et al. (2021) studied smart water management, and they worked on self-adjusting anode catalyst for it especially in the fuel cells which were anion exchange membrane ones. They used nanopore structure which were graded gradient to help in catalyst layer's design. They stated that moisture behaviour was found to self-regulating for the anode catalyst layer while proceeding with their research to get the desired outcomes.

Uddin and Jeong (2021) studied smart water management, and they worked on Bangladesh's urban pollution of rivers in the past forty years, to understand ecological risk, prospective public health, future forecasts as well as current policy towards their studied area. They found that Rupsha, Korotoa, Padma, Teesta as well as Pashur rivers were particularly in critical condition.

Choudhary et al. (2024) considered the Indian perspective and examined the automatic smart watering unit which were on the basis of sensors to be utilized for the paddy nurseries. They further developed automatic watering unit on the basis of sensors for irrigation purpose and they found the optimal parameters with the assistance of RSM model, which were 10 cm as spray height, 4.5 kg/cm2 as spray pressure as well as o.14 m/s conveyor speed.

Zhang et al. (2024) considered the concept fit for purpose, and they examined the system of smart water on the basis of it with a conceptual approach. They stated that in smart water system the gap in the method of general examination, was covered by them. They assessed the system on the basis of fit for purpose in the aspects of reliability, usefulness, as well as feasibility. They further claimed that the suggestions for not only the current system but for novel system, could be given with their proposed method.

Damman et al. (2023) considered connecting practice as well theory to advance in the direction of water smart society through their study to obtain practice as well as research-based definitions. In the direction of water smart society, they worked on leading not only the practitioners but also decision makers through their proposed strategic objectives in their study.

Kanyama et al. (2024) considered developing the enhanced efficiency of water by working on the networks of smart water metering to detect anomalies in them. They stated that the optimization of network becomes more important due to climate change, and they used Namibia's Windhoek's seventy-two months complete dataset, to work on an accurate analytical method to disclose varied kinds of anomaly frequently present in the above stated networks.

Cardell-Oliver et al. (2024) studied the databases of smart water meter, and they worked resolution as well activity known protection of privacy for their studied area. They stated that the sensitive info regarding activities of homes could be revealed through the databases of smart water meter, and they proposed a pipeline in regard to privacy protection with tuneable parameters to solve the problems of smart water meters.

Kanyama et al. (2024) studied the networks of the smart water metering, and they worked on a review paper where detection of anomalies could be done by applying machine learning in their studied area. They found that there is not enough systematic review of literatures while researching on detection of anomalies using machine learning in the above stated networks and they for effective networks of water, vigorous anomaly detection relying on machine learning, needs availability of data as a necessity.

Salman and Hasar (2023) studied the air pollution, waste, transportation as well as water smart applications with the assistance of techniques such as Internet of Things, in order to consider the aspects of environ in the concept of smart cities. They stated that leaks as well as losses of water could be avoided with the use of Internet of Things to make sure the urban water sustainability to complete their study.

Tavares et al. (2024) studied hybrid the systems of renewable water supply and they worked on optimising their smart feasibility with the technologies like digital twin, in order to target the cities which were isolated ones to use a multicriteria method. They discussed about applying machine learning in the systems of hybrid water energy and in Solver, they applied the method of generalized reduced gradient, to obtain their desired research results.

Abdi et al. (2024) studied different crude oils as well as smart water, in order to work on their interfacial tension. They stated that for utilizing smart water, the main mechanism is the ion hydration energy. They stated that the interacting system of crude oils or smart water's interfacial tension could be greatly affected by the potential of Hydrogen.

Durgun et al. (2024) studied sensors enabled by artificial intelligence in order to monitor the quality of real time water. They considered the systems of smart water in urban areas and worked on ultraviolet disinfection analysis in addition to identifying the contaminants in them. They further explained a new method which utilizes AI technologies incorporated in innovative sensor systems to assess the quality of water, in order to pay the way for more advanced systems to give solution to the most terrible environ challenges.

Newing et al. (2023) studied smart water metering, which could be called as a tool which is non-invasive in nature in order to infer the occupancy as well as type of dwelling and they covered the consequences of the gathered statistics of tourism along with neighbourhood level housing in their research.

Gumz and Fettermann (2024) studied smart meters, and they considered the perspective of the users in them, which also included their trends in the future in addition to present. They stated that to incorporate international smart meters, it is essential take the perspective of the users into consideration. They covered one thousand five hundred and fifty-five papers, and they performed complete bibliographic analysis on them to obtain the favourable results.

Mathews et al. (2023) studied water as well as energy management and they worked on a BIM integrated model for it, on the basis of Internet of Things, to be used in smart homes to create the design of a sample home while considering ASHRAE and NBC standards for incorporation.

Wu et al. (2020) studied the source management of urban water, and they worked on its quality forecast with the assistance of smart data. They stated that it was the first time that quality control of water was achieved by incorporating data driven technologies with a system of smart water supply and they also proposed two models to deal with prediction of biological indicators in the system of water supply for industrial ones, to obtain suitable outcomes.

Smart Agriculture and Irrigation

Morchid et al. (2024) studied agricultural water security, and they worked on a system of smart irrigation management on the basis of Internet of Things to improve it, with the assistance of inserted systems, cloud computing as well as telemetry data. They stated that their proposed system has 3 tier architecture which consisted of dashboard, ThingsBoard cloud, as well as Internet of Things devices. Here, Hypertext Transfer Protocol helped in transferring telemetry data to the ThingsBoard and then not only analysing but visualizing of data in real time could take place on it.

El-taibi et al. (2024) studied smart agriculture and they worked on improving the management of water in it by performing a study on smart irrigation system which on the basis of IoT as well as cloud technology. They stated that the utilization of water could be optimize with the help of cloud system, which also makes the irrigation efficient by connecting small farms. Further, the efficiency of water could be improved by the centralized data management system that has the ability to link smart farms.

Alves et al. (2023) studied smart farming, and they worked on developing a digital twin for it, as well as to save water, they further worked on irrigation management system. They created a simulation model and integrated an Internet of Things platform with it. They then not only examined the mentioned platform with their simulation model but also the various methos used to calculate irrigation prescriptions, in order to enhance the efficiency of the utilization of water for irrigation purposes.

Lephondo et al. (2024) considered the agriculture sector, and they worked on minimizing the use of water with the assistance of machine learning to see the consequences of smart irrigation system in the mentioned sector. The 4IR permitted system of smart irrigation was proposed by them to help the farmers in obtaining weather API, moisture of the soil, plugged sensors to forecast the time to irrigate or not, as well as various crop's capacity of water, in order to minimize the use of water in the sector.

Krisnawijaya et al. (2024) studied smart farming, and they covered the systems of data management development with the design of reference architecture. They performed an analysis of feature driven domain one, and system of data management's architecture modelling with the assistance of numerous architecture views to complete their research.

Singh et al. (2024) studied autonomous irrigation management, and they worked on a smart controller which was on the basis of soil moisture sensor for it, where their focus was on the recycled water in the hybrid bermudagrass present in California's southern coast, which led to their 3-year irrigation study.

Jimenez et al. (2021) studied irrigation systems of high scale, and they worked to allocate resources in it, with the method of smart water management. They stated that in a given irrigation district, water negotiation could be done by incorporating agent-based model, which were made of algorithms like satisfaction, negotiation as well as perception ones.

Khachatryan et al. (2019) studied smart landscape irrigation technologies, and they worked for it with willingness as well as preferences in order to achieve sustainable water management. They stated that compared to the traditional controllers, smart irrigation controllers on the basis of sensors were preferred. Similarly compared to offsite stations as well as over wired, onsite weather stations along with wireless sensors were preferred to see as the significant factors in savings of water bill annually when considering the preferences of consumers.

Bwambale et al. (2024) studied smart irrigation, and they worked on a predictive control framework, optimization, but also on modelling of their studied area. For scheduling the smart irrigation, they worked on the predictive control framework, and for future decisions on irrigation, model predictive control could be utilized and such framework consequents into efficient water use as well as high yields.

Li et al. (2024) studied climate based smart irrigation strategy and they worked on alleviating the consumption of water for agricultural activities along with taking care of food security. They found that the demand for irrigation as well as water requirement for crops are affected by climate change and their proposed smart irrigation could overcome such problems.

Sustainable Management

Vadruccio et al. (2024) studied management of water, and they worked on its efficiency with the assistance of IoT as well as smart metering. They evaluated how adopting smart meters in consuming water affects environment as well as economy. For the reference context, they considered the scenario of Italy, to develop an analytical model to obtain the desired results.

Glina et al. (2024) studied climate smart sustainable management, and they worked on soil water repellence in order to find out the significance of soil water repellence for their studied area, particularly for Poland's central region's fen peatland soils. They progressed with their examination of water repellence for the above stated soils, and they stated that the progression of the mush forming process is the basis organic soil's water repellence, as well as the decrease of water repellence could be noticed with the depletion of the organic matter.

Nakhaei et al. (2023) studied management of water, and they worked on a support system which was a smart sustainable decision one for their studied area to help the power plants, especially in the regions which were water stress ones. They not only incorporated multi criteria decision making but also PESTEL matrix methods, apart from AI based control as well as prediction sections, along with smart decision support system to be used for observation purpose, to accomplish their study.

Oberascher et al. (2021) studied water management in urban areas, and they worked on incorporating it with micro steps in order to create a solution which is on the basis of Internet of Things. They introduced the above-mentioned solution as smart rain barrel to take care of rainwater harvesting. For their case study, they considered alpine municipality, which was exposed to retrofitting in a hypothetical scenario, to accomplish their research.

Charmchi et al. (2021) studied hydro reservoirs, and they worked on smart supply side management in order to optimize their studied area with the assistance of a nexus concept on energy or water. They not only worked on a detailed new smart framework for the hydro reservoirs to optimize their management, but they also performed hydropower pinch analysis to obtain suitable outcomes.

Arepalli and Naik (2024) studied aqua ponds management, and they worked on an assessment framework which was about smart water quality on the basis of Internet of Things for their studied area with the assistance of their proposed improved Dilated Spatial temporal Convolution Neural Network to obtain suitable results.

Wang et al. (2024) considered not only advanced application but also fundamental mechanism for water harvesting by working on micro or nano architectonics along with smart materials. They analysed performance of water harvesting of various bionic systems and based on the various quantities as well as types of the imitated organisms, they categorized systematically the smart water harvester to accomplish their study.

Buriro et al. (2023) studied smart water flooding, and they worked on the advancements in them, by performing a complete study on the interaction of salinity as well as ions in the reservoirs of carbonate. They performed a critical analysis of the disagreement as well as agreement between the authors regarding smart water flooding.

RESEARCH GAPS

After going through several research and review papers as well as chapter, only the ones published in the recent years were considered further. There is a no doubt that in the recent years in the world as well as in India, researchers focused on advanced technologies to solve water problems in irrigation, agriculture, smart cities, and so on. However, there was a lack or absence of studies on the Indian policies regarding water problems which could help the country on the path of smart water management, in the recent years.

MAIN FOCUS OF THE CHAPTER

The chapter focuses on sustainable water management and the necessity of management of water in a sustainable way in the world. The chapter highlights the some of the current developments in the world in regard to sustainable water management in brief. Further, this book chapter especially focused on India's stand and how the Central Government is taking steps to lead the country on the path of sustainable water management. It highlighted a number of policies and initiatives taken by the Indian Government in the recent years to help the country in mitigating water related problems as well as in making the country progress in the context of sustainable water management. Furthermore, graphical analyses were performed to understand the future trends and necessary direction as well as steps to be taken to properly implement the objectives of sustainable water management in the long run.

ISSUES, CONTROVERSIES, and PROBLEMS

In this section, some major problems, controversies as well as issues are discussed which cannot be overlooked. Even though there have been a number of policies as well initiatives framed and executed to lead towards sustainable water management, proper implementation of these policies, initiatives and programmes were not there, which is why the ground reality and proposed plans had a big difference between them. Then, growing population and increasing urbanization along with migration of people from villages to cities over time, made the policies and existing facilities insufficient at a fast rate. Of course, India being a democratic country will have change in governments periodically which includes the Union Government and State Governments, this also means that the new government having different political beliefs as well as plans, might interrupt the earlier programmes or might cancel them or create new ones, which at the same time could be good and could also make the implementation of plans and schemes slow as well as not much effective with growing water concern. Then, more support from NGOs and public is also needed as waiting for the government to take care of all the water problems for a large as well as most populated country in the world is not practical.

SOME MAJOR POLICIES, PROGRAM AND INITIATIVES IN INDIA

In this section, a number of initiatives, programs as well as policies formed and implemented by the Government in India to lead the country in the direction of sustainable water management, are discussed. Mostly, the country's method towards sustainable water management, is a community focused one. Some of the major programs, policies as well as initiatives taken by India in the water sector are:

- Jal Jeevan Mission
- Pradhan Mantri Krishi Sinchayee Yojana
- Atal Bhujal Yojana
- Jal Shakti Abhiyan
- National Water Mission
- Ground Water Management and Regulation
- Namami Gange Programme
- National Hydrology Project
- River Basin Management
- AMRUT
- National Rural Drinking Water Program
- National Water Policy
- Smart Cities Misson
- Integrated Watershed Management Programme
- Mahatma Gandhi National Rural Employment and Guarantee Act

Jal Jeevan Mission

In the year 2019, Jal Jeevan Mission was initiated. The meaning of term is water is life and with that meaning, the mission has a target to have all households in the Indian rural areas to obtain piped supply of water by the year 2024. With the development of the local infrastructure, the mission stresses on community participation, decentralized management of water as well as sustainable supply of water in the country. The mission also wants to lessen the burden of human, particularly women who are mostly occupied with the collection of water by providing them piped supply of water to their homes, as well as by making sure people have access to safe drinking water in order to enhance their general health.

Figure 28. Satellite map of Jal Jeevan Mission centres in India

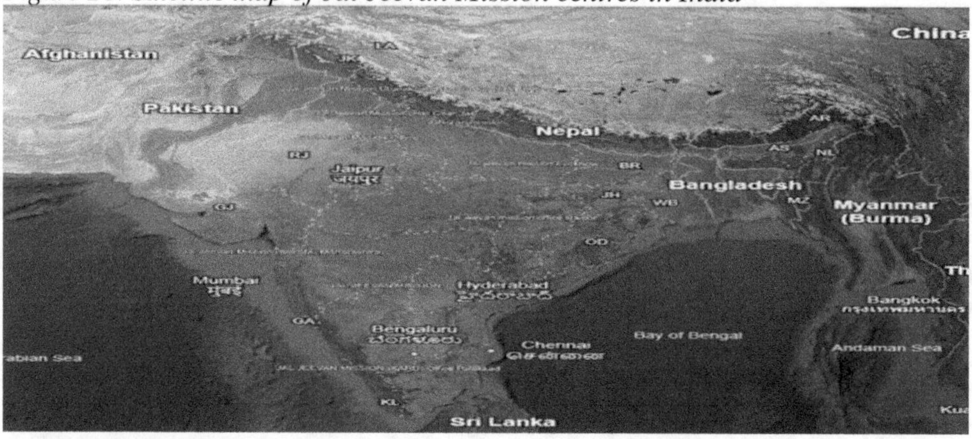

Source: Google Earth

Pradhan Mantri Krishi Sinchayee Yojana

This plan is often in short known as PMKSY, and the term literally means irrigation plan for the farmers. The plan was started in the year 2015 with the target to improve the efficiency of the use of water in the agriculture sector. The plan supports techniques like watershed management, micro irrigation which includes drip as well as sprinkler system of irrigation, in addition to rainwater harvesting in the country. This plan not only optimizes the use of water, but also targets to increase the food security assistance, productivity of agriculture sector, as well as enhance the livelihoods of the Indian farmers, which are consistent with sustainable water management path.

Figure 29. Satellite map of PMKSY office in New Delhi, India

Source: Google Earth

Atal Bhujal Yojana

The plan is often short knowns as ABHY, and in the year 2018, it was launched. The plan literally means groundwater plan named after India's former and late Prime Minister Atal Bihari Vajpayee. This plan's main focus is on management of groundwater in a sustainable manner in the country. The plan stresses on efficient practices of using water, demand side management, as well as initiatives to recharge groundwater which are community driven ones, particularly in areas which are water stressed ones. The plan further focuses on aquifers which are over exploited to make sure groundwater resources are available in a sustainable manner.

Figure 30. Satellite map of Atal Bhujal Yojana office in New Delhi, India

Source: Google Earth

Jal Shakti Abhiyan

This campaign is in short called as JSA and it literally means a campaign to strengthen water. The campaign was launched by the Ministry of Jal Shakti under the Government of India in the year 2019. The campaign main objectives were to recharge, conserve, reuse, as well as restore water resources in the country, to deal with the scarcity of water. The campaign particularly focuses on recharging as well as reusing of bore wells, conservation of water as well as rainwater harvesting, intensive afforestation, watershed development, in addition to renovation of conventional as well as other water bodies including tanks. The campaign took several measures to make sure water is conserved in an effective manner, by establishing rainwater harvesting cells, building bye-laws enforcement, water bodies in urban areas to be rejuvenated and so on. In two hundred and fifty-six water stressed districts in India, the campaign covered one thousand five hundred and ninety-two blocks out of the total two thousand eight hundred and thirty-six blocks. Due to pandemic, the campaign was affected, so the Ministry of Jal Shakti started the campaign named Catch the Rain on an annual basis, which led to "Jal Shakti Abhiyan: Catch The Rain" dealing with collection of rainwater and methods to harvest rainwater by underground tank, system of rooftop rainwater harvesting, in addition to wells.

Figure 31. Union Minister for Jal Shakti C.R. Paatil addressing the campaign in New Delhi, India

Sources: Ministry of Jal Shakti (GODL-India), Wikimedia

Figure 32. Satellite map of Ministry of Jal Shakti in New Delhi, India

Source: Google Earth

National Water Mission

This mission was identified as one of the 8 missions in 2008 under National Action Plan on Climate Change but got the Union Cabinet's approval by the 1st half of the year 2011. The mission targets to reduce wastage, water conservation, as well as making sure that there is equitable distribution across the different sectors in the country. It encourages use of water in an efficient manner with the assistance of public awareness campaigns, policy initiatives, in addition to technological intrusions. This mission is incorporating industrial sector, agricultural sector, as well as domestic consumption of water, as its target to ensure sustainable water practices and management. Further, there had been 5 goals under this mission which are: to examine the water resources being affected by the climate change; available of complete water data base in the public domain in the country; encouragement of basin level combined management of water resources; one fifth increase in the efficiency of utilizing water; in addition to encouragement of state as well as citizen actions for augmentation, conservation, as well as preservation of water, along with special attention to areas which are vulnerable in the country which also comprises of areas which are overexploited. The mission also runs several significant campaigns in the country such as Research & Development in the Water Sector, Jal Itihaas, Catch The Rain, Dialogues with District Magistrates, Bureau of Water Use Efficiency, in addition to Water Talk.

Figure 33. Satellite map of National Water Mission office in New Delhi, India

Source: Google Earth

Figure 34. Union Minister for Water Resources Shri H. Rawat chairing the first meeting of advisory board for National Water Mission in 2013 in New Delhi, India

Source: Ministry of Water Resources (GODL-India), Wikimedia

Ground Water Management and Regulation

This scheme is a Central Sector one, which starts from the year 2021 till the year 2026. In short, the scheme is often known as GWMR. The scheme incorporates many major activities like mapping of aquifers, to monitor the levels of the ground water, piezometers construction, outreach activities, groundwater resources examination, to monitor the quality of groundwater, mapping of spring shed as well as intrusions to rejuvenate aquifers, and control as well as regulation of extraction of groundwater. Further, under this scheme, Public Investment Board approved a project named National Aquifer Mapping and Management, from the year 2022 till the year 2026, which will be part of this scheme and include the seven thousand piezometers construction as well as the same number of Digital Water Level Recorders to be installed with the system of telemetry, construction of one thousand one hundred and thirty five wells in eleven states to generate data for mapping of aquifers, and three hundred thousand square kilometre area to be covered for heliborne surveys so that mapping in high resolution could be done for North Western India's arid parts.

Figure 35. Inauguration of conclave on aquifer mapping groundwater management in 2016 in New Delhi, India

Source: Ministry of Water Resources (GODL-India), Wikimedia

Namami Gange Programme

The programme is centred on the National River Ganga, which is symbolizes as one of the Goddess in the country from ancient times. The programme targets to conserve, effectively reduce the pollution as well as rejuvenate the national river. The Union Government approved this programme as a “Flagship Programme” in the mid of 2014 with a budget of over $ 3.2 billion USD at that time. The programme is considered to be an integrated conservation mission, with chief pillars as development of the river front, afforestation, infrastructure for the treatment of sewage, cleaning of the surface of river, biodiversity, Ganga Gram, to monitor the industrial effluent, as well as public awareness. The programme execution was divided into three parts, entry level activities to see some noticeable impact immediately, activities with medium term time frame which must be done within five years, and then activities with long term time frame which must be done within ten years.

Figure 36. German Ambassador Dr. Martin and the Secretary, Ministry of Water Resources, River Development and Ganga Rejuvination, Shri S. Shekar, signing an agreement to execute Namami Gange Programme in 2016 in New Delhi, India

Source: Ministry of Water Resources (GODL-India), Wikimedia

Figure 37. The Union Ministers addressing press conference on Namami Gange Programme in 2018 in New Delhi, India

Source: Ministry of Shipping (GODL-India), Wikimedia

National Hydrology Project

This is a national level project, which in the month of April 2016 got the approval of the Union Cabinet and is a scheme belonging to the Central Sector in India but also has the assistance of the World Bank. The budget of the project is more than $437 million USD, in which majority of the budget is for the project and a small portion for the National Water Informatics Centre's establishment. World Bank set to support $160 million USD and complete grant to the Indian states were given by the Central Government. The timeline of this project was 8 years, to be completed by this year 2024, but the project also has conditional extension till the end of September next year 2025, by the Department of Expenditure. There forty-eight agencies executing this project, where one forth are from the Central Government and rest are from the State Governments. Through this project, spatial information system enhancement, capacity building in addition to water resources must be strengthened, management applications as well as operation of water resources must be encouraged, and in situ system must be enhanced. Hence, this project will assess water resources, management of drought, reservoir operations, management of flood, and so on with the assistance of Decision Support Systems.

Figure 38. Satellite map of National Hydrology Project office in New Delhi, India

Source: Google Earth

River Basin Management

This scheme has 2 chief parts which are Brahmaputra Board as well as examination of the scheme named water resource development. The first part is being continued from the old plans, which was originally an Act of the Parliament in 1980 and became functional in 1982, this includes the measures to control the erosion of banks as well as floods in Barak Valley in addition to Brahmaputra river. The seven sisters' states and parts of West Bengal as well as Sikkim are affected by Brahmaputra Basin. The second part has two elements, which are Central Water Commission and then National Water Development Agency.

Figure 39. Satellite map of Central Water Commission centres in India

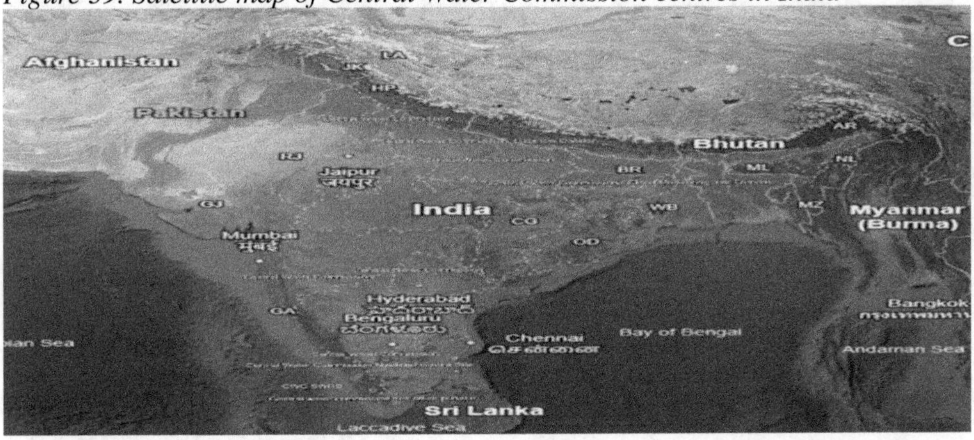

Source: Google Earth

Figure 40. Satellite map of National Water Development Agency centres in India

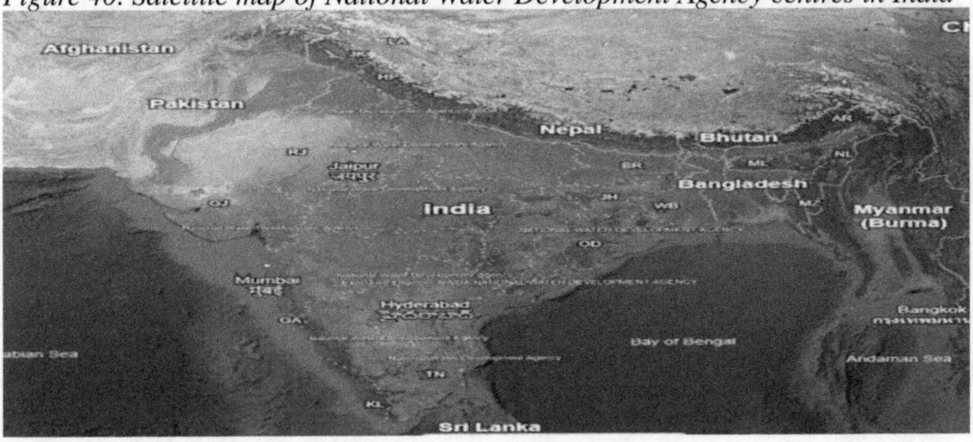

Source: Google Earth

Figure 41. Satellite map of Brahmaputra Board in Assam, India

Source: Google Earth

Figure 42. Satellite map of Brahmaputra river in India

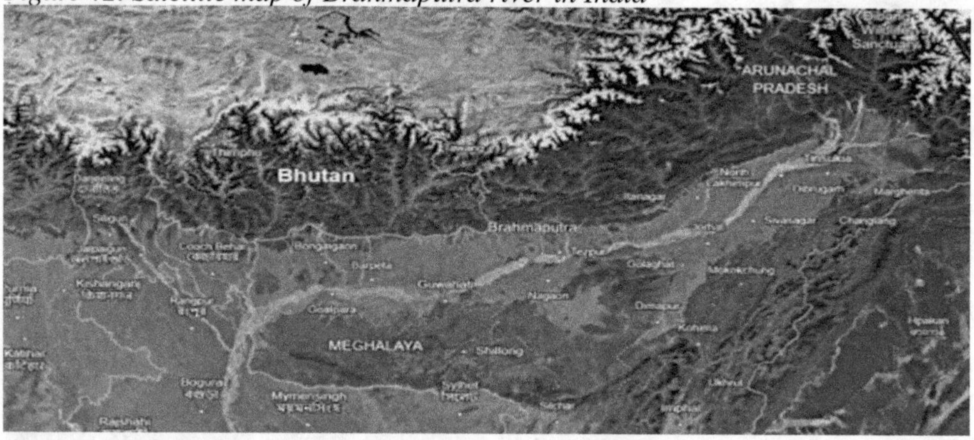

Source: Google Earth

Figure 43. Satellite map of Ganga River Basin Management and Studies centre in Kanpur, Uttar Pradesh, India

Source: Google Earth

Atal Mission for Rejuvenation and Urban Transformation

This mission is often called as AMRUT in short, it was launched by the Indian Government with the objective to give the general civic facilities such as sewerage, supply of water, parks as well as transport facility in urban areas. This mission covers five hundred Indian cities which comprise of both towns in addition to cities having a more than 100,000 people residing in them. The mission was from the financial year 2015-16 till the financial year 2019-20 with an outlay of over $ 5.9 billion USD. The ma parts of the project were septage, urban transport, system of water supply, drainage of storm water, capacity building, sewerage, green space along with parks, and so on. The mission covered sustainable water management in urban areas as one of the major parts of it, but of course, it was not limited to it. Then, AMRUT 2.0 scheme was launched by the beginning of October in the year 2021 and again for 5 financial years, which focussed on City Water Balance Plan development, in order to encourage water's circular economy in the country.

Figure 44. Satellite map of AMRUT centre in Lucknow, Uttar Pradesh, India

Source: Google Earth

Figure 45. Prime Minister Shri Modi disclosing the logo and tagline of AMRUT in 2015 in New Delhi, India

Source: Prime Minister's Office (GODL-India), Wikimedia

National Rural Drinking Water Programme

This programme was introduced in the country in the year 2009, by modifying the earlier programme named Accelerated Rural Water Supply Programme. In short, it is known as NRDWP and had the goal to give sufficient as well as safe water supply for domestic purposes including cooking, drinking, and so on to all the people in the villages or rural areas in the country in a sustainable manner. As per the audit report submitted by the Controller and Auditor General for the 5-year period from 2012 till the year 2017, the programme failed to complete its objectives and underperformed. From the latter half of the year 2019 onwards, the Union Government partnered with the State Governments in the country to execute JJM after including this former programme in order to provide drinking water to all the households in the villages via the connection of tap water till this year 2024.

Figure 46. National Review Meeting on NRDWP organized by Ministry of Drinking Water & Sanitation in 2016 in New Delhi, India

Source: Ministry of Drinking Water & Sanitation (GODL-India), Wikimedia

National Water Policy

The Ministry of Water Resources under the Government of India formed and implemented this policy for the first time in the year 1987 in the country. The policy was reviewed as well as updated in the year 2002, and then again in the year 2012. In the recent years, National Water Policy 2012 is being carried out and still in effect. The 2012 version of the policy says that enhanced technologies of the utilization of water must be started to detain the declining levels of groundwater in the areas which are overexploited ones in the country, a part of the flow of rivers must be kept separated for ecological purposes, all projects which are water resources one must be planned in a way that they could draw highest gains from the water resources as well as topography which are available in the given area, and so on.

Figure 47. Consultation meeting of Union Minister of Parliamentary Affairs & Water Resources Shri P.K. Bansal and NGOs on NWP in 2011

Source: Ministry of Water Resources (GODL-India), Wikimedia

Smart Cities Mission

This mission was launched the Government of India in the year 2015 with the goal of developing one hundred cities in the country to make them developed enough to be considered smart cities. There were many smart cities created and in progress due to the mission, but of course it can't cover the whole city in that manner in reality as it might sound on paper. However, in the chapter's context, this mission also included smart water system, by using smart and advanced technologies to detect any leakages in the supply system of water, illegal discharges or any contamination in the water supply. With this mission, the water sector was being developed in a smart and sustainable manner as well as managed in the urban areas in the country.

Figure 48. Launch of smart cities mission's projects in Pune by PM Shri Modi in 2016

Source: Prime Minister's Office (GODL-India), Wikimedia

Integrated Watershed Management Programme

The programme was launched in 2009 in the country to restore as well as mange the watersheds which were degraded in the villages, in order to develop the rural areas in a sustainable manner. However, later on, by 2015, this programme was merged with other water programmes under PMKSY.

Figure 49. National conference organized by the Indian government on IWMP in 2012

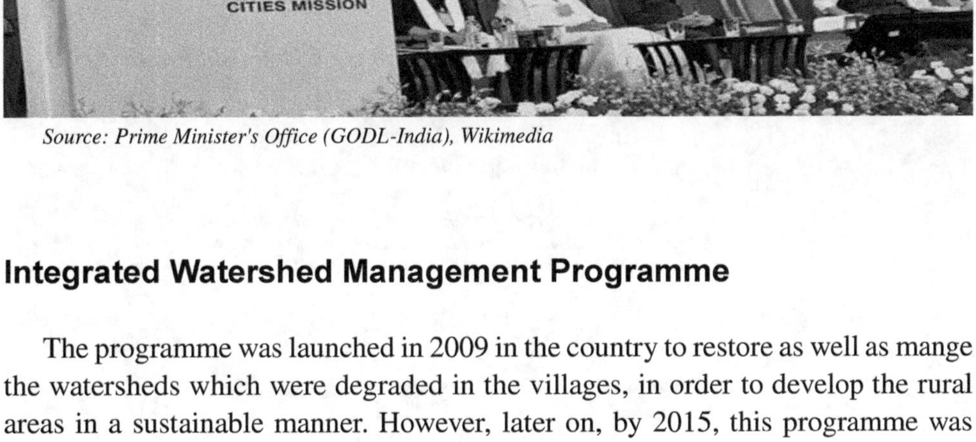

Source: Ministry of Rural Development (GODL-India), Wikimedia

Mahatma Gandhi National Rural Employment Guarantee Act

This act is also famous by the short form MGNREGA, which is mainly an employment plan for rural areas in India. It was passed in the year 2005, but it also includes facilities for managing as well as conserving water in villages. It provides assistance to building farm ponds, check dams, structures for water harvesting in villages to generate employment for the local people, but of course, the act is to create rural employment, so it is beyond the mentioned description, but in some ways, it does help in sustainable water management while creating rural employment.

Figure 50. Orientation workshop on MGNREGA in New Delhi, India, in 2010

Source: Ministry of Rural Development (GODL-India), Wikimedia

METHODOLGY

In this section, problems as well as challenges faced in the implementation of the above-mentioned programmes, policies as well as initiatives are discussed:

Jal Jeevan Mission: The major problems were the quality of water including fluoride as well as arsenic contaminants in the water, rural areas having insufficient infrastructure facilities, delays in distribution of funds and execution of projects under it.

PMKSY: The challenges faced under it were that the coordination between central government and state governments-based authorities were not having proper coordination, drought prone areas were having not enough available water, systems of micro irrigation were expected at a slow rate, and so on.

Atal Bhujal Yojana: The main problems were overextraction and over reliance on groundwater led to its depletion, local communities being inflexible with any changes in the old practices in agriculture, data on the levels of groundwater not accurate,

Jal Shakti Abhiyan: The major hurdles were absence of long-term thorough planning but focus on short term gains, pandemic also disrupted the mission greatly, behavioural change was hard to accomplish at the grassroots level, and so on.

National Water Mission: Some major obstacles in this mission were projects which were water saving ones faced execution delays in the country, coordination between different departments were missing, interstate disputes on the sharing of water resources, infrastructure to support measures for water conservation were insufficient, etc.

GWMR: Major problems under it were private borewells not checked by the authorities leading to groundwater overextraction, mechanisms to recharge groundwater were not enough to avoid the depletion of groundwater, and so on.

Nanami Gange Programme: Challenges under this programme were discharge of untreated sewage as well as industrial waste illegally in the river, engagement of communities was no way enough, encroachment in addition to dumping of solid waste, maintaining cleanliness of river for long time, etc.

National Hydrology Project: Obstacles faced under this project were sharing of real time data across regions in addition to states in India was hard to achieve, monitoring stations as well as hydrological data were inadequate considering their quality, pattern of rainfall being affected by climate change, etc.

River Basin Management: Some primary problems were management policies comprising integrated basin wide were missing, emissions by industries as well as untreated sewage being discharge in river basins causing pollution, disputes over water resources among neighbouring states in India, and so on.

AMRUT: Main challenges faced were approval of projects as well as their implementation being delayed in cities, less focus on holistic method and more focus on project oriented one and lack of considering local context as well needs.

NRDWP: Some hurdles in the course of this programme were rural areas having contamination of drinking water in many areas in the country which included arsenic, fluoride, nitrate, iron, etc., infrastructure supporting water supply not receiving inadequate maintenance, providing supply of water to last mile network or tribal places or remote areas were hard to achieve.

NWP: Major problems were the pricing policy of water in various regions or states were different, proper implementation at grassroot level was difficult, exploitation and depletion of water sources, ineffective conservation measures of water and so on.

Smart Cities Mission: The primary problems were technologies for smart water management had lower priority compared to others in the project, technological as well as monetary limitations for smart water technologies and infrastructure, real time monitoring to catch leakages was hard to implement, etc.

IWMP: Some noteworthy obstacles included local communities were inadequately participating or lacking necessary awareness about activities related to watershed, technical limitations at the grassroot level to implement the programme, and delays.

MGNREGA: The challenges in this context were lack of efficiency while executing projects which were water-based ones, delays in the distribution of funds as well as start of the projects, technical limitations for infrastructure which were related to water sector, corruption and so on.

RESULTS AND DISCUSSION

In this section data from different sources were collected which includes World Bank and so on, to perform graphical analysis.

Table 1. Growth of world population

World Population	In Billions
Year 2020	7.887
Year 2024 (live)	8.183
Year 2030 (expected)	8.569
Year 2050 (expected)	9.664

Source: Worldometer

Table 2. India's population growth

Population of India	In Billions
Year 2020	1.4026
Year 2024 (live)	1.4548
Year 2030 (expected)	1.5251
Year 2050 (expected)	1.6796

Source: Worldometer

Table 3. Comparison of share of water resources and population by India in the world

Share of India in the World	In (%)
Population Share in 2024	18
Water Resources Share	4

Source: Worldometer, and Press Information Bureau, Ministry of Jal Shakti

Table 4. Comparison of treated and untreated sewage in 23 big cities in India

Treatment of Sewage in 23 Big Indian Cities	In (%)
Treated Domestic & Industrial Sewage Before Discharging in Water	31
Untreated Domestic & Industrial Sewage Before Discharging in Water	69

Source: UN India

Table 5. Share of land by water related disasters in India

Major Disasters Related to Water	Share of Land in India (%)
Drought Affected Area in Feb 2024	26.04
Drought Affected Area in Jan 2022	11.83
Flood Prone Areas	12.16

Source: Ministry of Home Affairs, Government of India, and Statista

Table 6. Comparison of average potential of rivers and usable potential in India

Major River Basins in India	Average Potential (billion cubic metres)	Usable (billion cubic metres)
Ganga	525.02	250
Brahmaputra	537.24	24
Total	1869.35	690.1

Source: Central Water Commission 2010, Government of India, and The Asian Development Research Institute

Table 7. India's per capita availability of water over the years and future predictions

Per Capita Water Ability in India	In cubic meters
Year 2001	1816
Year 2011	1545
Year 2021	1486
Year 2031 (Expected)	1367
Year 2050 (Calculated)	1247

Source: World Bank, Worldometer, and Author's Calculation

India's population in 2021 = 1.41 billion (World Bank)
India's per capita water availability in 2021 = 1,486
India's population in 2050 = 1.6796 billion (Worldometer)
Hence, India's per capita water availability = (1.41 x 1486) / 1.6796 = 1,247.475 = 1,247

Table 8. Share of households or people in India having tap water connection or improved access to water

Share of People/Households in India	In (%)
People with basic water access in 2022	93.3
Tap water connection in rural households in 2024	74.9
Tap water connection in rural households in 11 cities/UT in 2024	100
Tap water functional connection in WB rural houses in 2024	44.7
Improved access to water in India in 2022	97.5
Improved access to water through pipe system in India in 2022	44.8

Source: Statista

Table 9. Prediction of demand of water in India in the future

Water Demand in India	In (Billion Cubic Meter)
Irrigation by MOWR in 2010	688
Irrigation by MOWR in 2050	1072
Irrigation by NCIWRD in 2010 (low)	543
Irrigation by NCIWRD in 2010 (high)	557
Irrigation by NCIWRD in 2050 (low)	628
Irrigation by NCIWRD in 2050 (high)	807

continued on following page

Table 9. Continued

Water Demand in India	In (Billion Cubic Meter)
Total by MOWR in 2010	813
Total by MOWR in 2050	1447
Total by NCIWRD in 2010 (low)	694
Total by NCIWRD in 2010 (high)	710
Total by NCIWRD in 2050 (low)	973
Total by NCIWRD in 2050 (high)	1180

Source: Basin Planning Directorate, CWC, XI Plan Document, Government of India

Table 10. Growth of tapped water connection in India's rural households

Tapped Water Connection in Rural Households in India	In Millions
Year 2014-15	1.710138
Year 2017-18	4.250807

Source: format C-36 of IMIS, Ministry of Drinking Water and Sanitation, Government of India, (via Lok Sabha Questions, Sansad Portal)

Table 11. Progress of NRDWP in rural areas in India in the year 2017

NRDWP Progress in Rural India in 2017	In (%)
Targeted habitations completely covered	76.86
Targeted habitations partially covered	18.83
Targeted habitants affected by quality	4.32

Source: IMIS, Ministry of Drinking Water and Sanitation, Government of India, (via Lok Sabha Questions, Sansad Portal)

Table 12. Share of Indians access to improved sources of water

Status of People Access to Improved Water Sources in India	In (%)
Share of India's population to improved drinking water source by World Bank	94
Share of India's population to improved drinking water source by IMIS in 2017	94.38
Total rural population getting complete access to more than 40 LPCD in 2017	71.01
Total rural population getting partial access to more than 40 LPCD in 2017	23.37

Source: World Bank, and IMIS, Ministry of Drinking Water and Sanitation, Government of India, (via Lok Sabha Questions, Sansad Portal)

Table 13. Reliance on groundwater in India

India's Reliance on Groundwater	In (%)
Agriculture sector for Irrigation on groundwater in 2012	60
Drinking water supplies on groundwater in 2012	85

Source: World Bank (2012)

Table 14. Status of water-based problems in the world

Global Status on Water Related Problems	In (%)
Climate Change events related to water	90
Climate Change events not related to water	10
People access to unsafe as well as unimproved drinking water	12
People without any kind of sanitation	30

Source: World Sustainable Development Summit 2022, TERI, and World Bank (2023), Climate Change and Water (Infographic)

Table 15. Global share of sectors utilizing water

Share of Sectors Using Water	In (%)
Agriculture in the World	70
Industry in the world	19
Municipalities in the world	11

Source: UN World Water Development Report 2024, UNESCO, and CityChangers

Table 16. Comparison of status of cities facing and not facing water insecurity in the world

Status of Cities in the World	In (%)
Cities facing water insecurity	25
Cities not facing water insecurity	75

Source: World Bank (2023), Water Resource Management (Infographic)

Table 17. Growth of smart water management's market in the world

International Market Growth of Smart Water Management	In Millions (USD)
Year 2023	6,300
Year 2033 (expected)	21,400

Source: GlobeNewswire

Figure 51. Growth of world population

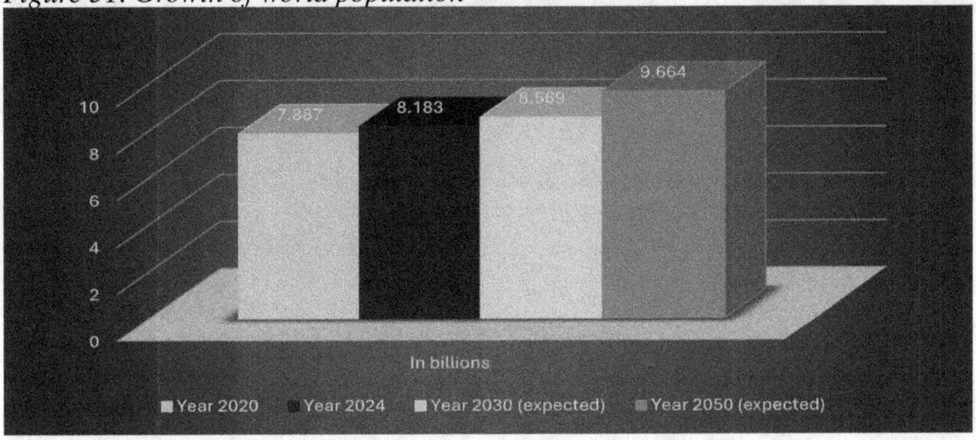

Source: Worldometer

Figure 52. India's population growth

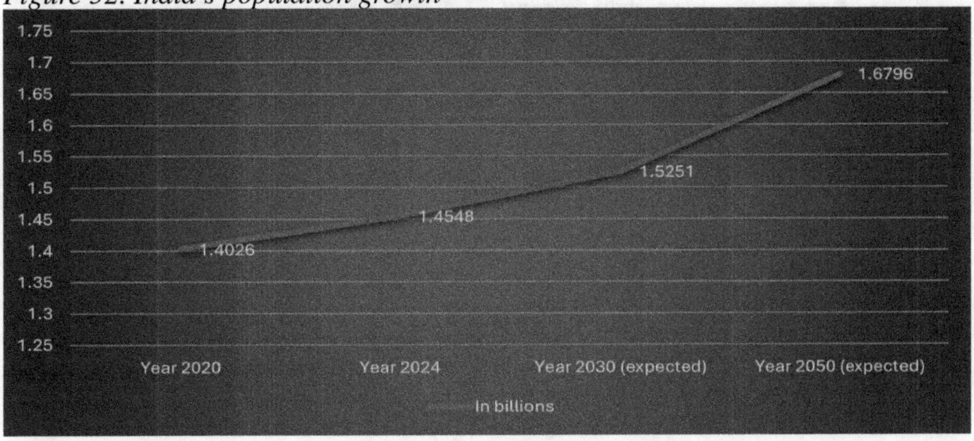

Source: Worldometer

Figure 53. Comparison of share of water resources and population by India in the world

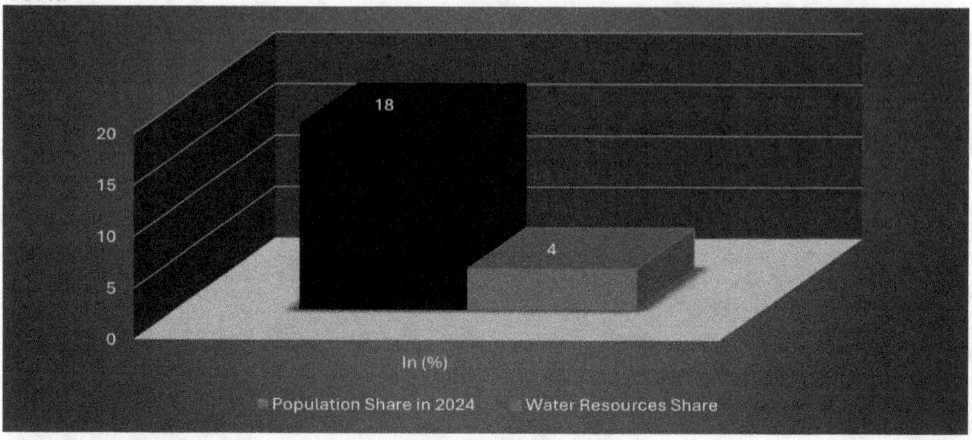

Source: Worldometer, and Press Information Bureau, Ministry of Jal Shakti

Figure 54. Comparison of treated and untreated sewage in 23 big cities in India

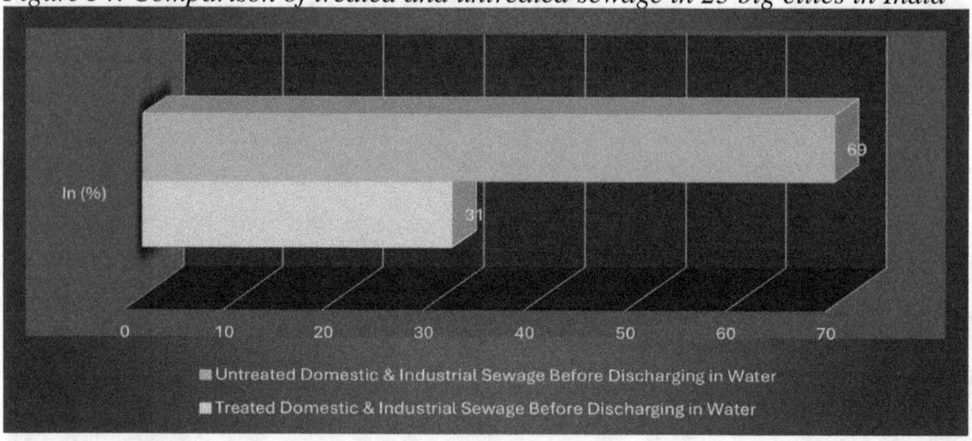

Source: UN India

Figure 55. Share of land by water related disasters in India

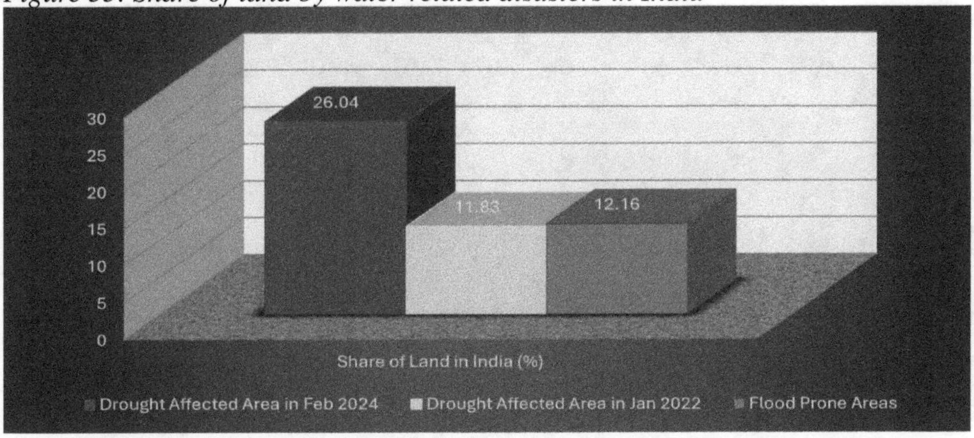

Source: Ministry of Home Affairs, Government of India, and Statista

Figure 56. Comparison of average potential of rivers and usable potential in India

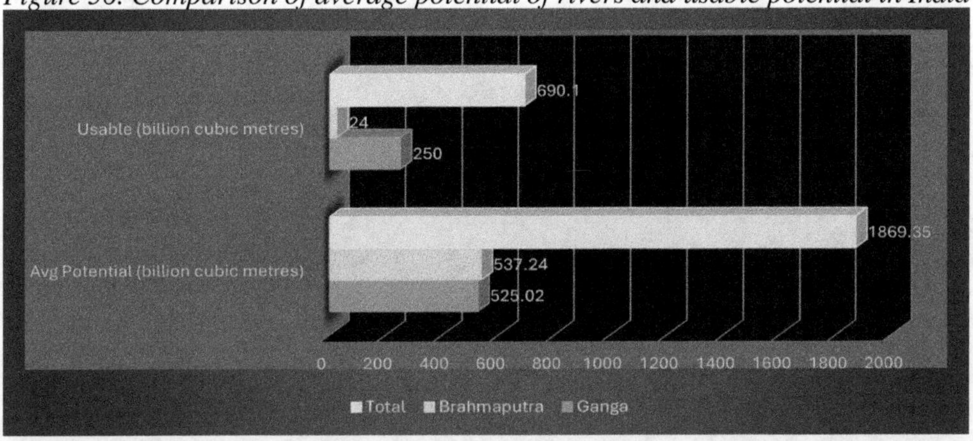

Source: Central Water Commission 2010, Government of India, and The Asian Development Research Institute

Figure 57. India's per capita availability of water over the years and future predictions

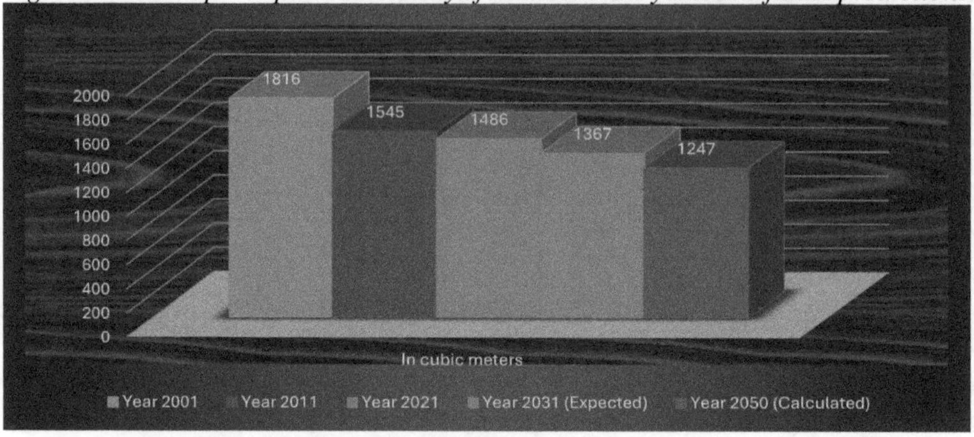

Source: World Bank, Worldometer, and Author's Calculation

Figure 58. Share of households or people in India having tap water connection or improved access to water

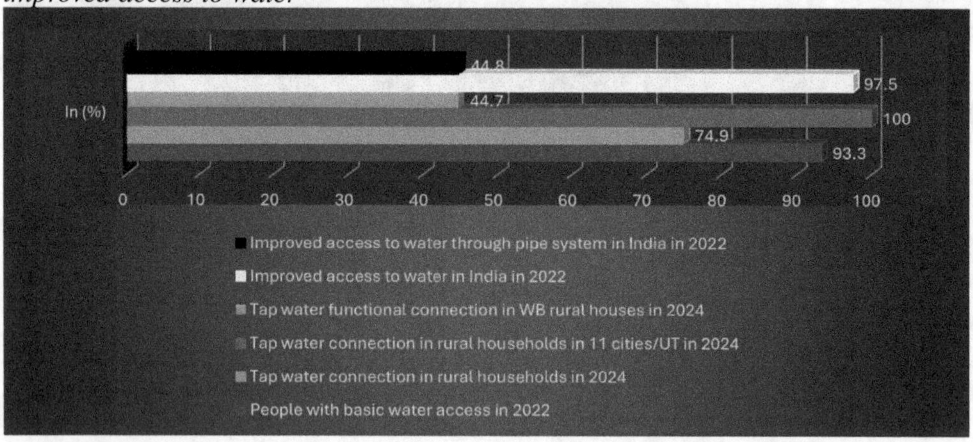

Source: Statista

Figure 59. Prediction of demand of water in India in the future

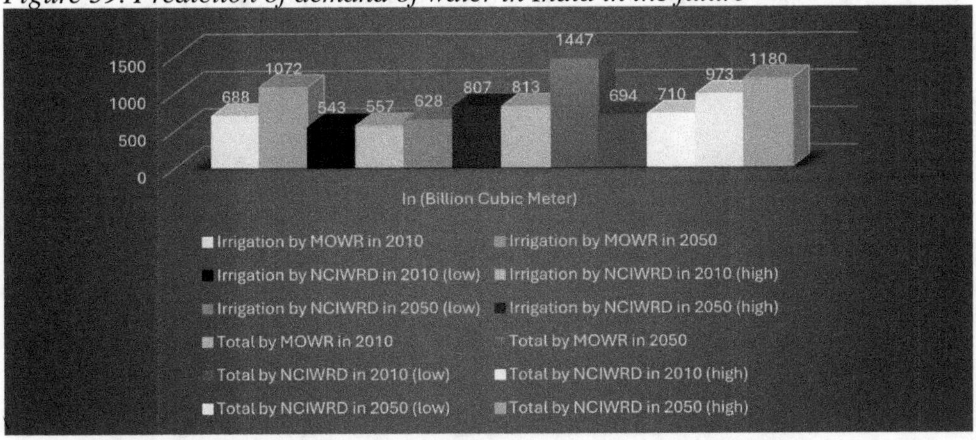

Source: Basin Planning Directorate, CWC, XI Plan Document, Government of India

Figure 60. Growth of tapped water connection in India's rural households

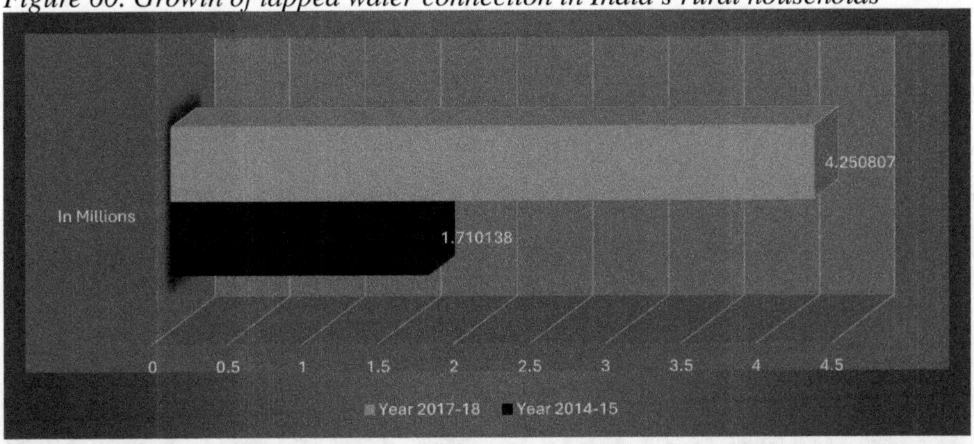

Source: format C-36 of IMIS, Ministry of Drinking Water and Sanitation, Government of India, (via Lok Sabha Questions, Sansad Portal)

Figure 61. Progress of NRDWP in rural areas in India in the year 2017

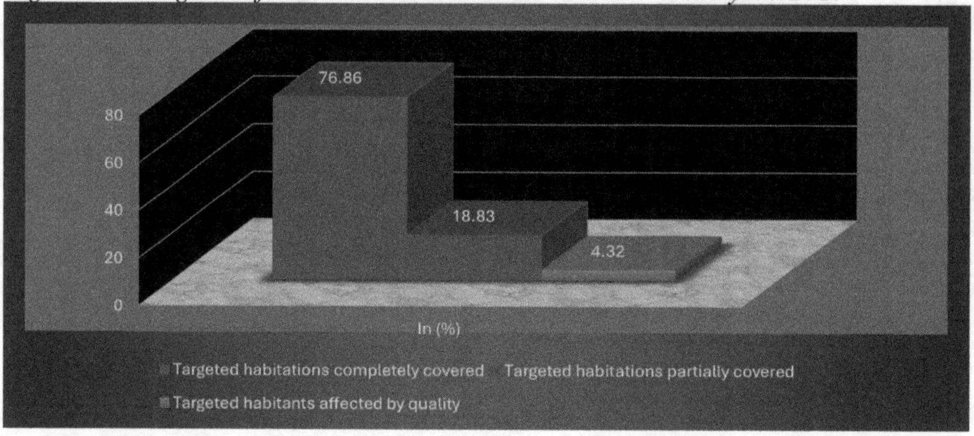

Source: IMIS, Ministry of Drinking Water and Sanitation, Government of India, (via Lok Sabha Questions, Sansad Portal)

Figure 62. Share of Indians access to improved sources of water

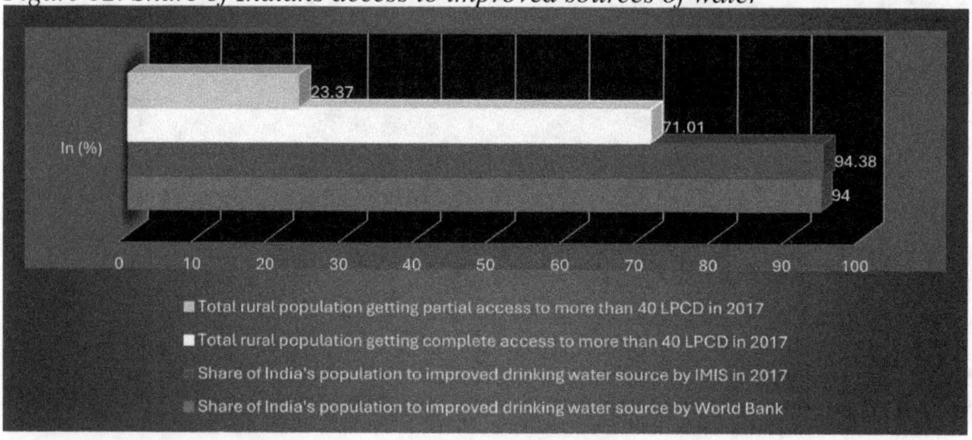

Source: World Bank, and IMIS, Ministry of Drinking Water and Sanitation, Government of India, (via Lok Sabha Questions, Sansad Portal)

Figure 63. Reliance on groundwater in India

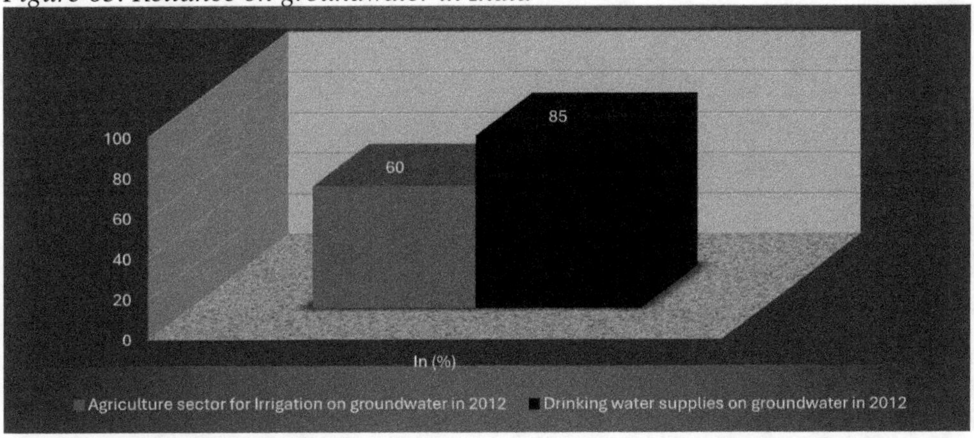

Source: World Bank (2012)

Figure 64. Status of water-based problems in the world

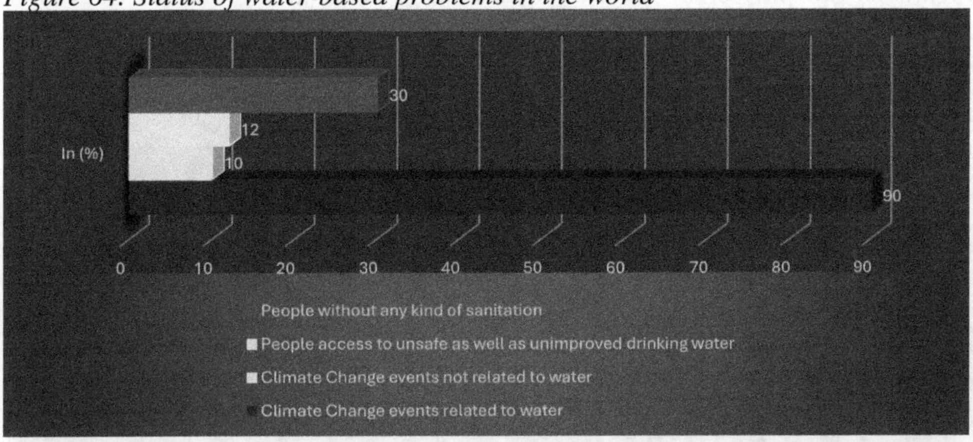

Source: World Sustainable Development Summit 2022, TERI, and World Bank (2023), Climate Change and Water (Infographic)

Figure 65. Global share of sectors utilizing water

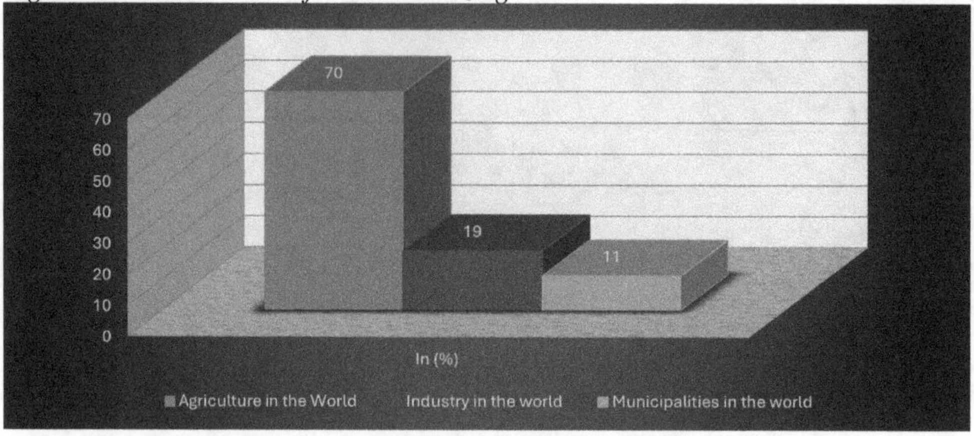

Source: UN World Water Development Report 2024, UNESCO, and CityChangers

Figure 66. Comparison of status of cities facing and not facing water insecurity in the world

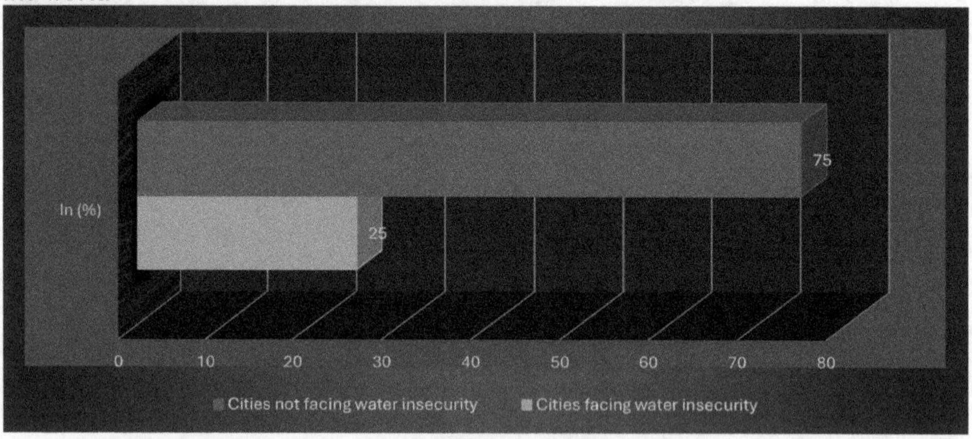

Source: World Bank (2023), Water Resource Management (Infographic)

Figure 67. Growth of smart water management's market in the world

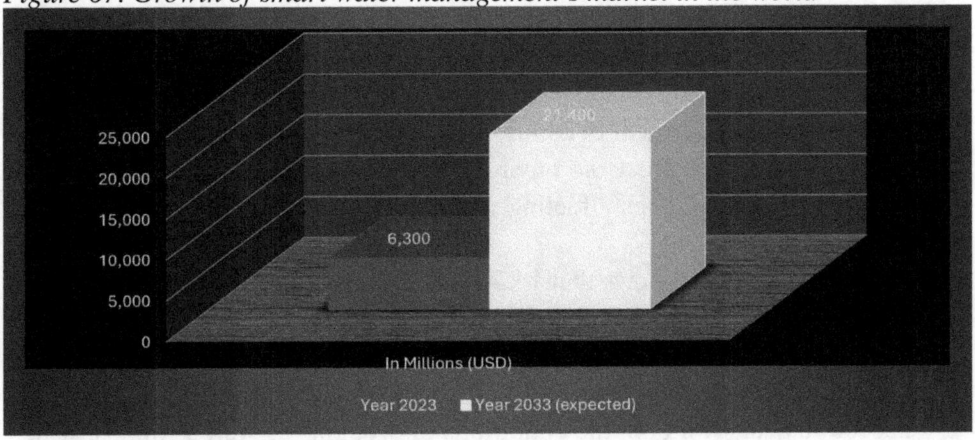

Source: GlobeNewswire

Major Advantages and Challenges of Sustainable Water Management

In this subsection, challenges in addition to advantages along with progress of sustainable water management is understood by analysing the above graphs:

- From Fig. 51, it is clear that the world population is expected to increase rapidly with time, but the water resources won't increase but decrease due to climate change, which will worsen the problems of water related once. This means that the current sustainable water management level will become outdated or lack impact with growing water demand and depleting resources.
- Based on Fig. 52, India is also showing a positive trend with the passage of time, means that the country's population will grow significantly and increasing the demand of water resources, but the country's water resources are barely enough at present, and if not taken appropriate measures then the country's water sources will not only deplete at an alarming rate, putting more stress on the available water sources.
- According to Fig. 53, it clearly shows that India has only 4% water resources in the world, but that water resources are supposed to take care of 18% of the world's population, despite the fact that 18% is 1.44 billion people roughly meaning compared to other countries in the world, people in India has to face scarcity or water resources distribution will be less on average, this problem will worsen by 2030 or by 2050 even more.

- From Fig. 54, only 31% of the sewage coming from industries as well as from domestic ones are treated before releasing it into in water sources, meaning that a big portion of the released sewage is not being treated before discharge. Even though flowing water naturally cleans itself, but discharging 69% of sewage without treatment in 23 big cities in India will deteriorate the existing water sources and affect the environment as well as compromise with the quality of water sources affecting native fishes, plants and microbes in the water sources.
- Based on Fig. 55, it shows that 14.21% of India's land became more prone to drought in just 2 years, making the number reach 26.04% of the land in the country affected by it. Then over 12% of the land is prone to floods, meaning that the current management of water is not enough, and some more steps are needed to deal with both the conditions of droughts as well as floods, which might worsen in the future.
- According to Fig. 56, the average potential of the Indian rivers is high on average, but around one third of their potential could be used, meaning a small portion of the limited water resources are available for Indians to manage agriculture, industries as well as for domestic purposes. This means that there is a need to increase the usable potential of Indian river to manage the water resources properly as well as to meet the significant growing demand for water in the future.
- From Fig. 57, the per capita water availability in cubic meters of water in India were 1816 in in 2001, but it kept decreasing significantly with time, and from 2011 onwards making India enter the list of water stressed countries. Based on calculations, the per capita water availability in India might reach 1247 by 2050. Hence, the current rate is making matters worse for the people in the country in the future. With growing population, depleting resources, it is important for India to take suitable measures to manage the water resources in a sustainable manner, either by minimizing the wastages, leaks, or by increasing the usable potential of water sources or by controlling the population of the country or by incorporating all of them.
- Based on Fig. 58, the tap water connection in the rural households in eleven cities or union territories in India is 100% by 2024, but functional tap water connections in the rural areas West Bengal is 44.7%, meaning the rural areas around major cities or around union territories in the country have way better tap water connections than other rural areas in the country. In 2022, 93.3% of the people in the country have access to basic access to water or minimum water access, 74.9% of the rural households have tap water access, indicating that the government policies and initiatives have worked on world's most populated country to have minimum water access as well as more and more

tap water connectivity. The figure also shows the progress of the government policies with the improved access to water as well as improved access to water through pipe system in the country.

- According to Fig. 59, the Ministry of Water Resources as well as National Commission on Integrated Water Resources Development in India predicted that the demand for irrigation as well as total water demand will rise drastically from the year 2010 to the year 2050. The main reason might be because of the growing population of the country, which is at a fast rate over the years, which will automatically increase the demand for everything including drinking water, food, industrial items, which will ultimately increase the water demand in all sectors, but that will worsen the water problems in the country as the water resources are limited and they are being depleted over the years. Hence, current progress of policies or initiatives in the country might not be sufficient for the future.
- From Fig. 60, the tapped water connection was over 1.7 million in rural areas in the country in the financial year 2014-15 but by the financial year 2017-18, it reached over 4.2 million, indicating that there was rapid increase in the tapped water supply connectivity in the rural areas in India, also showing the progress of government policies as well as programmes to manage water properly.
- Based on Fig. 61, the targeted habitants who were completely covered in the rural areas in India was 76.86% in 2017 under National Rural Drinking Water Programme, whereas 18.83% of them were partially covered by the government programme, 4.32% were facing with the quality of water supply. It is safe to say that the government programme was helpful in covering a vast number of people in rural areas in the country and shows hope for further improvement.
- According to Fig. 62, majority of the rural area population received 40 LPCD of water supply at the minimum in India, which was close to three quarters of the rural population, along with the vast majority of the country receiving improved access to water, which shows the progress of the government policies and programmes, but at the same time it is far from the enough for many people to have comfortable daily lives involving regular chores related to water.
- From Fig. 63, the agriculture sector as well as the drinking water supplies are heavily dependent on the groundwater in India, but the groundwater is also being depleted with time, which creates more water problems in the future as the demand for water from these sectors will increase in the future with the growing population taken into account. This will create stress on the current management of water resources.

- Based on Fig, 64, 12% of the world's population doesn't have access to safe and improved access to drinking water, whereas almost one third of the world's population lacking any kind of sanitation, indicating that the sustainable management of water in the world at present is far from the target in reality. Then, 90% of the events are water related ones which are occurring due to climate change in the world, means that sustainable water management is the need of the hour in the world and if not taken seriously, then the current scenario will worsen at a significant rate in the future.
- According to Fig. 65, agriculture in the world consumes 70% of the used water, then 19% by industries, and municipalities consuming 11%. However, the global population is increasing at a significant rate, meaning that food production through agriculture sector will need more productivity, but with rapid development, the industries will also increase and improve their productivity, then municipalities in the world will have facilitate the increasing population's demand for water, ultimately demand for water increasing by all these sectors in the future, but given that the water resources available to utilize is limited and depleting at a fast rate, it is wise to improve the current management of water resources in the world, also controlling the population growth, and reducing wastages as well as mismanagement of water resources, along with improved techniques to have better sustainable water management in the world.
- From Fig. 66, 25% of the cities in the world are facing water insecurity, but in the future with the growing population and more urbanization as well as more migration of people from villages to cities, the water insecurity problem will increase in cities in the world, particularly in the developing countries like India, many others. The developed countries will also face this problem but not at the scale of developing countries which are improving their rate of development at a fast rate over the years.
- Based on Fig. 67, the market of Smart Water Management in the world is expected to grow at a rapid rate from 2023 till the year 2033, showing the demand for advanced technologies to improve the water sector in the world, but this will also lead to sustainable water management, meaning that it will be wise to invest in smart as well as advanced technologies to improve the water sector leading to smart as well as sustainable water management globally.

LIMITATIONS

The chapter mostly focused on the number of policies and initiatives led by the Indian Government to walk on the path of sustainable water management in brief. However, the chapter couldn't focus on one specific policy or initiative in depth. Then, this book chapter was unable to focus on other similar countries having policies and indicatives which are similar to some extent to India to develop in the direction of sustainable water management, in that detail. There might be some other policies as well as initiatives implemented by India, that might not be covered in the book chapter.

SOLUTIONS AND RECOMMENDATIONS

The Union Government in India along with State Governments have implemented various schemes, programmes and policies throughout the country, which it devised from time to time, over the decades. Of course, considerable progress has been made in the direction of sustainable water management, but India is one of the water stressed countries in the world from the 2nd decade of the 21st century and the depletion of water resources as well as growing population of the country has made the sustainable water management progress far from enough in reality. It is advisable and recommended that the implementation of the various policies, schemes and programmes must be improved on the ground level by the government, and awareness among people to be increased to help the country in improving the current scenario as well as to increase the progress of sustainable water management of the country at significant rate. Also, update of policies, initiatives, programmes are essential from time to time in the future as well, so the goals of sustainable development in the water sector in country could be meet with the increasing population and demand for water resources, despite the fact that water resources are limited and being depleted. Then, new advanced technologies could be implemented more on a large scale in the future throughout the country, to optimize the management of water resources in a sustainable manner. Of course, these things are being carried out at present in different parts of the country, but they are no way near the satisfactory level. Furthermore, NGOs and general public in India are also advised to participate more actively instead on just relying on government programmes to take care of water related problems or incorporate sustainable water management properly throughout the country, otherwise it is not practical for the most populated country in the world to implement sustainable water management at the grassroot level throughout the country to conserve water resources and avoid its depletion.

FUTURE RESEARCH DIRECTIONS

The chapter has discussed various measures and policies taken in the world to move on the path of sustainable water management, where many countries were covered, and special focus was given to India. In the future, research could be performed to make the advanced technologies more cost effective, so that there mass incorporation on a large scale could be done, way more than the current level. Then, research could be done on specific policies as well as programmes in depth, related to water in the future to understand the progress and benefits to the communities along with any modifications needed could be suggested through the studies by the researchers.

CONCLUSION

The chapter explained the concept of sustainable water as well as sustainable water management and covered various countries in the world as cases to show the development of the world in the direction of sustainable water management. However, special focus was given to India, where many government policies, programmes and initiatives were covered by this chapter to highlight the progress of India on the path of sustainable water management. Various problems, and challenges in those discussed polices, initiatives as well as programmes led by the Central Government in collaboration with the State Governments in India, were talked about. Then, data from various available sources were taken into account to perform graphical analysis to showcase the need as well as demand of sustainable water management in India and in the world as well as to show the progress of India in this direction. A lot of progress has been made by the Government of India in the direction of sustainable water management, but the current growing population, increasing demand for water resources, as well as depletion of the limited water resources in the country, made these progress far away from the satisfactory level of sustainable water management in reality. So, necessary steps must be taken by the country to overcome this hurdle and deal with future problems which might occur at the current rate. Then, many parts of the world are suffering with water related problems and with climate change, the situation is expected to become worse along with the increasing global population, so more focus on sustainable water management must be given as well as successful implementation of devised polices and initiatives to improve the situation.

ACKNOWLEDGEMENT

This research received no specific grant from any funding agency in the public, commercial, or not-for-profit sectors.

REFERENCES

Abdi, A., Awarke, M., Malayeri, M. R., & Riazi, M. (2024). Interfacial tension of smart water and various crude oils. *Fuel*, *356*, 129563. DOI: 10.1016/j.fuel.2023.129563

Alves, R. G., Maia, R. F., & Lima, F. (2023). Development of a Digital Twin for smart farming: Irrigation management system for water saving. *Journal of Cleaner Production*, *388*, 135920. DOI: 10.1016/j.jclepro.2023.135920

Arepalli, P. G., & Naik, K. J. (2024). An IoT based smart water quality assessment framework for aqua-ponds management using Dilated Spatial-temporal Convolution Neural Network (DSTCNN). *Aquacultural Engineering*, *104*, 102373. DOI: 10.1016/j.aquaeng.2023.102373

Buriro, M.A., Wei, M., Bai, B., & Yao, Y. (2023). Advances in smart water flooding: A comprehensive study on the interplay of ions, salinity in carbonate reservoirs. *Journal of Molecular Liquids*, Volume 390, Part B, 123140.

Bwambale, E., Abagale, F. K., & Anornu, G. K. (2024). Towards a modelling, optimization and predictive control framework for smart irrigation. *Heliyon*, *10*(18), e38095. DOI: 10.1016/j.heliyon.2024.e38095 PMID: 39364235

Cardell-Oliver, R., Cominola, A., & Hong, J. (2024). Activity and resolution aware privacy protection for smart water meter databases. *Internet of Things : Engineering Cyber Physical Human Systems*, *25*, 101130. DOI: 10.1016/j.iot.2024.101130

Choudhary, V., Machavaram, R., Singh, N., & Nagar, H. (2024). Assessment of sensor-based automatic smart watering unit for paddy nurseries under Indian perspective. *Smart Agricultural Technology*, *8*, 100518. DOI: 10.1016/j.atech.2024.100518

Damman, S., Schmuck, A., Oliveira, R., Koop, S. H. A., Almeida, M. C., Alegre, H., & Ugarelli, R. M. (2023). Towards a water-smart society: Progress in linking theory and practice. *Utilities Policy*, *85*, 101674. DOI: 10.1016/j.jup.2023.101674

DiCarlo, M., Berglund, E. Z., Kaza, N., Grieshop, A., Shealy, L., & Behr, A. (2023). Customer complaint management and smart technology adoption by community water systems. *Utilities Policy*, *80*, 101465. DOI: 10.1016/j.jup.2022.101465

Durgun, Y. (2024). Real-time water quality monitoring using AI-enabled sensors: Detection of contaminants and UV disinfection analysis in smart urban water systems. *Journal of King Saud University. Science*, *36*(9), 103409. DOI: 10.1016/j.jksus.2024.103409

Et-taibi, B., Abid, M. R., Boufounas, E.-M., Morchid, A., Bourhnane, S., Abu Hamed, T., & Benhassou, D. (2024). Enhancing water management in smart agriculture: A cloud and IoT-Based smart irrigation system. *Results in Engineering*, *22*, 102283. DOI: 10.1016/j.rineng.2024.102283

Figueiredo, I., Esteves, P., & Cabrita, P. (2021). Water wise – a digital water solution for smart cities and water management entities. *Procedia Computer Science*, *181*, 897–904. DOI: 10.1016/j.procs.2021.01.245

Fong, B., Housh, M., Hong, G. Y., & Wang, J. M. (2024). Pipeline Management Technologies for Sustainable Water Supply in a Smart City Environment. *Encyclopedia of Sustainable Technologies (Second Edition)*, Volume 3, Pages 671-679.

Glina, B., Yetunde, F. M., Mendyk, L., & Piernik, A. (2024). Soil water repellency and its importance for the climate-smart sustainable management of fen peatland soils in Central Poland. *Geoderma Regional*, *39*, e00867. DOI: 10.1016/j.geodrs.2024.e00867

Gumz, J., & Fettermann, D. C. (2024). User's perspective in smart meter research: State-of-the-art and future trends. *Energy and Building*, *308*, 114025. DOI: 10.1016/j.enbuild.2024.114025

Hu, X., Li, Y., Sun, Y., Zhang, C., Ding, W., & Deng, J. (2022). Water permits trading framework for urban water demand management based on smart metering. *Journal of Environmental Management*, *304*, 114208. DOI: 10.1016/j.jenvman.2021.114208 PMID: 34891052

Jimenez, A.-F., Cardenas, P.-F., & Jimenez, F. (2021). Smart water management approach for resource allocation in High-Scale irrigation systems. *Agricultural Water Management*, *256*, 107088. DOI: 10.1016/j.agwat.2021.107088

Kanyama, M. N., Shava, F. B., Gamundani, A. M., & Hartmann, A. (2024). Anomalies identification in Smart Water Metering Networks: Fostering improved water efficiency. *Physics and Chemistry of the Earth Parts A/B/C*, *134*, 103592. DOI: 10.1016/j.pce.2024.103592

Kanyama, M. N., Shava, F. B., Gamundani, A. M., & Hartmann, A. (2024). Machine learning applications for anomaly detection in Smart Water Metering Networks: A systematic review. *Physics and Chemistry of the Earth Parts A/B/C*, *134*, 103558. DOI: 10.1016/j.pce.2024.103558

Kavya, M., Mathew, A., Shekar, P. R., & P, S. (2023). Short term water demand forecast modelling using artificial intelligence for smart water management. *Sustainable Cities and Society*, *95*, 104610. DOI: 10.1016/j.scs.2023.104610

Khachatryan, H., Suh, D. H., Xu, W., Useche, P., & Dukes, M. D. (2019). Towards sustainable water management: Preferences and willingness to pay for smart landscape irrigation technologies. *Land Use Policy*, *85*, 33–41. DOI: 10.1016/j.landusepol.2019.03.014

Krisnawijaya, N. N. K., Tekinerdogan, B., Catal, C., & van der Tol, R. (2024). Reference architecture design for developing data management systems in smart farming. *Ecological Informatics*, *81*, 102613. DOI: 10.1016/j.ecoinf.2024.102613

Kumar, R., Singh, R., Agrawal, A., & A. (2024). Chapter 5 - Global scenario of water footprints in smart cities. *Current Directions in Water Scarcity Research*, Volume 8, Pages 63-68.

Lephondo, I., Telukdarie, A., Munien, I., Onkonkwo, U., & Vermeulen, A. (2024). The Outcomes of Smart Irrigation System using Machine Learning to minimize water usage within the Agriculture Sector. *Procedia Computer Science*, *237*, 525–532. DOI: 10.1016/j.procs.2024.05.136

Li, M., Zhou, S., Shen, S., Wang, J., Yang, Y., Wu, Y., Chen, F., & Lei, Y. (2024). Climate-smart irrigation strategy can mitigate agricultural water consumption while ensuring food security under a changing climate. *Agricultural Water Management*, *292*, 108663. DOI: 10.1016/j.agwat.2023.108663

Mathews, M. E., Shaji, A. E., Anand, N., Andrushia, A. D., Chin, S. C., & Lubloy, E. (2023). *IoT-based BIM integrated model for energy and water management in smart homes*. Intelligent Edge Computing for Cyber Physical Applications, Intelligent Data-Centric Systems. DOI: 10.1016/B978-0-323-99412-5.00009-5

Morchid, A., Jebabra, R., Khalid, H. M., El Alami, R., Qjidaa, H., & Jamil, M. Q. (2024). IoT-based smart irrigation management system to enhance agricultural water security using embedded systems, telemetry data, and cloud computing. *Results in Engineering*, *23*, 102829. DOI: 10.1016/j.rineng.2024.102829

Nakhaei, M., Ahmadi, A., Gheibi, M., Chahkandi, B., Hajiaghaei-Keshteli, M., & Behzadian, K. (2023). A smart Sustainable decision Support system for water management oF power plants in water stress regions. *Expert Systems with Applications*, *230*, 120752. DOI: 10.1016/j.eswa.2023.120752

Newing, A., Hibbert, O., Van-Alwon, J., Ellaway, S., & Smith, A. (2023). Smart water metering as a non-invasive tool to infer dwelling type and occupancy – Implications for the collection of neighbourhood-level housing and tourism statistics. *Computers, Environment and Urban Systems*, *105*, 102028. DOI: 10.1016/j.compenvurbsys.2023.102028

Oberascher, M., Kinzel, C., Kastlunger, U., Kleidorfer, M., Zingerle, C., Rauch, W., & Sitzenfrei, R. (2021). Integrated urban water management with micro storages developed as an IoT-based solution – The smart rain barrel. *Environmental Modelling & Software*, *139*, 105028. DOI: 10.1016/j.envsoft.2021.105028

Oberascher, M., Rauch, W., & Sitzenfrei, R. (2022). Towards a smart water city: A comprehensive review of applications, data requirements, and communication technologies for integrated management. *Sustainable Cities and Society*, *76*, 103442. DOI: 10.1016/j.scs.2021.103442

Salman, M. Y., & Hasar, H. (2023). Review on environmental aspects in smart city concept: Water, waste, air pollution and transportation smart applications using IoT techniques. *Sustainable Cities and Society*, *94*, 104567. DOI: 10.1016/j.scs.2023.104567

Singh, A. (2023). The Current Scenario of Sustainable Planning and Development of the Indian Built Environment. In Cobîrzan, N., Muntean, R., & Felseghi, R. (Eds.), *Circular Economy Implementation for Sustainability in the Built Environment* (pp. 106–138). IGI Global., DOI: 10.4018/978-1-6684-8238-4.ch005

Singh, A. (2023). The Current Trends and Need of Infrastructure Projects in the Country India. In Popescu, C., Yu, P., & Wei, Y. (Eds.), *Achieving the Sustainable Development Goals Through Infrastructure Development* (pp. 235–261). IGI Global., DOI: 10.4018/979-8-3693-0794-6.ch010

Singh, A. (2023). The Need and Development of Dams in India as Well as in the World for a Better Future. In Lousada, S. (Ed.), *Geoinformatics in Support of Urban Politics and the Development of Civil Engineering* (pp. 156–180). IGI Global., DOI: 10.4018/978-1-6684-6449-6.ch007

Singh, A. (2023). The Present and Future Role of Artificial Intelligence in the Promotion of Sustainable Practices in the Construction Sector. In Kahraman, C., Sari, I. U., Oztaysi, B., Cebi, S., Cevik Onar, S., & Tolga, A. Ç. (Eds.), *Intelligent and Fuzzy Systems. INFUS 2023. Lecture Notes in Networks and Systems* (Vol. 759). Springer., DOI: 10.1007/978-3-031-39777-6_2

Singh, A. (2023). The Recent Trend of Artificial Neural Network in the Field of Civil Engineering. In Vasant, P. (Eds.), *Intelligent Computing and Optimization. ICO 2023. Lecture Notes in Networks and Systems* (Vol. 855). Springer., DOI: 10.1007/978-3-031-50158-6_32

Singh, A., Verdi, A., Haver, D., Sapkota, A., & Iradukunda, J. C. (2024). Using a soil moisture sensor-based smart controller for autonomous irrigation management of hybrid bermudagrass with recycled water in coastal Southern California. *Agricultural Water Management*, *299*, 108906. DOI: 10.1016/j.agwat.2024.108906

Singh, M., & Ahmed, S. (2021). IoT based smart water management systems: A systematic review. *Materials Today: Proceedings*, *46*(Part 11), 5211–5218. DOI: 10.1016/j.matpr.2020.08.588

Sundaresan, S., Kumar, K. S., Kumar, T. A., Ashok, V., & Julie, E. G. (2021). *Blockchain architecture for intelligent water management system in smart cities*. Blockchain for Smart Cities. DOI: 10.1016/B978-0-12-824446-3.00006-5

Sushma, N., Suresh, H. N., Mohana Lakshmi, J., & Santhosh Kumar, K. B. (2024). Experimental investigation on wireless integrated smart system for energy and water resource management in Indian smart cities. *Results in Engineering*, *23*, 102687. DOI: 10.1016/j.rineng.2024.102687

Tavares, M., Perez-Sanchez, M., Carravetta, A., Coronado-Hernandez, O. E., Lopez-Jimenez, P. A., & Ramos, H. M. (2024). Smart Feasibility Optimization of Hybrid Renewable Water Supply Systems by Digital Twin Technologies: A Multicriteria Approach Applied to Isolated Cities. *Sustainable Cities and Society*, *115*, 105834. DOI: 10.1016/j.scs.2024.105834

Tayerani Charmchi, A. S., Ifaei, P., & Yoo, C. (2021). Smart supply-side management of optimal hydro reservoirs using the water/energy nexus concept: A hydropower pinch analysis. *Applied Energy*, *281*, 116136. DOI: 10.1016/j.apenergy.2020.116136

Uddin, M. J., & Jeong, Y.-K. (2021). Urban river pollution in Bangladesh during last 40 years: Potential public health and ecological risk, present policy, and future prospects toward smart water management. *Heliyon*, *7*(2), e06107. DOI: 10.1016/j.heliyon.2021.e06107 PMID: 33659727

Vadruccio, R., Salvadori, G., & Tumino, A. (2024). Smart metering and Internet of Things for efficient water management. *Procedia Computer Science*, *239*, 1498–1505. DOI: 10.1016/j.procs.2024.06.324

Wang, X., Wei, H., Luo, N., Luo, H., Zhou, X., Qin, B., Mei, Y., & Zhang, Y. (2024). Smart Materials and Micro/Nanoarchitectonics for Water Harvesting: From Fundamental Mechanism to Advanced Application. *Composites. Part A, Applied Science and Manufacturing*, *184*, 108241. DOI: 10.1016/j.compositesa.2024.108241

Wu, D., Wang, H., & Seidu, R. (2020). Smart data driven quality prediction for urban water source management. *Future Generation Computer Systems*, *107*, 418–432. DOI: 10.1016/j.future.2020.02.022

Wu, Z. Y., Chew, A., Meng, X., Cai, J., Pok, J., Kalfarisi, R., Lai, K. C., Hew, S. F., & Wong, J. J. (2023). High Fidelity Digital Twin-Based Anomaly Detection and Localization for Smart Water Grid Operation Management. *Sustainable Cities and Society*, *91*, 104446. DOI: 10.1016/j.scs.2023.104446

Zhang, J., Liu, Y., Zhu, W., Pei, Y., Yin, Y., Qin, Y., Li, X., & Guiver, M. D. (2021). Self-adjusting anode catalyst layer for smart water management in anion exchange membrane fuel cells. *Cell Reports. Physical Science*, *2*(3), 100377. DOI: 10.1016/j.xcrp.2021.100377

Chapter 12
Understanding Mobility Patterns in Machico's Low-Density Rural Areas Through Stated Preference Surveys

Raul Costa Alves
https://orcid.org/0000-0002-3680-6626
Universidad de Extremadura, Spain

Sérgio António Neves Lousada
https://orcid.org/0000-0002-8429-2164
University of Madeira, Portugal

José Cabezas
https://orcid.org/0000-0002-7209-3731
Universidad de Extremadura, Spain

ABSTRACT

This research investigates the mobility patterns of residents in Machico, a low-density rural area in Madeira Island, Portugal. The study examines how socioeconomic, demographic, and infrastructural factors influence transportation choices. Using a stated preference approach, data was collected through structured interviews. The findings reveal that service frequency and travel time are the most critical factors in choosing transportation modes. The study recommends enhancing service frequency, ensuring reliability, offering flexible solutions like on-demand services and ridesharing, and maintaining affordable fares to improve public transportation in

DOI: 10.4018/979-8-3693-2448-6.ch012

the region. These findings have implications for similar rural contexts worldwide.

1. INTRODUCTION

The municipality of Machico, nestled on the western edge of Madeira Island, Portugal, presents a unique landscape where the sea meets the mountains, culminating at Ponta de São Lourenço (Alves, 2023a). Encompassing 68.31 km2 and five distinct parishes (Figure 1), Machico exemplifies the challenges faced by low-density rural areas globally.

Figure 1. Municipality of Machico and its respective parishes. (Source: Alves, 2022)

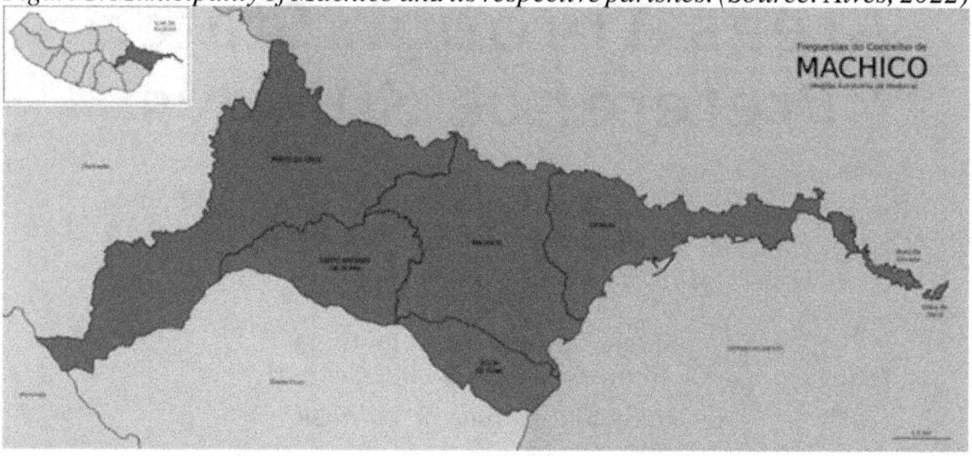

These challenges, amplified by rugged terrain, dispersed housing (Figure 2), and limited public transportation, necessitate a deep understanding of mobility patterns to foster inclusive and sustainable development. The insights gleaned from this study in Machico have the potential to inform transportation policies not only within the region but also in similar rural contexts worldwide, contributing to a broader dialogue on sustainable rural mobility.

Figure 2. Housing dispersion in the Porto da Cruz Parish (Source: Authors)

This research endeavours to analyse the intricate mobility patterns of Machico's residents, examining how socioeconomic factors (family income, employment, education), demographic factors (age, gender, family composition), and infrastructural factors (road quality, public transportation availability, active mobility infrastructure) shape their transportation choices (Figure 3). Furthermore, the study aims to identify the primary barriers to mobility and propose actionable recommendations to enhance accessibility and elevate the quality of public transportation in the region.

Figure 3. Infrastructure of the rural area of the Machico parish (Source: Authors)

To achieve these objectives, the authors employ a stated preference research approach, a methodology designed to capture individual preferences and behaviours across diverse hypothetical transportation scenarios. Data collection was conducted through structured interviews facilitated by tablets to ensure accuracy and efficiency. The subsequent analysis involved both qualitative (interviews) and quantitative (statistics) methods, providing a comprehensive understanding of the multifaceted nature of rural mobility in Machico.

The results of the study reveal that service frequency and travel time are paramount in shaping the transportation mode choices of Machico's residents. Enhancing the frequency of transportation services not only augments user satisfaction but also has the potential to significantly increase the utilization of public transportation. A strategic approach to achieve this without escalating operational costs could involve the implementation of flexible and demand-responsive transportation schedules.

The reliability of services emerges as another pivotal factor. Punctual and dependable transportation services align with residents' expectations and can be ensured through the integration of real-time monitoring systems and continuous feedback mechanisms.

Flexible transportation solutions, such as on-demand transportation services and shared ride programs, are particularly vital in low-density areas where the operation of regular buses may not be economically viable. The introduction of innovative technologies, such as mobile applications for scheduling on-demand transportation, can offer the agility needed to better cater to residents' diverse demands.

While less influential than frequency and travel time, affordable fare policies remain indispensable. Subsidies and reduced fares for vulnerable groups, including the elderly and low-income individuals, can significantly enhance accessibility to public transportation and foster social inclusion. Striking a balance between affordability and maintaining service quality and frequency is a complex challenge that necessitates well-conceived and sustainable transportation policies.

2. LITERATURE REVIEW

The following section delves into the existing literature on rural mobility, exploring its unique characteristics, challenges, and the factors that influence transportation choices in these contexts. The section also examines the stated preference methodology and its application in rural mobility research, providing a foundation for the subsequent methodological approach in this study.

2.1. Characteristics of Rural Mobility

Mobility in low-density rural areas has garnered increasing attention due to the unique challenges these regions face compared to their urban counterparts. While urban areas often benefit from well-developed public transportation systems, rural areas grapple with infrequent service, vast distances between key destinations, and a pronounced reliance on private vehicles. Nutley (2003) underscores that the scarcity of public transportation in rural settings can lead to social and economic isolation, particularly for elderly residents and those with limited incomes. This lack of adequate accessibility translates to difficulties in accessing essential services such as healthcare, education, and employment, thereby exacerbating social vulnerabilities.

These conclusions are echoed in studies that reveal the detrimental impact of limited mobility on the quality of life in rural areas. Farrington and Farrington (2005) delve into how the absence of reliable public transportation curtails social participation and inclusion, especially in more remote communities. Their research underscores the imperative of tailoring transportation policy planning to the specific needs of rural populations. However, their analysis lacks a comprehensive framework for understanding the multifaceted nature of rural mobility, which encompasses not only public transportation but also private vehicle usage, active transportation modes, and the influence of socioeconomic and demographic factors.

Nowadays the challenges of rural mobility are further compounded by the digital divide, where limited internet connectivity and digital literacy can hinder the adoption of technology-driven solutions (Esteban-Navarro et al., 2020). This digital divide can create additional barriers for rural residents, particularly the elderly and those with lower incomes, in accessing and utilizing innovative transportation services.

2.2. Specific Challenges of Rural Mobility

Low Frequency and Service Coverage

Public transportation in rural areas is often characterized by limited-service frequency and inadequate coverage. The "Local Development Strategy for the period 2014-2020" (ELD, 2014) for Machico highlights the insufficient provision of basic services, including public transportation, in the region's rural areas. This constraint not only hinders access to essential services and employment opportunities but also fosters a reliance on private vehicles, further marginalizing those without access to personal transportation. Rural territories often suffer from scarce and inadequate access to basic services, amenities, and opportunities, resulting in territorial and socioeconomic marginalization (Vitale Brovarone & Cotella, 2020).

Long Distances between Points of Interest

Rural areas typically exhibit greater distances between residences, workplaces, schools, and healthcare facilities compared to urban areas. The rugged terrain, as observed in Machico, can lead to circuitous routes with sharp curves, amplifying travel times between destinations (Figure 4).

Figure 4. Rural area of the Porto da Cruz parish (Source: Authors)

Moreover, the lower population density results in a wider dispersion of points of interest, contributing to the perception of extended distances. This necessitates long-distance travel, rendering transportation both costly and time-intensive, and undermining the efficiency and viability of public transportation systems.

Dependence on Individual Transportation

The dearth of public transportation alternatives often compels rural residents to depend on private vehicles. The "Integrated Strategic Transport Plan for the Autonomous Region of Madeira (PIETRAM) 2014-2020" report documents a substantial rise in car usage on Madeira Island, partly attributed to enhanced road accessibility through new infrastructure developments. However, this increased reliance on private vehicles, while potentially convenient for some, poses significant challenges for those without access to cars, such as the elderly, young adults, and individuals with lower incomes. This dependence not only restricts their mobility but also heightens the risk of social exclusion (Figure 5). The low density and scattered housing fabric of small towns and villages make it difficult to operate traditional public transport services (Vitale Brovarone & Cotella, 2020).

Figure 5. Rural area of the Caniçal parish (Source: Authors)

Proposed Solutions

Nutley (2003) advocates for tailored transportation policies for rural areas, proposing the implementation of on-demand services and the integration of diverse transportation modes. This approach recognizes the unique needs and constraints of rural communities and seeks to provide flexible and adaptable solutions. Farrington and Farrington (2005) suggest harnessing information and communication technologies (ICTs) to develop intelligent mobility solutions that cater to the specific requirements of rural populations. However, the effectiveness of ICT-based solutions may be contingent on factors such as internet connectivity and digital literacy in rural areas, necessitating a nuanced approach to implementation.

Also, the increasing trend of urbanization and the concentration of economic opportunities in urban centres can lead to a decline in rural populations and exacerbate the challenges of providing sustainable transportation services in these areas (Berg et al., 2019). The outmigration of younger generations can result in an aging population with specific mobility needs, further straining existing transportation systems.

Proposed Solutions

Recent advancements in autonomous vehicle technology and ride-sharing platforms offer promising solutions for addressing the challenges of rural mobility (Eichholz, 2023). These technologies have the potential to provide flexible, on-demand transportation services in low-density areas, reducing reliance on private vehicles and improving accessibility for residents.

The integration of renewable energy sources, such as solar and wind power, into rural transportation systems can contribute to environmental sustainability and reduce reliance on fossil fuels (Standar et al., 2021). The adoption of electric vehicles and the development of charging infrastructure in rural areas can further support the transition to a cleaner and more sustainable transportation system.

2.3. Factors Influencing Mobility

Mobility in rural areas is influenced by a complex interplay of socioeconomic, demographic, and infrastructural factors. Farrington and Farrington (2005) identify age, gender, income, and vehicle ownership as pivotal determinants of transportation choices in rural settings. Furthermore, the quality and accessibility of road infrastructure play a crucial role in shaping residents' mobility patterns.

Socioeconomic Factors

Income levels significantly influence the ability to own and maintain a private vehicle, as well as the affordability of transportation services. Low-income families often face limited transportation options, hindering their access to essential services and opportunities. Employment location and the necessity for daily commutes also factor into transportation decisions, particularly in rural areas where job opportunities may be dispersed. Additionally, the accessibility of educational institutions is paramount for families with school-age children, impacting their choice of residence and transportation modes.

Demographic Factors

The elderly and young adults encounter distinct mobility challenges. Elderly individuals may experience physical limitations or health conditions that restrict their mobility options, while young people may be dependent on their parents or guardians for transportation due to age restrictions or lack of financial resources. Gender also plays a role in shaping mobility needs, with women often requiring flexible transportation arrangements to balance work and family responsibilities.

Family composition, including the number of children and the presence of individuals with disabilities, further influences transportation requirements and the feasibility of utilizing different modes of transport.

Infrastructure Factors

The maintenance and quality of road infrastructure are crucial for rural mobility. Poor road conditions (Figure 6) can significantly hinder mobility, especially in areas with challenging topography like Machico. Figure 6 displays a color-coded map of the Machico and Água de Pena parishes, featuring 5 levels of classification: Level 1, indicated in red, represents poor condition; Level 2, in orange, denotes fair condition; Level 3, in yellow, reflects average condition; and the two best conditions are shown in green, with light green for good condition and dark green for excellent condition.

Figure 6. Condition of roads with color-coded classification in Machico and Água de Pena Parishes (Source: Authors)

The frequency, scheduling, and coverage of public transportation services are equally critical determinants of mobility. Inadequate or unreliable services can lead to an overreliance on private vehicles, exacerbating the challenges faced by those without access to personal transportation. The availability of active mobility infrastructure, such as dedicated bike lanes and pedestrian walkways, can encourage sustainable transportation choices and promote healthier lifestyles in rural communities. However, the development and maintenance of such infrastructure may require substantial investments and community support.

The COVID-19 pandemic has had a profound impact on mobility patterns globally, including in rural areas (Cattivelli & Rusciano, 2020). The shift towards remote work and online services has altered travel behaviours and highlighted the importance of digital connectivity in rural communities.

Rural Youth Mobility and Job Choice

The COVID-19 pandemic has exacerbated the challenges faced by rural youth, potentially leading to increased out-migration as economic opportunities dwindle and uncertainties rise. The pandemic's disruption of global and local economies has disproportionately affected rural areas, where pre-existing vulnerabilities and limited access to resources have been further exposed. The resulting economic hardship and lack of prospects may compel young people to seek better opportunities elsewhere, contributing to the ongoing trend of youth out-migration from rural areas. (Phillipson et al., 2020)

The declining attractiveness of the agricultural sector among young people is a global trend with far-reaching implications for the future of food production and rural sustainability. The allure of higher wages, better working conditions, and diverse career paths in urban areas often outweighs the perceived hardships and uncertainties associated with agricultural work. This phenomenon, coupled with the increasing accessibility of non-agricultural jobs in rural areas, further contributes to the exodus of young people from the agricultural sector. (Calicioglu et al., 2019).

Aging Workforce in the Agricultural Sector:

The aging of the agricultural workforce is a pressing concern that transcends national borders, affecting both developed and developing countries alike. The physical demands and inherent risks associated with agricultural labour can pose significant challenges for older workers, potentially leading to decreased productivity and increased vulnerability to accidents and health issues. The lack of young people entering the sector to replace retiring farmers further exacerbates this problem, raising concerns about the long-term sustainability of agriculture and food security. (Kim et al., 2014; Szabo et al., 2021).

The decision of young people to leave agriculture is often driven by a complex interplay of factors, including economic considerations, social perceptions, and individual aspirations. The perceived low wages, lack of social prestige, and limited opportunities for advancement in the agricultural sector can deter young people from pursuing farming as a career. Additionally, the allure of urban areas, with their diverse employment opportunities and modern amenities, can further incentivize youth out-migration. (Tocco et al., 2013).

Factors influencing mode choice:

The study found that individual characteristics and lifestyle, the journey type, and the perceived service performance of each transport mode tend to influence the choice of transport (Youssef et al., 2021).

2.4. Stated Preference Surveys

The stated preference (SP) research methodology has emerged as a valuable tool for understanding transportation behaviours and preferences, particularly in contexts where real-world choices are limited or constrained. According to Louviere, Hensher, and Swait (2000), this approach enables researchers to evaluate hypothetical scenarios, offering valuable insights into how individuals might respond to changes in transportation conditions, such as the introduction of new services, modifications to existing infrastructure, or variations in pricing.

Importance and Application

SP research is especially pertinent in rural mobility studies, where transportation conditions are often more dynamic and unpredictable than in urban areas. This methodology allows for the creation and assessment of scenarios that do not currently exist but could be implemented in the future, aiding in the identification of optimal solutions that align with the preferences and needs of rural residents.

Advantages

The SP methodology offers several advantages in the context of rural mobility research. It enables the evaluation of hypothetical scenarios, facilitating the exploration of innovative and unconventional transportation solutions that may not be readily apparent in real-world settings. By considering a multitude of factors, such as cost, travel time, convenience, and comfort, SP research allows for a comprehensive understanding of the trade-off's individuals are willing to make when choosing transportation modes. This information is invaluable for policymakers and transportation planners seeking to design effective and appealing services. Additionally, SP research allows for the analysis of preferences across different population segments, ensuring that the diverse needs of rural residents, including the elderly, young adults, and low-income individuals, are considered.

Complexity of Mobility Decisions

In rural areas, mobility decisions are often complex and influenced by a myriad of factors beyond simply cost and travel time. SP methodology excels at capturing this complexity by incorporating variables such as convenience, reliability, safety, and accessibility to specific destinations. For instance, Shergold and Parkhurst (2012) highlight the significant impact of road maintenance on rural mobility, demonstrating how poor road conditions can severely restrict travel and diminish quality of life. By utilizing hypothetical scenarios, SP research can assess how infrastructure improvements or enhancements to public transportation frequency could influence residents' mobility choices, providing crucial data for evidence-based decision-making.

The use of advanced statistical techniques, such as machine learning algorithms, can enhance the predictive power of stated preference models and provide more accurate insights into individual preferences and behaviours (Zantalis et al., 2022). These techniques can help identify complex patterns and interactions between distinct factors influencing transportation choices.

2.5. Applications in Rural Contexts

The application of the stated preference methodology in rural contexts is instrumental in identifying and prioritizing mobility solutions that are most likely to resonate with local needs and preferences. In Machico, characterized by its rugged topography and dispersed housing patterns, this methodology is particularly relevant for gaining a nuanced understanding of residents' transportation behaviours and aspirations.

Local Contextualization

The unique geographical features of Machico, including winding roads and mountainous terrain, pose distinct challenges for traditional public transportation models. The dispersed nature of housing necessitates flexible and adaptable transportation solutions that can cater to the diverse needs of residents scattered across the region (Alves, 2023b). Moreover, ensuring accessibility to essential services, such as healthcare and education, remains a persistent concern in rural areas like Machico. The limited frequency and coverage of existing public transportation services underscore the need for innovative approaches that can bridge the gaps in service provision.

Proposed Solutions

On-demand transportation services, which can be tailored to the specific needs and schedules of residents, offer a promising avenue for addressing the transportation challenges in low-density areas. Leveraging technology, such as mobile applications for booking and tracking rides, can enhance the convenience and accessibility of these services. Adjusting bus routes to better serve remote or underserved areas, while ensuring convenient schedules, can also significantly improve mobility options for rural residents. Collaboration between local authorities and transportation operators is essential for developing efficient and well-coordinated routes that maximize coverage and minimize travel times. Investing in active transportation infrastructure, such as bike lanes and walking trails, can promote sustainable modes of transport and contribute to healthier lifestyles in rural communities. Furthermore, incentivizing the use of bicycles and walking through public awareness campaigns or financial incentives can further encourage active mobility and reduce reliance on private vehicles.

The successful implementation of rural mobility solutions often requires a collaborative approach involving various stakeholders, including local communities, government agencies, transportation providers, and technology companies (Szymanska & Koloszko-Chomentowska, 2022). Effective communication, coordination, and shared decision-making are crucial for overcoming implementation challenges and ensuring the long-term sustainability of these solutions.

2.5.1. Case Studies Of Innovative Rural Transportation Solutions

The Yangsan City, South Korea

In South Korea, the city of Yangsan has implemented a "city-type bus" service to address the transportation needs of its rural areas, characterized by an aging population and declining ridership. This service utilizes smaller buses with flexible routes and schedules, adapting to the demand and travel patterns of the local population. The city-type bus service has shown increased ridership compared to traditional large buses, demonstrating its potential for improving accessibility in rural areas. However, challenges such as ensuring cost-effectiveness and balancing fixed routes with demand-responsive operations remain. This case study underscores the importance of tailoring transportation solutions to the specific needs and characteristics of rural communities and highlights the potential benefits of flexible and demand-responsive services. (Roh & Kim, 2022)

Pertunmaa Municipality, Finland

In Finland, the municipality of Pertunmaa is grappling with the challenges of an aging and shrinking population, impacting the viability of its rural communities and the provision of services. The lack of public transport, coupled with the reliance on private cars, creates significant barriers to accessing essential services for older residents who can no longer drive. The municipality is exploring innovative solutions such as densifying the municipal centre with accessible housing options, improving walkability, and leveraging digital services to enhance service delivery. This case study underscores the importance of a multi-faceted approach to addressing rural mobility challenges, encompassing urban planning, transportation solutions, and digital inclusion, to ensure the well-being and independence of older residents in shrinking rural communities. (Verma & Taegen, 2019)

Lohmar Municipality, Germany

In Germany, the rural municipality of Lohmar is actively seeking to improve its mobility offerings and promote sustainable transportation through a Smart City project. A case study conducted in Lohmar investigated the acceptance and usage behaviour of Shared Mobility Services, including car-sharing, bike-sharing, and e-scooter sharing. The study revealed a prominent level of interest in these services, particularly e-scooter sharing, but also identified challenges related to limited availability and the prevalence of private car ownership. The findings emphasize the need for comprehensive and accessible Shared Mobility Services, coupled with effective promotion and education, to encourage their adoption and contribute to a more sustainable transportation system in rural areas like Machico. (Schaefer et al., 2022)

Connecting to Machico

These diverse case studies provide valuable insights for addressing the mobility challenges in Machico's low-density rural areas. The successful implementation of on-demand transportation services, integrated mobility platforms, and even autonomous vehicles in other rural contexts demonstrates the potential for adopting similar approaches in Machico. The lessons learned from these cases highlight the importance of:

- Community engagement to understand the specific needs and preferences of residents.
- Flexible scheduling and routing to accommodate the dispersed population and varying demand patterns.

- Technological integration to provide seamless and user-friendly mobility solutions.
- Addressing safety concerns and public perception when introducing modern technologies such as autonomous vehicles.
- Tailoring transportation solutions to the specific demographic and geographic characteristics of the community.
- Considering the economic feasibility and sustainability of the proposed solutions.

By learning from these international experiences and adapting them to the unique context of Machico, it is possible to develop and implement effective and sustainable transportation solutions that enhance accessibility, improve quality of life, and promote the overall well-being of residents in Machico's low-density rural areas.

2.6. Comparison of Studies

Comparing previous studies provides valuable insights into the effectiveness of different approaches and solutions for rural mobility. Nutley (2003) focuses on the importance of specific transportation policies and on-demand transportation services, concluding that the lack of adequate public transportation can lead to social and economic isolation. Farrington and Farrington (2005) emphasize using ICTs to develop intelligent mobility solutions, noting that the absence of adequate public transportation restricts social participation and inclusion. However, these studies do not specifically address the unique challenges posed by the rugged terrain and dispersed settlements often found in rural areas like Machico.

Shergold and Parkhurst (2012) expand on this by highlighting the role of sustainable transport solutions, such as integrated transport systems and active travel options, in addressing rural transport issues. They emphasize the need for a comprehensive approach that considers not only the availability of transport but also its affordability, accessibility, and environmental impact. However, their research primarily focuses on broad policy recommendations and does not delve into the specific preferences and needs of rural residents in different contexts.

Table 1 summarizes the contributions and methodologies of each study, providing a clear comparison of their approaches and findings regarding rural mobility.

Table 1. Comparative Analysis of Rural Mobility Studies: Methodologies and Key Findings (Source: Authors)

Nutley	**2003**	**General, focusing on rural areas**	**Qualitative analysis and literature review**	**The lack of adequate public transportation can lead to social and economic isolation, particularly for the elderly and low-income individuals. Proposed solutions include on-demand services and integration of different transportation modes.**
Farrington & Farrington	2005	General, focusing on rural areas	Literature review and policy analysis	Insufficient public transportation limits social participation and inclusion, recommending the use of ICTs for intelligent mobility solutions. However, it does not fully address the multifaceted nature of rural mobility.
Shergold & Parkhurst	2012	General, focusing on rural areas	Policy analysis and literature review	Highlights the importance of sustainable transport solutions, such as integrated transport systems and active travel options. Emphasizes the need for a comprehensive approach considering availability, accessibility, and environmental impact.
Louviere, Hensher & Swait	2000	General, focusing on research methods	Stated Preference (SP) methodology	SP methodology allows the evaluation of hypothetical scenarios, providing insights into how people might respond to changes in transportation conditions, such as new services or infrastructure modifications.

2.7. Relevance of the Study

This study aims to bridge the gaps identified in previous research by focusing on the specific context of Machico, a region characterized by unique geographical and demographic features. By employing a stated preference research methodology, this study seeks to gather detailed data on the preferences and mobility needs of rural residents, allowing for the development of tailored solutions that address the specific challenges faced by the community.

The study's findings have the potential to inform public policy formulation and contribute to improving accessibility and quality of life for rural residents. By applying advanced methodologies and considering Machico's specific context, this study aims to contribute to creating more effective and sustainable mobility solutions.

Contributions:

- Detailed Understanding of Preferences: The application of stated preference research enables a nuanced understanding of residents' preferences, providing valuable data for evidence-based transportation policy formulation.

- Evaluation of Hypothetical Scenarios: By evaluating hypothetical scenarios, the study can predict the acceptance and effectiveness of new transportation solutions before they are implemented, saving resources, and minimizing potential disruptions.
- Context-Specific Solutions: Focusing on Machico's unique context allows for the development of mobility solutions that are tailored to the region's specific characteristics, such as its topography and dispersed housing patterns.
- Informed Public Policies: The data obtained from this study can inform the development of more effective public policies that promote social inclusion, enhance accessibility, and improve the overall quality of life for rural residents.

This study contributes to the growing body of research on rural mobility by focusing on the specific context of Machico and employing a stated preference methodology to gather detailed data on residents' preferences and needs. The findings of this study can inform the development of evidence-based policies and contribute to improving accessibility and quality of life for rural residents not only in Machico but also in similar contexts worldwide.

3. METHODOLOGY

The study adopted a stated preference (SP) research approach to investigate mobility patterns in Machico's low-density rural areas. The SP methodology was selected for its capacity to capture the complexities of transportation decisions in rural settings, where choices are often limited and influenced by unique geographical challenges (Ahern et al., 2008).

3.1. Research Design and Objectives

The research design was structured to fulfil three primary objectives:

1. Explore Transportation Preferences: The study aimed to understand the relative significance of numerous factors, including cost, travel time, service frequency, and convenience, in residents' transportation choices.
2. Identify Barriers to Mobility: The research sought to uncover the challenges residents encounter in accessing transportation, encompassing infrastructure limitations, socioeconomic factors, and service gaps.

3. Recommend Improvements: The study intended to propose actionable recommendations to enhance the accessibility, quality, and sustainability of transportation services in Machico.

The following steps were undertaken to achieve these objectives (Figure 7):

1. Literature Review: A comprehensive review of literature on rural mobility, stated preferences, and relevant research methodologies was conducted. This review served to inform the research design and subsequent data analysis.
2. Development of Hypothetical Scenarios: Key attributes of transportation services pertinent to Machico residents were identified. Hypothetical scenarios were then created, systematically varying these attributes to elicit preferences and understand trade-offs.
3. Questionnaire Design: A structured questionnaire was developed, incorporating both sociodemographic questions and stated preference questions based on the hypothetical scenarios.
4. Data Collection: Face-to-face interviews were conducted with a representative sample of residents using tablet computers. During these interviews, participants were presented with the hypothetical scenarios, and their choices were recorded.
5. Sampling: A stratified sampling approach was employed to ensure the generalizability of the findings. The population was stratified based on parish and age group, with participants randomly selected within each stratum.

Figure 7. Research Methodology Flowchart (Source: Authors)

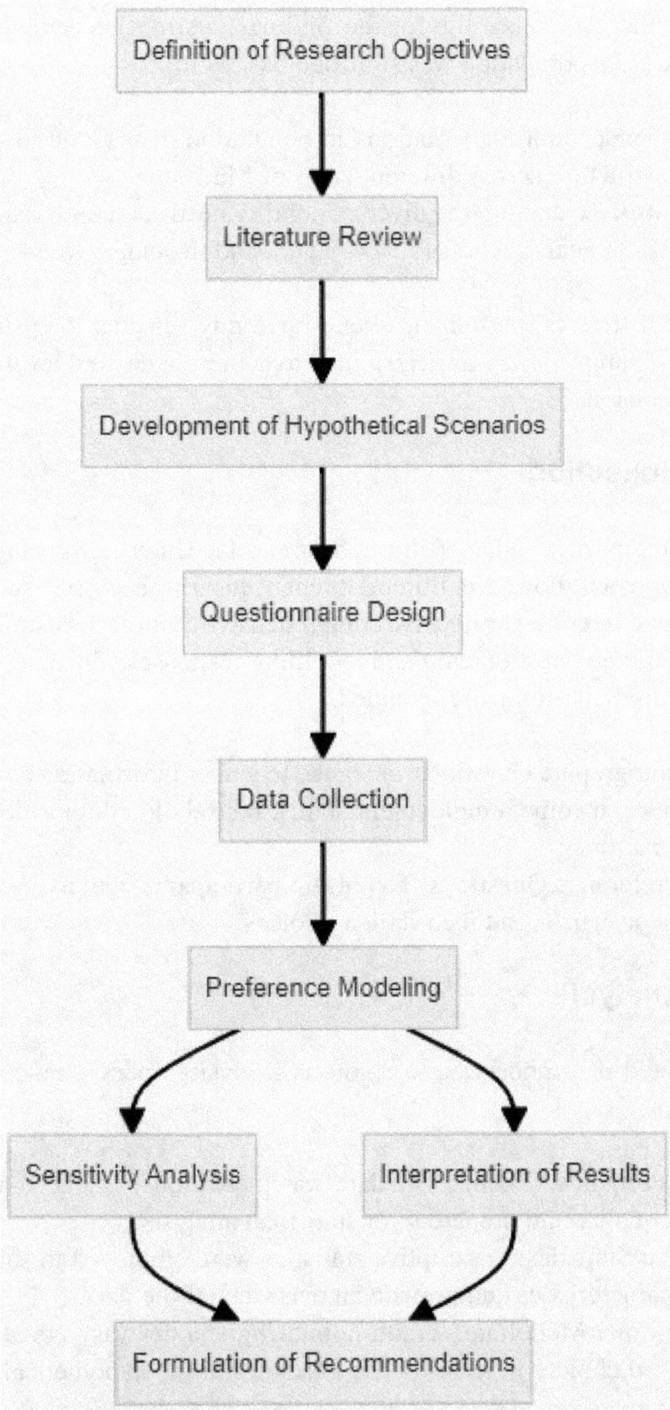

3.2. Sampling

To ensure the generalizability of the findings, a stratified sampling approach was employed. The population was stratified based on:

- **Parish:** To account for variations in population density, access to services, and infrastructure across different areas of Machico.
- **Age Group:** To capture the diverse mobility needs and preferences of young adults (18-34 years), adults (35-64 years), and the elderly (65+ years).

Within each stratum, participants were randomly selected to ensure representativeness. The sample size was determined to achieve a desired level of statistical power and precision.

3.3. Data Collection

Data collection was conducted through face-to-face interviews using tablet computers. This approach allowed real-time data entry, ensuring accuracy, and minimizing the risk of errors. Interviewers receive comprehensive training to ensure consistency in administering the questionnaire and recording responses (Zhang et al, 2008).

The structured questionnaire included:

- Sociodemographic Questions: designed to gather information on participants' age, gender, income, employment status, household composition, and other relevant factors.
- Stated Preference Questions: To present participants with hypothetical transportation scenarios and record their choices.

3.4. Data Analysis

The collected data underwent a rigorous analysis process, encompassing the following steps:

1. Data Cleaning and Coding: The data was meticulously cleaned and coded to ensure its quality and prepare it for statistical analysis.
2. Descriptive Statistics: Descriptive statistics were employed to summarize the sample characteristics and provide an overview of the data.
3. Discrete Choice Modelling: A multinomial logit model was estimated to explain the observed choices in terms of the attributes of the hypothetical scenarios.

4. Model Validation and Sensitivity Analysis: The model underwent validation, and sensitivity analysis was performed. This was done to assess the model's robustness and evaluate the impact of changes in transportation attributes on residents' choices.

The results of this analysis were instrumental in:

- Identifying the key factors influencing transportation preferences.
- Pinpointing barriers to mobility.
- Formulating evidence-based recommendations for improving transportation services in Machico.

4. RESULTS

The stated preference (SP) survey revealed valuable insights into the mobility patterns of Machico residents, particularly regarding their choice of transportation modes. The results demonstrate that service frequency and travel time are the most influential attributes in respondents' decision-making, aligning with previous studies like Farrington and Farrington (2005), which identified frequency and travel time as crucial factors in public transportation choices in rural areas.

4.1. Importance of Service Frequency

Service frequency emerged as the predominant factor influencing the choice of transportation mode. Respondents showed a clear preference for services with higher frequency, indicating that the availability of public transportation at regular intervals is fundamental to their satisfaction and usage. This preference aligns with the findings of Nutley (2003), who emphasized the importance of adequate service frequency in preventing social and economic isolation in rural areas.

4.2. Travel Time as a Decisive Factor

Travel time also plays a crucial role in mode choice. The survey results revealed that respondents value the speed and efficiency of transportation, opting for modes that minimize travel time. This finding is consistent with the literature on rural mobility, which highlights the challenges posed by long distances between points of interest in rural areas (ELD, 2014).

4.3. Influence of Travel Cost

While travel cost is a relevant factor, its influence is less significant compared to service frequency and travel time. However, the research indicates that affordable fare policies are important for encouraging public transportation use, especially among vulnerable groups like the elderly and low-income individuals. This finding supports the argument for inclusive fare policies to improve accessibility and promote social inclusion in rural transportation.

4.4. Preference for Flexible Transportation Solutions

The survey also revealed a growing demand for flexible transportation solutions, such as on-demand services and ridesharing programs. These options are particularly appealing in low-density areas where regular bus operation may not be economically viable. Table 1 presents the respondents' preferences for several types of transportation services.

Table 2. Preferences for Types of Transportation Services: Percentage Distribution (Source: Authors)

Type of Service	Percentage of Preference
Regular Public Transportation	45%
On-Demand Transportation	28%
Ridesharing (Carpooling)	17%
Individual Transportation (Own Car)	10%

4.5. Spatial Distribution of Mobility Patterns

The spatial analysis of mobility patterns in Machico revealed significant variations between different parishes. More densely populated areas, such as Machico and Caniçal, exhibit higher public transportation usage, while more rural and dispersed areas, like Porto da Cruz and Santo António da Serra, show greater reliance on private vehicles. This variation in public transportation usage can be attributed to the higher concentration of services and transportation infrastructure in more urban areas, as discussed by Nutley (2003).

It is important to note that this chapter presents only the results of the stated preference survey. A more in-depth discussion of these results, including comparisons with other studies and their implications for public policy, will be conducted in Chapter 5.

5. ANALYSIS AND DISCUSSIONS

The subsequent section provides an in-depth analysis and discussion of the survey results, exploring the implications for public transportation policies in Machico. The section also addresses the study's limitations and suggests avenues for future research, concluding with practical recommendations to enhance mobility in the region.

5.1. In-depth Analysis of Results

The stated preference survey in Machico has unveiled a complex rural mobility landscape, where the frequency of public transportation services emerges as the most influential factor in transportation mode choice (Figure 8).

Figure 8. Rural areas of Ribeira Seca in the Parish of Machico (Source: Authors)

The preference for frequent services aligns with the broader literature on rural mobility, emphasizing the critical role of accessible and available public transportation in meeting the needs of rural populations (Nutley, 2003; Farrington & Farrington, 2005; Berg & Ihlström, 2019). The desire for frequent services stems from the challenges of long distances and limited alternatives in rural areas. Infrequent services can lead to extended waiting times and travel durations, negatively impacting residents' quality of life and potentially forcing reliance on private vehicles, which may not be economically feasible for all.

Travel time also emerged as a key determinant in mode choice, reflecting the need for efficient travel options, especially given the considerable distances between destinations in rural settings. While travel cost is a factor, its influence is less pronounced compared to service frequency and travel time. However, the research underscores the importance of affordable fare policies, particularly for vulnerable groups, to ensure inclusivity and accessibility in public transportation. The demand for flexible transportation solutions like on-demand services and ridesharing further highlights the need for adaptable services that can cater to the diverse needs and schedules of rural residents.

The spatial analysis of mobility patterns in Machico reveals disparities between parishes, with more urbanized areas exhibiting higher public transportation usage compared to rural and dispersed areas that rely more on private vehicles. This variation can be attributed to the greater concentration of services and infrastructure in urban areas, as noted by Nutley (2003). The role of digitalization in enhancing rural accessibility, as suggested by Vitale Brovarone & Cotella (2020), could be further explored in this context. The role of digitalization in enhancing rural accessibility, as suggested by Vitale Brovarone & Cotella (2020), could be further explored in this context.

5.2. Implications for Public Transportation Policies

The findings of this study corroborate existing research on rural mobility, particularly regarding the importance of service frequency, travel time, and affordability. The emphasis on service frequency and travel time aligns with previous studies like Farrington and Farrington (2005), which also identified these as crucial factors in public transportation choices in rural areas. The need for affordable fare policies echoes the concerns raised in earlier research about social inclusion and accessibility in rural transportation.

However, the study also provides novel insights into the growing demand for flexible transportation solutions in rural areas, highlighting the potential of on-demand services and ridesharing to address the unique challenges of low-density contexts. This aligns with recent research emphasizing the importance of innovative and adaptable solutions to meet the evolving mobility needs of rural communities (Eichholz, 2023; Szymanska & Koloszko-Chomentowska, 2022). The spatial analysis of mobility patterns adds another layer of understanding, revealing the disparities in transportation usage across different parishes within Machico, which resonates with findings from other studies on the impact of urbanization and population density on rural mobility (Berg et al., 2019; Verma & Taegen, 2019).

In addition, the study's findings resonate with recent research on the impact of the COVID-19 pandemic on rural mobility patterns (Phillipson et al., 2020). The pandemic has highlighted the vulnerabilities of rural transportation systems and the need for resilient and adaptable solutions that can address the evolving needs of rural communities in times of crisis. The study's emphasis on flexible transportation options and the role of digitalization aligns with the recommendations from recent literature on post-pandemic rural mobility.

The observed preferences and patterns can be explained by a combination of factors, including:

- **Desire for Convenience and Efficiency**: The preference for frequent services and shorter travel times reflects the desire for convenient and efficient transportation options, which are particularly important in rural areas where distances are greater, and alternatives are limited.
- **Need for Adaptability**: The demand for flexible solutions underscores the need for transportation services that can cater to the diverse needs and schedules of rural residents, who may face challenges with traditional fixed-route services.
- **Influence of Infrastructure and Services**: The spatial disparities in transportation usage highlight the impact of infrastructure and service availability on mobility choices, emphasizing the need for targeted investments and interventions in underserved areas.
- **Changing Demographics and Lifestyles**: The findings also reflect the changing demographics and lifestyles in rural areas, with an increasing number of residents seeking flexible and technology-driven solutions to meet their mobility needs.

5.3. Study Limitations and Suggestions for Future Research

This study, while providing valuable insights, has certain limitations. The reliance on hypothetical scenarios in the stated preference survey may not fully capture actual behaviour. Additionally, the sample, although stratified, may not be perfectly representative of the entire population of Machico. Future research could employ longitudinal studies to track the evolution of mobility patterns and evaluate the long-term impact of implemented policies. Expanding the research to other rural areas with similar characteristics would allow for comparisons and the identification of generalizable patterns. Further investigation into the environmental impact of different transportation options and the exploration of innovative solutions for sustainable mobility in rural contexts would also be valuable contributions.

5.4. Practical Recommendations

The study's findings have significant implications for transportation planning and policymaking in Machico and other rural areas. Policymakers and planners should consider the following:

- Prioritize Service Frequency: Increasing the frequency of public transportation services, particularly during peak hours and in areas with high demand, can significantly improve accessibility and reduce reliance on private vehicles.
- Develop Flexible Transportation Solutions: Implementing on-demand services, ridesharing programs, and other flexible options can address the unique challenges of low-density areas and cater to the diverse needs of rural residents.
- Implement Affordable Fare Policies: Ensuring that public transportation is affordable for all, particularly vulnerable groups, is crucial for promoting social inclusion and equitable access to opportunities.
- Tailor Solutions to Local Contexts: Transportation policies and solutions should be tailored to the specific needs and characteristics of different areas within the region, considering factors such as population density, land use patterns, and existing infrastructure (Figure 9).
- Embrace Digital Technologies: Integrating digital technologies, such as mobile applications and real-time information systems, can enhance the convenience, efficiency, and accessibility of rural transportation services.
- Build Resilience: The COVID-19 pandemic has underscored the need for resilient transportation systems that can adapt to unforeseen disruptions.
- Public transport design: It was suggested that public transport should be designed to meet the required level of service of the customer to encourage them using it (Youssef et al., 2021).

Policymakers should consider strategies for enhancing the resilience of rural transportation, such as diversifying transportation options and promoting sustainable modes of travel.

Figure 9. Access roads with considerable slopes in the parish of Machico (Source: Authors)

6. CONCLUSIONS AND RECOMMENDATIONS

This study investigated mobility patterns in low-density rural areas, focusing on the Machico region of Madeira Island, Portugal. Utilizing a stated preference research methodology, the authors identified key factors influencing residents' transportation choices and proposed recommendations to enhance mobility within the region. The study's findings contribute significantly to the understanding of rural mobility challenges and offer insights for policymakers and transportation planners seeking to improve transportation services in similar contexts worldwide.

Key Findings

The research revealed that service frequency and travel time are the most influential factors in residents' choice of transportation mode, aligning with previous studies like Farrington and Farrington (2005), which identified frequency and travel time as crucial factors in public transportation choices in rural areas. The availability and reliability of public transportation services are crucial for user satisfaction.

Increasing service frequency and ensuring punctual and reliable services can lead to greater satisfaction and increased use of public transportation.

Flexible transportation solutions, such as on-demand services and ridesharing programs, are essential in low-density areas where regular bus operation may not be economically viable. Implementing these solutions can fill gaps left by traditional services and reduce the need for personal vehicles, contributing to more sustainable and inclusive transportation options.

While travel cost is a factor, its influence is less significant compared to service frequency and travel time. However, affordable fare policies are essential, especially for vulnerable groups like the elderly and low-income individuals. Subsidies or reduced fares can improve accessibility to public transportation and support social inclusion, aligning with the principles of social justice in transportation planning (Farrington & Farrington, 2005).

Recommendations and Implementation Plan

The study's findings support the following recommendations and a phased implementation plan to enhance mobility in Machico:

Phase 1: Short-term Improvements (0-1 year)

- Increase Service Frequency: Conduct a comprehensive analysis of current demand patterns and peak hours to optimize bus schedules and increase frequency where needed. This is expected to result in increased ridership, reduced travel times, and improved user satisfaction. The economic impact could include increased revenue for public transportation providers and potential cost savings for residents who rely on public transportation.
- Awareness Campaigns: Launch targeted campaigns to inform residents about existing public transportation options, schedules, and benefits, utilizing various communication channels (e.g., social media, local newspapers, community events). Increased awareness is expected to lead to higher ridership and a shift away from private vehicle use, potentially reducing traffic congestion and associated costs.

Phase 2: Mid-term Enhancements (1-3 years)

- Develop On-Demand Transportation: Partner with technology companies to design and implement a user-friendly mobile application for on-demand transportation services in less populated areas. This could improve accessibility for residents in remote areas, reduce transportation costs for individuals, and potentially create new business opportunities for local transportation providers.

- Pilot Ridesharing Programs: Initiate pilot programs with ridesharing companies or develop community-based carpooling initiatives to assess their feasibility and effectiveness in the Machico context. Successful ridesharing programs could reduce the number of vehicles on the road, decrease traffic congestion, and lower transportation costs for participants.
- Fare Policy Review: Conduct a thorough review of current fare policies and explore options for reduced fares or subsidies for vulnerable groups, ensuring affordability and social inclusion. This could increase ridership among low-income individuals and the elderly, promoting social equity and reducing transportation-related financial burdens.

Phase 3: Long-term Investments (3-5 years)

- Road Infrastructure Improvements: Develop a comprehensive plan for road maintenance and upgrades, prioritizing areas with poor conditions or limited access to public transportation. Improved road infrastructure can enhance safety, reduce travel times, and lower vehicle operating costs, contributing to economic development and improved quality of life.
- Continuous Monitoring and Evaluation: Establish a system for ongoing monitoring and evaluation of transportation services and infrastructure, using key performance indicators (KPIs) to track progress and identify areas for further improvement. This data-driven approach can ensure the effectiveness and sustainability of transportation policies and investments over time.

Implications for Public Policies

The study's findings underscore the need for public policies that prioritize rural mobility. By investing in public transportation infrastructure, promoting flexible transportation solutions, and ensuring affordability, policymakers can create a more equitable and sustainable transportation system in Machico. The phased implementation plan outlined above provides a roadmap for translating research findings into actionable policies that can improve the lives of rural residents.

Future Research

This study serves as a foundation for future research on rural mobility. Longitudinal studies could assess how mobility patterns evolve over time and evaluate the long-term impact of implemented policies. Expanding the research to other rural areas with similar characteristics would allow for comparisons and the identification of generalizable patterns. Additionally, further research could investigate the environmental impact of different transportation options and explore innovative solutions to promote sustainable mobility in rural contexts.

Community Engagement

To ensure the success and sustainability of the proposed recommendations, it is crucial to actively engage the community in the planning and decision-making process. This can be achieved through public consultations, focus groups, and participatory workshops, where residents can voice their opinions, concerns, and ideas. By incorporating community feedback, transportation policies and solutions can be tailored to the specific needs and preferences of the local population, leading to greater acceptance and utilization of public transportation services.

REFERENCES

Ahern, A. A., & Tapley, N. (2008). The use of stated preference techniques to model modal choices on interurban trips in Ireland. *Transportation Research Part A, Policy and Practice*, *42*(1), 15–27. DOI: 10.1016/j.tra.2007.06.005

Alves, R., Lousada, S., Cabezas, J., & Gómez, J. (2022). The Importance of the Strategic Urban Rehabilitation Plan in the Sustainable Development of the Municipality of Machico. *Sustainability (Basel)*, *14*(24), 16816. DOI: 10.3390/su142416816

Alves, R., Lousada, S., Cabezas, J., & Gómez, J. M. N. (2023a). Flood Risk and Vulnerability: An Approach to Urban Floods in Ribeira de Machico in the Autonomous Region of Madeira. In Lousada, S. (Ed.), *Geoinformatics in Support of Urban Politics and the Development of Civil Engineering* (pp. 104–131). IGI Global., DOI: 10.4018/978-1-6684-6449-6.ch005

Alves, R., Lousada, S., Cabezas, J., & Gómez, J. M. N. (2023b). Local Housing Strategy: Analysis of Importance and Implementation in Machico Municipality, Madeira. *Land (Basel)*, *12*(9), 1778. DOI: 10.3390/land12091778

Berg, J., & Ihlström, J. (2019). The Importance of Public Transport for Mobility and Everyday Activities among Rural Residents. *Social Sciences (Basel, Switzerland)*, *2019*(8), 58. DOI: 10.3390/socsci8020058

Calicioglu, O., Flammini, A., Bracco, S., Bellù, L., & Sims, R. (2019). The Future Challenges of Food and Agriculture: An Integrated Analysis of Trends and Solutions. *Sustainability (Basel)*, *2019*(11), 222. DOI: 10.3390/su11010222

Cattivelli, V., & Rusciano, V. (2020). Social Innovation and Food Provisioning during Covid-19: The Case of Urban–Rural Initiatives in the Province of Naples. *Sustainability (Basel)*, *2020*(12), 4444. DOI: 10.3390/su12114444

Eichholz, L. (2023). The Implications of a Co-Created Software Solution for Mobility in Rural Areas. *Smart Cities*, *2023*(6), 2706–2721. DOI: 10.3390/smartcities6050122

ELD. (2014). Estratégia Local de Desenvolvimento para o periodo 2014-2020.

Esteban-Navarro, M.-Á., García-Madurga, M.-Á., Morte-Nadal, T., & Nogales-Bocio, A.-I. (2020). The Rural Digital Divide in the Face of the COVID-19 Pandemic in Europe—Recommendations from a Scoping Review. *Informatics (MDPI)*, *2020*(7), 54. DOI: 10.3390/informatics7040054

Farrington, J., & Farrington, C. (2005). Rural accessibility, social inclusion, and social justice: Towards conceptualisation. *Journal of Transport Geography*, *13*(1), 1–12. DOI: 10.1016/j.jtrangeo.2004.10.002

Kim, Y., Kim, D., Park, K., & Kim, D. (2014). A Study on Analysis of Industrial Injury Characteristics of Aging Workers in Agriculture. *Journal of Ergonomics Society of Korea*, *33*(6), 477–486. DOI: 10.5143/JESK.2014.33.6.477

Louviere, J. J., Hensher, D. A., & Swait, J. D. (2000). *Stated choice methods: analysis and applications*. Cambridge University Press. DOI: 10.1017/CBO9780511753831

Nutley, S. (2003). Transport and Social Exclusion: An Overview of the Literature. *Transport Policy*, *10*(1), 1–11.

Phillipson, J., Gorton, M., Turner, R., Shucksmith, M., Aitken-McDermott, K., Areal, F., Cowie, P., Hubbard, C., Maioli, S., McAreavey, R., Souza-Monteiro, D., Newbery, R., Panzone, L., Rowe, F., & Shortall, S. (2020). The COVID-19 Pandemic and Its Implications for Rural Economies. *Sustainability (Basel)*, *2020*(12), 3973. DOI: 10.3390/su12103973

PIETRAM. (2016). Plano Integrado Estratégico de Transportes da Região Autónoma da Madeira (PIETRAM) – 2014/2020.

Roh, C.-G., & Kim, J. (2022). What Are More Efficient Transportation Services in a Rural Area? A Case Study in Yangsan City, South Korea. *International Journal of Environmental Research and Public Health*, *2022*(19), 11263. DOI: 10.3390/ijerph191811263 PMID: 36141536

Schaefer, C., Stelter, A., Holl-Supra, S., Weber, S., & Niehaves, B. (2022). The Acceptance and Use Behavior of Shared Mobility Services in a Rural Municipality. *Smart Cities*, *2022*(5), 1229–1240. DOI: 10.3390/smartcities5040062

Shergold, I., & Parkhurst, G. (2012). Understanding and Addressing Rural Transport Issues: The Role of Sustainable Transport Solutions. *Transport Reviews*, *32*(6), 745–763.

Standar, A., Kozera, A., & Satoła, Ł. (2021). The Importance of Local Investments Co-Financed by the European Union in the Field of Renewable Energy Sources in Rural Areas of Poland. *Energies*, *2021*(14), 450. DOI: 10.3390/en14020450

Szabo, S., Apipoonanon, C., Pramanik, M., Leeson, K., & Singh, D. R. (2021). Perceptions of an ageing agricultural workforce and farmers' productivity strategies: Evidence from Prachinburi Province, Thailand. *Outlook on Agriculture*, *50*(3), 294–304. DOI: 10.1177/00307270211025053

Szymanska, E., & Koloszko-Chomentowska, Z. (2022). Sustainable Innovative Mobility Solutions Preferred by Inhabitants of Rural Areas—The Case of Lithuania and Poland. *Sustainability (Basel)*, *2022*(14), 6603. DOI: 10.3390/su14116603

Tocco, B., Bailey, A., & Davidova, S. (2013). Determinants to Leave Agriculture and Change Occupational Sector: Evidence from an Enlarged Eu. Factor Markets Working Paper, 32.

Verma, I., & Taegen, J. (2019). Access to Services in Rural Areas from the Point of View of Older Population—A Case Study in Finland. *International Journal of Environmental Research and Public Health*, *2019*(16), 4854. DOI: 10.3390/ijerph16234854 PMID: 31810348

Vitale Brovarone, E., & Cotella, G. (2020). Improving Rural Accessibility: A Multilayer Approach. *Sustainability (Basel)*, *2020*(12), 2876. DOI: 10.3390/su12072876

Youssef, Z., Alshuwaikhat, H., & Reza, I. (2021). Modeling the Modal Shift towards a More Sustainable Transport by Stated Preference in Riyadh, Saudi Arabia. *Sustainability (Basel)*, *2021*(13), 337. DOI: 10.3390/su13010337

Zantalis, F., Koulouras, G., Karabetsos, S., & Kandris, D. (2019). A Review of Machine Learning and IoT in Smart Transportation. *Future Internet*, *2019*(11), 94. DOI: 10.3390/fi11040094

Zhang, J., Fujiwara, A., & Thein, S. (2008). Capturing Travelers' Stated Mode Choice Preferences Under Influence of Income in Yangon City, Myanmar. *Journal of Transportation Systems Engineering and Information Technology*, *2008*(8), 49–62. DOI: 10.1016/S1570-6672(08)60032-X

KEY TERMS AND DEFINITIONS

Machico: A city located on the island of Madeira, Portugal, facing significant housing challenges due to increased demand and geographical limitations.

Rural Accessibility: Refers to the ease with which residents of rural areas can access services, employment, and other essential activities. Accessibility is a critical factor for social inclusion and economic development in these areas.

Sustainable Mobility: Refers to transportation practices that minimize environmental impact, promote efficiency, and are economically viable. In rural areas, this can include the use of public transportation, shared rides, and other alternative means of transport.

Rural Transport Planning: Involves the creation and implementation of policies and strategies to improve transportation in rural areas, focusing on efficiency, accessibility, and sustainability.

On-Demand Transport: A transportation system that provides services according to users' needs, often through advance bookings. This type of service is particularly useful in low-density population areas where regular transport is not feasible.

Road Infrastructure: Includes all the roads, bridges, and other structures necessary for efficient and safe transportation. The maintenance and improvement of road infrastructure are crucial to ensuring access and mobility in rural areas.

Compilation of References

Abas, D. N., Zakaria, R., Aminuddin, E., Abdullah, R., Khan, J. S., & Sahamir, S. R. (2018). An outlook on strategic framework development needs of roles and involvement of joint property management in high rise green buildings in Malaysia. *Chemical Engineering Transactions*, *63*, 427–432. DOI: 10.3303/CET1863072

Abdalla, H., Fattah, K. P., Abdallah, M., & Al-Tamimi, A. K. (2021). Environmental footprint and economics of a full-scale 3d-printed house. *Sustainability (Basel)*, *13*(21), 11978. DOI: 10.3390/su132111978

Abdi, A., Awarke, M., Malayeri, M. R., & Riazi, M. (2024). Interfacial tension of smart water and various crude oils. *Fuel*, *356*, 129563. DOI: 10.1016/j.fuel.2023.129563

Acolin, A., & Reina, V. (2022). Housing cost burden and life satisfaction. *Journal of Housing and the Built Environment*, *37*(4), 1–27. DOI: 10.1007/s10901-021-09921-1 PMID: 36776144

Ahern, A. A., & Tapley, N. (2008). The use of stated preference techniques to model modal choices on interurban trips in Ireland. *Transportation Research Part A, Policy and Practice*, *42*(1), 15–27. DOI: 10.1016/j.tra.2007.06.005

Ahmad, T., Aibinu, A. A., & Stephan, A. (2019). Managing green building development–a review of current state of research and future directions. *Building and Environment*, *155*, 83–104. DOI: 10.1016/j.buildenv.2019.03.034

Ahvenniemi, H., Huovila, A., Pinto-Seppä, I., & Airaksinen, M. (2017). What are the differences between sustainable and smart cities? *Cities (London, England)*, *60*, 234–245. DOI: 10.1016/j.cities.2016.09.009

Akadiri, P. O., Chinyio, E. A., & Olomolaiye, P. O. (2012). Design of a sustainable building: A conceptual framework for implementing sustainability in the building sector. *Buildings (Basel, Switzerland)*, *2*(2), 126–152. DOI: 10.3390/buildings2020126

Akbarnezhad, A., Huan, M., Mesgari, S., & Castel, A. (2015). Recycling of geopolymer concrete. *Construction & Building Materials, 101*, 152–158. DOI: 10.1016/j.conbuildmat.2015.10.037

Akhanova, G., Nadeem, A., Kim, J. R., & Azhar, S. (2020). A multi-criteria decision-making framework for building sustainability assessment in Kazakhstan. *Sustainable Cities and Society, 52*, 101842. DOI: 10.1016/j.scs.2019.101842

Alam, F., & Jin, Y. (2023). The utilisation of small wind turbines in built-up areas: Prospects and challenges. *Wind, 3*(4), 418–439. DOI: 10.3390/wind3040024

Albitar, M., Ali, M. M., & Visintin, P. (2017). Experimental study on fly ash and lead smelter slag-based geopolymer concrete columns. *Construction & Building Materials, 141*, 104–112. DOI: 10.1016/j.conbuildmat.2017.03.014

Allen, D., & Shonnard, D. R. (2015). *Sustainable engineering: Concepts, design and case studies*. Prentice Hall.

Ally, A. N., Blanche, M. M., Nana, U. J. P., Grâce, M. M., François, N., & Pettang, C. (2021). Recovery of mining wastes in building materials: A review. *Open Journal of Civil Engineering, 11*(4), 379–397. DOI: 10.4236/ojce.2021.114022

Almeida, J., Ribeiro, A. B., Silva, A. S., & Faria, P. (2020). Overview of mining residues incorporation in construction materials and barriers for full-scale application. *Journal of Building Engineering, 29*, 101215. DOI: 10.1016/j.jobe.2020.101215

Al-Naje, F. Q., Abed, A. H., & Al-Taie, A. J. (2020). A review of sustainable materials to improve geotechnical properties of soils. *Al-Nahrain Journal for Engineering Sciences, 23*(3), 289–305. DOI: 10.29194/NJES.23030289

Alqaisi, R., Le, T. M., & Khabbaz, H. (2020). Applications of recycled sustainable materials and by-products in soil stabilization. In Recent Thoughts in Geoenvironmental Engineering: Proceedings of the 3rd GeoMEast International Congress and Exhibition, Egypt 2019 on Sustainable Civil Infrastructures–The Official International Congress of the Soil-Structure Interaction Group in Egypt (SSIGE) (pp. 91-117). Springer International Publishing. DOI: 10.1007/978-3-030-34199-2_7

Altomonte, S., & Schiavon, S. (2013). Occupant satisfaction in leed and non-leed certified buildings. *Building and Environment, 68*, 66–76. DOI: 10.1016/j.buildenv.2013.06.008

Alves, R., Lousada, S., Cabezas, J., & Gómez, J. M. N. (2023b). Port of Machico, Madeira: Assessment of the Consequences of Overtopping Using AHP Methodology. In S. Lousada (Ed.), Geoinformatics in Support of Urban Politics and the Development of Civil Engineering (pp. 77-103). IGI Global. https://doi.org/DOI: 10.4018/978-1-6684-6449-6.ch004

Alves, R. G., Maia, R. F., & Lima, F. (2023). Development of a Digital Twin for smart farming: Irrigation management system for water saving. *Journal of Cleaner Production*, *388*, 135920. DOI: 10.1016/j.jclepro.2023.135920

Alves, R., Lousada, S., Cabezas, J., & Gómez, J. M. N. (2022). The importance of the strategic urban rehabilitation plan in the sustainable development of the municipality of Machico. *Sustainability (Basel)*, *14*(24), 16816. DOI: 10.3390/su142416816

Alves, R., Lousada, S., Cabezas, J., & Gómez, J. M. N. (2023). Local housing strategy: Analysis of importance and implementation in Machico municipality, Madeira. *Land (Basel)*, *12*(9), 1778. DOI: 10.3390/land12091778

Alves, R., Lousada, S., Cabezas, J., & Gómez, J. M. N. (2023c). Modeling of Flood Risk: Vulnerability and Adaptation of Machico City Center. In Lousada, S. (Ed.), *Geoinformatics in Support of Urban Politics and the Development of Civil Engineering* (pp. 25–49). IGI Global., DOI: 10.4018/978-1-6684-6449-6.ch002

Alves, R., Silva, A., & Lousada, S. (2022a). The impact of local accommodation in the sustainable development of Machico. In Aranzadi, T. R. (Ed.), *Planeamiento Ecológico En Las Iniciativas de Desarrollo* (pp. 305–322). Academic Press.

Amalfi, R. L., Vakili-Farahani, F., & Thome, J. R. (2016). Flow boiling and frictional pressure gradients in plate heat exchangers. Part 2: Comparison of literature methods to database and new prediction methods. *International Journal of Refrigeration*, *61*, 185–203. DOI: 10.1016/j.ijrefrig.2015.07.009

Amasyali, K., & El-Gohary, N. M. (2018). A review of data-driven building energy consumption prediction studies. *Renewable & Sustainable Energy Reviews*, *81*, 1192–1205. DOI: 10.1016/j.rser.2017.04.095

Anacker, K. B. (2019). Introduction: Housing affordability and affordable housing. *International Journal of Housing Policy*, *19*(1), 1–16. DOI: 10.1080/19491247.2018.1560544

Antunes, P., Travanca, R., Rodrigues, H., Melo, J., Jara, J., Varum, H., & André, P. (2012). Dynamic Structural Health Monitoring of Slender Structures Using Optical Sensors. *Sensors (Basel)*, *12*(5), 5. Advance online publication. DOI: 10.3390/s120506629 PMID: 22778661

APEC. (2016). *Review of APEC Green Building Project: Final Report.* Nathan Associates Inc.

Aram, F., García, E. H., Solgi, E., & Mansournia, S. (2019). Urban green space cooling effect in cities. *Heliyon*, *5*(4), e01339. Advance online publication. DOI: 10.1016/j.heliyon.2019.e01339 PMID: 31008380

Arepalli, P. G., & Naik, K. J. (2024). An IoT based smart water quality assessment framework for aqua-ponds management using Dilated Spatial-temporal Convolution Neural Network (DSTCNN). *Aquacultural Engineering*, *104*, 102373. DOI: 10.1016/j.aquaeng.2023.102373

Arsenyeva, O. P., Tovazhnyanskyy, L. L., Kapustenko, P. O., & Khavin, G. L. (2011). The generalized correlation for friction factor in crisscross flow channels of plate heat exchangers. *Chemical Engineering Transactions*, *25*, 399–404. DOI: 10.3303/CET1125067

Arsenyeva, O., Klemeš, J. J., Plankovskyy, S., & Kapustenko, P. (2022). The influence of plate corrugation geometry on heat and mass transfer performance of plate heat exchangers for condensation of steam in the presence of air. *Thermal Science and Engineering Progress*, *30*, 101248. DOI: 10.1016/j.tsep.2022.101248

Asadpoure, A., Mohammadi, S., & Vafai, A. (2006). Modeling crack in orthotropic media using a coupled finite element and partition of unity methods. *Finite Elements in Analysis and Design*, *42*(13), 1165–1175. DOI: 10.1016/j.finel.2006.05.001

Aslam, M. S., Huang, B., & Cui, L. (2020). Review of construction and demolition waste management in China and USA. *Journal of Environmental Management*, *264*, 110445. DOI: 10.1016/j.jenvman.2020.110445 PMID: 32217323

Asyera, E., Gamal, A., & Pandjaitan, T. H. (2020). Occupant behavioral change for energy efficiency in office buildings. *AIP Conference Proceedings, 2255(1*, 020451. DOI:DOI: 10.1063/5.0020451

Atanda, J. O. (2019). Developing a social sustainability assessment framework. *Sustainable Cities and Society*, *44*, 237–252. DOI: 10.1016/j.scs.2018.09.023

Aung, T., Liana, S.R., Htet, A., & Bhaumik, A.. (2023). Implementing green facades: A step towardssustainable smart buildings. *Journal of Smart Cities and Society, 2*, 41-51. htttps://.DOI: 10.3233/SCS-230014

Averfalk, H., Ingvarsson, P., Persson, U., Gong, M., & Werner, S. (2017). Large heat pumps in Swedish district heating systems. *Renewable & Sustainable Energy Reviews*, *79*, 1275–1284. DOI: 10.1016/j.rser.2017.05.135

Averfalk, H., & Werner, S. (2017). Essential improvements in future district heating systems. *Energy Procedia, 116*, 217–225. DOI: 10.1016/j.egypro.2017.05.069

Babaev, V., Sukhonos, M., Starostina, A., & Beletsky, I. (2016). Improving the processes of cost management in the construction and energy projects. *Eastern-European Journal of Enterprise Technologies*, *4*(3), 10–17. DOI: 10.15587/1729-4061.2016.75515

Badi, S., & Murtagh, N. (2019). Green supply chain management in construction: A systematic literature review and future research agenda. *Journal of Cleaner Production*, *223*, 312–322. DOI: 10.1016/j.jclepro.2019.03.132

Bahramian, M., & Yetilmezsoy, K. (2020). Life cycle assessment of the building industry: An overview of two decades of research (1995–2018). *Energy and Building*, *219*, 109917. DOI: 10.1016/j.enbuild.2020.109917

Bakar, N. N. A., Hassan, M. Y., Abdullah, H., Rahman, H. A., Abdullah, M. P., Hussin, F., & Bandi, M. (2015). Energy efficiency index as an indicator for measuring building energy performance: A review. *Renewable & Sustainable Energy Reviews*, *44*, 1–11. DOI: 10.1016/j.rser.2014.12.018

Bamber, N., Turner, I., Arulnathan, V., Li, Y., Zargar Ershadi, S., Smart, A., & Pelletier, N. (2020). Comparing sources and analysis of uncertainty in consequential and attributional life cycle assessment: Review of current practice and recommendations. *The International Journal of Life Cycle Assessment*, *25*(1), 168–180. DOI: 10.1007/s11367-019-01663-1

Banerjee, G. and Ganguly, S. (2018) Water resources and environmental sustainability, IBP publication.

Barah, B.C. (2015) 'Traditional water harvesting systems: An ecological economics Survey' New age international publisher.

Barrett, P. (2023). Intersections between housing affordability and meanings of home: A review. *Kotuitui*, *18*(1), 27–44. DOI: 10.1080/1177083X.2022.2090969

Basten, V., Latief, Y., Berawi, M.A., Budiman, R., & Riswanto. (2017). Evaluation of green building rating tools based on existing green building achievement in Indonesia using Life Cycle Assessment Method. *AIP Conferences Proceedings, 1818*, 020008, .DOI: 10.1063/1.4976872

Bedon, C., & Fasan, M. (2019). Reliability of Field Experiments, Analytical Methods and Pedestrian's Perception Scales for the Vibration Serviceability Assessment of an In-Service Glass Walkway. *Applied Sciences (Basel, Switzerland)*, *9*(9), 9. Advance online publication. DOI: 10.3390/app9091936

Berger, J. (2019). Signaling can increase consumers' willingness to pay for green products. Theoretical model and experimental evidence. *Journal of Consumer Behaviour, 18*(3), 233–246. DOI: 10.1002/cb.1760

Berg, J., & Ihlström, J. (2019). The Importance of Public Transport for Mobility and Everyday Activities among Rural Residents. *Social Sciences (Basel, Switzerland), 2019*(8), 58. DOI: 10.3390/socsci8020058

Berglund, E. Z., Monroe, J. G., Ahmed, I., Noghabaei, M., Do, J., Pesantez, J. E., Khaksar Fasaee, M. A., Bardaka, E., Han, K., Proestos, G. T., & Levis, J. (2020). Smart infrastructure: A vision for the role of the civil engineering profession in smart cities. *Journal of Infrastructure Systems, 26*(2), 03120001. DOI: 10.1061/(ASCE)IS.1943-555X.0000549

Beshr, A. A. E.-W. (2015). Structural Deformation Monitoring and Analysis of Highway Bridge Using Accurate Geodetic Techniques. *Engineering, 7*(8), 8. Advance online publication. DOI: 10.4236/eng.2015.78045

Bhuiyan C., Wolfgang-Albert, F. and Jain S. K. (2019) Water security and sustainability: Proceeding of Down to Earth-2019, Springer publication.

Bibri, S. E., Krogstie, J., & Kärrholm, M. (2020). Compact city planning and development: Emerging practices and strategies for achieving the goals of sustainability. Developments in the built environment, 4, 100021.

Biletskyi, I., Plankovskyy, S., Tsegelnyk, Y., Teliura, N., & Osinska, M. (2023). Examining the possibility of heat recovery during sanitation of sewage collectors. In Arsenyeva, O. (Eds.), *Smart Technologies in Urban Engineering* (pp. 350–362). Springer., DOI: 10.1007/978-3-031-46877-3_31

Binsacca, R. (2010, September 02). *Interview With Green Construction Consultant.* From Builder: https://www.builderonline.com/building/building-science/interview-with-green-construction-consultant_o

Biradkar, A. H. (2020). *Basic civil and environmental engineering*. Nirali Parkashan.

Birdie, J. S., and Birdie, G.S. (2014) Water supply & Sanitary engineering-including environmental engineering and pollution control, Dhanpat Rai books publication.

Bjørn, A., Chandrakumar, C., Boulay, A. M., Doka, G., Fang, K., Gondran, N., Hauschild, M. Z., Kerkhof, A., King, H., Margni, M., McLaren, S., Mueller, C., Owsianiak, M., Peters, G., Roos, S., Sala, S., Sandin, G., Sim, S., Vargas-Gonzalez, M., & Ryberg, M. (2020). Review of life-cycle based methods for absolute environmental sustainability assessment and their applications. *Environmental Research Letters, 15*(8), 083001. DOI: 10.1088/1748-9326/ab89d7

Blackburne, L., Gharehbaghi, K., & Hosseinian-Far, A. (2021). The knock-on effects of green buildings: High-rise construction design implications. *International Journal of Structural Integrity*, *13*(1), 57–77. DOI: 10.1108/IJSI-06-2021-0062

Bonfante, M. C., Raspini, J. P., Fernandes, I. B., Fernandes, S., Campos, L. M., & Alarcon, O. E. (2021). Achieving Sustainable Development Goals in rare earth magnets production: A review on state of the art and SWOT analysis. *Renewable & Sustainable Energy Reviews*, *137*, 110616. DOI: 10.1016/j.rser.2020.110616

Borowska-Stefańska, M., & Wiśniewski, S. (2022). The role of road transportation in the flood evacuation process. In Oxford Research Encyclopedia of Natural Hazard Science. DOI: 10.1093/acrefore/9780199389407.013.440

Bourdeau, M., Zhai, X., Nefzaoui, E., Guo, X., & Chatellier, P. (2019). Modeling and forecasting building energy consumption: A review of data-driven techniques. *Sustainable Cities and Society*, *48*, 101533. DOI: 10.1016/j.scs.2019.101533

Brauer, R. L. (2022). *Safety and health for engineers*. John Wiley & Sons.

Buchwald, A., Dombrowski, K., & Weil, M. (2005). The influence of calcium content on the performance of geopolymeric binder especially the resistance against acids. *Proceedings of the world geopolymer*, 35-9.

Bujor, O., Prodan, I. C., & Popa, A. (2021). Energy Geostructures: An Innovative Renewable Energy Technology for Cooling and Heating of Buildings. Environmental and Human Impact of Buildings: An Energetics Perspective, 183-204. DOI: 10.1007/978-3-030-57418-5_7

Bulkeley, H. (2022). Climate changed urban futures: environmental politics in the anthropocene city. In *Trajectories in environmental politics* (pp. 263–281). Routledge. DOI: 10.4324/9781003213321-14

Bulkeley, H., & Betsill, M. M. (2013). Revisiting the urban politics of climate change. *Environmental Politics*, *22*(1), 136–154. DOI: 10.1080/09644016.2013.755797

Buriro, M.A., Wei, M., Bai, B., & Yao, Y. (2023). Advances in smart water flooding: A comprehensive study on the interplay of ions, salinity in carbonate reservoirs. *Journal of Molecular Liquids*, Volume 390, Part B, 123140.

Bwambale, E., Abagale, F. K., & Anornu, G. K. (2024). Towards a modelling, optimization and predictive control framework for smart irrigation. *Heliyon*, *10*(18), e38095. DOI: 10.1016/j.heliyon.2024.e38095 PMID: 39364235

Calicioglu, O., Flammini, A., Bracco, S., Bellù, L., & Sims, R. (2019). The Future Challenges of Food and Agriculture: An Integrated Analysis of Trends and Solutions. *Sustainability (Basel)*, *2019*(11), 222. DOI: 10.3390/su11010222

Campisi, T., Acampa, G., Marino, G., & Tesoriere, G. (2020). Cycling master plans in Italy: The I-BIM feasibility tool for cost and safety assessments. *Sustainability (Basel)*, *12*(11), 4723. DOI: 10.3390/su12114723

Cannavale, A., Zampini, G., Carlucci, F., Pugliese, M., Martellotta, F., Ayr, U., Maiorano, V., Ortica, F., Fiorito, F., & Latterini, L. (2022). Energy and daylighting performance of building integrated spirooxazine photochromic films. *Solar Energy*, *242*(7), 424–434. DOI: 10.1016/j.solener.2021.10.058

Cao, M., Li, R., & Wang, Z. (2024). Research on integrated art design of building exterior finishing with photovoltaic application technology. *E3S Web of Conferences*, *512*, 01004. .DOI: 10.1051/e3sconf/202451201004

Cardell-Oliver, R., Cominola, A., & Hong, J. (2024). Activity and resolution aware privacy protection for smart water meter databases. *Internet of Things : Engineering Cyber Physical Human Systems*, *25*, 101130. DOI: 10.1016/j.iot.2024.101130

Castellani, V., Beylot, A., & Sala, S. (2019). Environmental impacts of household consumption in Europe: Comparing process-based LCA and environmentally extended input-output analysis. *Journal of Cleaner Production*, *240*, 117966. DOI: 10.1016/j.jclepro.2019.117966 PMID: 31839696

Catanho, A. (1997). Contribution to the Geological and Geotechnical Characterization of the Different Volcanic Complexes of Madeira Island.

Cattivelli, V., & Rusciano, V. (2020). Social Innovation and Food Provisioning during Covid-19: The Case of Urban–Rural Initiatives in the Province of Naples. *Sustainability (Basel)*, *2020*(12), 4444. DOI: 10.3390/su12114444

Cecconet, D., Raček, J., Callegari, A., & Hlavínek, P. (2019). Energy recovery from wastewater: A study on heating and cooling of a multipurpose building with sewage-reclaimed heat energy. *Sustainability (Basel)*, *12*(1), 116. DOI: 10.3390/su12010116

Chadwick, A., Morfett, J., & Borthwick, M. (2021). *Hydraulics in civil and environmental engineering*. Crc Press. DOI: 10.1201/9781003026839

Chakroborty, P., & Das, A. (2017). *Principles of transportation engineering*. PHI Learning Pvt. Ltd.

Chang, I., Lee, M., & Cho, G. C. (2019). Global CO2 emission-related geotechnical engineering hazards and the mission for sustainable geotechnical engineering. *Energies*, *12*(13), 2567. DOI: 10.3390/en12132567

Chang, I., Lee, M., Tran, A. T. P., Lee, S., Kwon, Y. M., Im, J., & Cho, G. C. (2020). Review on biopolymer-based soil treatment (BPST) technology in geotechnical engineering practices. *Transportation Geotechnics*, *24*, 100385. DOI: 10.1016/j.trgeo.2020.100385

Chan, S., Boran, I., Van Asselt, H., Ellinger, P., Garcia, M., Hale, T., Hermwille, L., Liti Mbeva, K., Mert, A., Roger, C. B., Weinfurter, A., Widerberg, O., Bynoe, P., Chengo, V., Cherkaoui, A., Edwards, T., Gütschow, M., Hsu, A., Hultman, N., & Shrivastava, M. K. (2021). Climate ambition and sustainable development for a new decade: A catalytic framework. *Global Policy*, *12*(3), 245–259. DOI: 10.1111/1758-5899.12932

Charlesworth, S. M., & Booth, C. A. (Eds.). (2016). *Sustainable surface water management: a handbook for SUDS*. John Wiley & Sons. DOI: 10.1002/9781118897690

Chavarry-Vallejos, C. M., Gutiérrez, A. A. V., Bancayán, C., Oré, L. J. C. R., Rodríguez, J. S. T., Andrés, Í., & Serrano, J. E. (2023). Use Of Mining Tailings in The Manufacture of Ecological Bricks to Encourage the Construction of Sustainable Housing. *International Journal (Toronto, Ont.)*, *10*(3), 2555–2567.

Chelleri, L., Waters, J. J., Olazabal, M., & Minucci, G. (2015). Resilience trade-offs: Addressing multiple scales and temporal aspects of urban resilience. *Environment and Urbanization*, *27*(1), 181–198. DOI: 10.1177/0956247814550780

Cheng, J. C., Lu, Q., & Deng, Y. (2016). Analytical review and evaluation of civil information modeling. *Automation in Construction*, *67*, 31–47. DOI: 10.1016/j.autcon.2016.02.006

Chen, J., Zhao, W., Luo, H., Li, P., & Wen, Y. (2008). A real-time system for monitoring distant dynamic displacement of structures. *Geo-Spatial Information Science*, *11*(1), 49–55. DOI: 10.1007/s11806-007-0158-1

Chin, D. (2020). *Water resource engineering*. Pearson Publication.

Choi, J. O., Chen, X. B., & Kim, T. W. (2019). Opportunities and challenges of modular methods in dense urban environment. *International Journal of Construction Management*, *19*(2), 93–105. DOI: 10.1080/15623599.2017.1382093

Choudhary, V., Machavaram, R., Singh, N., & Nagar, H. (2024). Assessment of sensor-based automatic smart watering unit for paddy nurseries under Indian perspective. *Smart Agricultural Technology*, *8*, 100518. DOI: 10.1016/j.atech.2024.100518

Choudhury, M., & Rizos, C. (2010). Slow structural deformation monitoring using Locata—A trial at Tumut Pond Dam. *Journal of Applied Geodesy*, *4*(4), 177–187. DOI: 10.1515/jag.2010.018

Chung, S. S., & Lo, C. W. (2003). Evaluating sustainability in waste management: The case of construction and demolition, chemical and clinical wastes in Hong Kong. *Resources, Conservation and Recycling*, *37*(2), 119–145. DOI: 10.1016/S0921-3449(02)00075-7

Clemente, D., Calheiros-Cabral, T., Rosa-Santos, P., & Taveira-Pinto, F. (2021). Hydraulic and Structural Assessment of a Rubble-Mound Breakwater with a Hybrid Wave Energy Converter. *Journal of Marine Science and Engineering*, *9*(9), 9. Advance online publication. DOI: 10.3390/jmse9090922

CML. (2012). Área de Reabilitação Urbana de Lisboa. Available online: https://www.lisboa.pt/temas/urbanismo/reabilitacao-urbana/area-de-reabilitacao. (accessed on 17 Maio 2024).

CMM. (2005). Plano Diretor Municipal de Machico. Available online: https://www.cm-machico.pt/municipal/machico-atua/urbanismo/pdm-plano-diretor-municipal. (accessed on 10 Janeiro 2024).

CMM. (2016). Área de Reabilitação Urbana de Machico. Available online: https://www.cm-machico.pt/municipal/machico-atua/urbanismo/aru-area-de-reabilitacao-urbana. (accessed on 10 Janeiro 2024).

CMM. (2019). Plano Estratégico de Reabilitação Urbana de Machico e do Porto da Cruz. Available online: https://www.cm-machico.pt/municipal/machico-atua/urbanismo/peru-plano-estrategico-de-reabilitacao-urbana. (accessed on 10 Janeiro 2024).

CMM. (2022). *Relatório da Estratégia Local de Habitação*. Câmara Municipal de Machico.

Coban, H. H., Lewicki, W., Miśkiewicz, R., & Drożdż, W. (2022). The Economic Dimension of Using the Integration of Highway Sound Screens with Solar Panels in the Process of Generating Green Energy. *Energies*, *16*(1), 178. DOI: 10.3390/en16010178

Cobîrzan, N., Muntean, R., Thalmaier, G., & Felseghi, R. A. (2022). Recycling of mining waste in the production of masonry units. *Materials (Basel)*, *15*(2), 594. DOI: 10.3390/ma15020594 PMID: 35057312

Coelho, A., & De Brito, J. (2012). Influence of construction and demolition waste management on the environmental impact of buildings. *Waste Management (New York, N.Y.)*, *32*(3), 532–541. DOI: 10.1016/j.wasman.2011.11.011 PMID: 22182407

Colangelo, F., Farina, I., Travaglioni, M., Salzano, C., Cioffi, R., & Petrillo, A. (2021). Innovative materials in Italy for eco-friendly and sustainable buildings. *Materials (Basel)*, *14*(8), 2048. DOI: 10.3390/ma14082048 PMID: 33921708

Cole, L. B. (2019). Green building literacy: A framework for advancing green building education. *International Journal of STEM Education*, *6*(1), 1–13. DOI: 10.1186/s40594-019-0171-6

Corominas, L., Byrne, D. M., Guest, J. S., Hospido, A., Roux, P., Shaw, A., & Short, M. D. (2020). The application of life cycle assessment (LCA) to wastewater treatment: A best practice guide and critical review. *Water Research*, *184*, 116058. DOI: 10.1016/j.watres.2020.116058 PMID: 32771688

Cossel, A. (2019). Analytical solutions for thermo-mechanical soil structure interaction in energy piles (Doctoral dissertation).

Coughlan, M., Long, M., & Doherty, P. (2020). Geological and geotechnical constraints in the Irish Sea for offshore renewable energy. *Journal of Maps*, *16*(2), 420–431. DOI: 10.1080/17445647.2020.1758811

Cox, W. (2022). *Demographia International Housing Affordability*. Urban Reform Institute & Frontier Centre for Public Policy.

Cunha, R. P., & Bourne-Webb, P. J. (2022). A critical review on the current knowledge of geothermal energy piles to sustainably climatize buildings. *Renewable & Sustainable Energy Reviews*, *158*, 112072. DOI: 10.1016/j.rser.2022.112072

Dacquay, C., Holländer, H. M., Kavgic, M., Maghoul, P., Liu, H., & Fujii, H. (2020). Evaluation of an integrated sewage pipe with ground heat exchanger for long-term efficiency estimation. *Geothermics*, *86*, 101796. DOI: 10.1016/j.geothermics.2019.101796

Dai, H. N., Zheng, Z., & Zhang, Y. (2019). Blockchain for Internet of Things: A survey. *IEEE Internet of Things Journal*, *6*(5), 8076–8094. DOI: 10.1109/JIOT.2019.2920987

Dalal, P. H., Patil, M., Wanare, R., Dave, T. N., & Iyer, K. K. (2021). Recent innovations and practices in geotechnical engineering for sustainable infrastructure development. Geohazard Mitigation: Select Proceedings of VCDRR 2021, 431-445. DOI: 10.1007/978-981-16-6140-2_34

Damman, S., Schmuck, A., Oliveira, R., Koop, S. H. A., Almeida, M. C., Alegre, H., & Ugarelli, R. M. (2023). Towards a water-smart society: Progress in linking theory and practice. *Utilities Policy*, *85*, 101674. DOI: 10.1016/j.jup.2023.101674

Darko, A., Chan, A. P., Huo, X., & Owusu-Manu, D. G. (2019). A scientometric analysis and visualization of global green building research. *Building and Environment*, *149*, 501–511. DOI: 10.1016/j.buildenv.2018.12.059

Davies, S. R., Halpern, M., Horst, M., Kirby, D. A., & Lewenstein, B. (2019). Science stories as culture: Experience, identity, narrative and emotion in public communication of science. *JCOM, Journal of Science Communication*, *18*(5), A01. Advance online publication. DOI: 10.22323/2.18050201

de Macedo, L. S. V., Picavet, M. E. B., de Oliveira, J. A. P., & Shih, W. Y. (2021). Urban green and blue infrastructure: A critical analysis of research on developing countries. *Journal of Cleaner Production*, *313*, 127898. DOI: 10.1016/j.jclepro.2021.127898

Dehghani, M. H., Karri, R. R., Tyagi, I., & Scholz, M. (Eds.). (2023). *Water, the Environment, and the Sustainable Development Goals*. Elsevier.

Deka, M. (2018) Water Sustainability, Notion press publication.

Delil, A. A. M., Baturkin, V., Gorbenko, G., Gakal, P., & Ruzaykin, V. (2002). Modeling of a miniature loop heat pipe with a flat evaporator. *SAE Technical Paper* 2002-01-2506. https://doi.org/DOI: 10.4271/2002-01-2506

DiCarlo, M., Berglund, E. Z., Kaza, N., Grieshop, A., Shealy, L., & Behr, A. (2023). Customer complaint management and smart technology adoption by community water systems. *Utilities Policy*, *80*, 101465. DOI: 10.1016/j.jup.2022.101465

Direção Regional de Estatística da Madeira (DREM). (2022a). Estatísticas da Construção e Habitação da Região Autónoma da Madeira 2022. Funchal: DREM. Available online: https://estatistica.madeira.gov.pt/ (accessed on 27 Maio 2023).

Direção Regional de Estatística da Madeira (DREM). (2022b). Madeira em Números da Região Autónoma da Madeira 2022. Funchal: DREM. Available online: https://estatistica.madeira.gov.pt/ (accessed on 27 Maio 2024).

Dorairaj, D., & Osman, N. (2021). Present practices and emerging opportunities in bioengineering for slope stabilization in Malaysia: An overview. *PeerJ*, *9*, e10477. DOI: 10.7717/peerj.10477 PMID: 33520435

Do, T. A., Hoang, T. T., Bui-Tien, T., Hoang, H. V., Do, T. D., & Nguyen, P. A. (2020). Evaluation of heat of hydration, temperature evolution and thermal cracking risk in high-strength concrete at early ages. *Case Studies in Thermal Engineering, 21*, 100658. DOI: 10.1016/j.csite.2020.100658

Drechsler, M., & Graham, A. (2005, October). Geopolymers-an innovative materials technology bringing resource sustainability to construction and mining industries. In *Proceedings of the IQA Annual Conference* (pp. 12-15).

Dulguun, E. (2023). Implementation of cutting-edge technologies in facility management challenges and opportunities in the digital era. *Journal Dimensie Management and Public Sector, 4*(4), 24–33. DOI: 10.48173/jdmps.v4i4.243

Duman, A. C., Erden, H. S., Gönül, Ö., & Güler, Ö. (2021). A home energy management system with an integrated smart thermostat for demand response in smart grids. *Sustainable Cities and Society, 65*, 102639. DOI: 10.1016/j.scs.2020.102639

Durgun, Y. (2024). Real-time water quality monitoring using AI-enabled sensors: Detection of contaminants and UV disinfection analysis in smart urban water systems. *Journal of King Saud University. Science, 36*(9), 103409. DOI: 10.1016/j.jksus.2024.103409

Dürrenmatt, D. J., & Wanner, O. (2014). A mathematical model to predict the effect of heat recovery on the wastewater temperature in sewers. *Water Research, 48*, 548–558. DOI: 10.1016/j.watres.2013.10.017 PMID: 24216228

Eichholz, L. (2023). The Implications of a Co-Created Software Solution for Mobility in Rural Areas. *Smart Cities, 2023*(6), 2706–2721. DOI: 10.3390/smartcities6050122

ELD. (2014). Estratégia Local de Desenvolvimento para o periodo 2014-2020.

Elkhapery, B., Kianmehr, P., & Doczy, R. (2021). Benefits of retrofitting school buildings in accordance to LEED v4. *Journal of Building Engineering, 33*, 101798. DOI: 10.1016/j.jobe.2020.101798

Eslamnoor, N., & Vural, S. (2021). User behavior and energy efficiency in north cyprus university dormitories: A case study. *Journal of Housing and the Built Environment, 37*(3), 1159–1178. DOI: 10.1007/s10901-021-09878-1

Esteban-Navarro, M.-Á., García-Madurga, M.-Á., Morte-Nadal, T., & Nogales-Bocio, A.-I. (2020). The Rural Digital Divide in the Face of the COVID-19 Pandemic in Europe—Recommendations from a Scoping Review. *Informatics (MDPI), 2020*(7), 54. DOI: 10.3390/informatics7040054

Et-taibi, B., Abid, M. R., Boufounas, E.-M., Morchid, A., Bourhnane, S., Abu Hamed, T., & Benhassou, D. (2024). Enhancing water management in smart agriculture: A cloud and IoT-Based smart irrigation system. *Results in Engineering*, *22*, 102283. DOI: 10.1016/j.rineng.2024.102283

Fan, Y., & Xia, X. (2018). Energy-efficiency building retrofit planning for green building compliance Author links open overlay panel. *Building and Environment*, *136*, 312–321. DOI: 10.1016/j.buildenv.2018.03.044

Farhadi, A., & Iqbal, M. T. (2024). Optimal sizing and techno-economic analysis of a hybrid power system for postville. *European Journal of Engineering and Technology Research*, *9*(2), 5–12. DOI: 10.24018/ejeng.2024.9.2.3127

Farhat, O., Faraj, J., Hachem, F., Castelain, C., & Khaled, M. (2022). A recent review on waste heat recovery methodologies and applications: Comprehensive review, critical analysis and potential recommendations. *Cleaner Engineering and Technology*, *6*, 100387. DOI: 10.1016/j.clet.2021.100387

Farrington, J., & Farrington, C. (2005). Rural accessibility, social inclusion, and social justice: Towards conceptualisation. *Journal of Transport Geography*, *13*(1), 1–12. DOI: 10.1016/j.jtrangeo.2004.10.002

Fauzi, R. T., Lavoie, P., Sorelli, L., Heidari, M. D., & Amor, B. (2019). Exploring the current challenges and opportunities of life cycle sustainability assessment. *Sustainability (Basel)*, *11*(3), 636. DOI: 10.3390/su11030636

Fei, W., Opoku, A., Agyekum, K., Oppon, J. A., Ahmed, V., Chen, C., & Lok, K. L. (2021). The critical role of the construction industry in achieving the sustainable development goals (SDGs): Delivering projects for the common good. *Sustainability (Basel)*, *13*(16), 9112. DOI: 10.3390/su13169112

Fet, J. (2024, June 10). Civil vs. Transportation Engineering: Infrastructure Insights. Site Name, i.e. Jain University. https://set.jainuniversity.ac.in/blogs/civil-engineering-vs-transportation-engineering-infrastructure

Fields, N. L., Miller, V. J., Cronley, C., Hyun, K. K., Mattingly, S. P., Khademi, S., Nargesi, S. R. R., & Williams, J. (2020). Interprofessional collaboration to promote transportation equity for environmental justice populations: A mixed methods study of civil engineers, transportation planners, and social workers' perspectives. *Transportation Research Interdisciplinary Perspectives*, *5*, 100110. DOI: 10.1016/j.trip.2020.100110

Figueiredo, I., Esteves, P., & Cabrita, P. (2021). Water wise – a digital water solution for smart cities and water management entities. *Procedia Computer Science*, *181*, 897–904. DOI: 10.1016/j.procs.2021.01.245

Flor, G., Sánchez-Fernández, M., Flor-Blanco, G., & de Sanjosé Blasco, J. J. (2023). Application of Geomatic Techniques for the Assessment of Anthropogenic Changes in the Urban Beaches of "La Magdalena" (Santander, Spain). *Remote Sensing (Basel)*, *15*(1), 1. Advance online publication. DOI: 10.3390/rs15010256

Fong, B., Housh, M., Hong, G. Y., & Wang, J. M. (2024). Pipeline Management Technologies for Sustainable Water Supply in a Smart City Environment. *Encyclopedia of Sustainable Technologies (Second Edition)*, Volume 3, Pages 671-679.

Forrest, R., & Hirayama, Y. (2015). The financialisation of the social project: Embedded liberalism, neoliberalism and home ownership. *Urban Studies (Edinburgh, Scotland)*, *52*(2), 233–244. DOI: 10.1177/0042098014528394

Freemark, Y., & Steil, J. (2021). Local power and the location of subsidized renters in comparative perspective: Public support for low- and moderate-income households in the United States, France, and the United Kingdom. *Housing Studies*, *0*, 1–29. DOI: 10.1080/02673037.2021.1910628

Freire, R., & Taveira-Pinto, F. (2011). *ANÁLISE DO COMPORTAMENTO HIDROMORFOLÓGICO DO QUEBRAMAR DA AGUDA. 5.as Jornadas de Hidráulica.* Recursos Hídricos e Ambiente.

Fu, C., Lu, L., & Pirabi, M. (2023). Advancing green finance: A review of sustainable development. *Digital Economy and Sustainable Development*, *1*(1), 44265. DOI: 10.1007/s44265-023-00020-3

Gao, H., Koch, C., & Wu, Y. (2019). Building information modelling based building energy modelling: A review. *Applied Energy*, *238*, 320–343. DOI: 10.1016/j.apenergy.2019.01.032

García, B. (2013). Ports: Definition and study of types, sizes and business models. *Journal of Industrial Engineering and Management*, *6*(4). Advance online publication. DOI: 10.3926/jiem.770

García-Mayor, C., Martí, P., Castaño, M., & Bernabeu-Bautista, Á. (2020). The unexploited potential of converting rail tracks to greenways: The Spanish Vias Verdes. *Sustainability (Basel)*, *12*(3), 881. DOI: 10.3390/su12030881

Garg, S. K. (2018) Environmental engineering swage waste disposal and air pollution engineering, Khanna publication.

Ghadir, P., Zamanian, M., Mahbubi-Motlagh, N., Saberian, M., Li, J., & Ranjbar, N. (2021). Shear strength and life cycle assessment of volcanic ash-based geopolymer and cement stabilized soil: A comparative study. *Transportation Geotechnics, 31*, 100639. DOI: 10.1016/j.trgeo.2021.100639

Ghosh, B., & Karmakar, S. (2024). 3d printing technology and future of construction: A review. *IOP Conference Series. Earth and Environmental Science, 1326*(1), 012001. DOI: 10.1088/1755-1315/1326/1/012001

Giupponi, L., Borgonovo, G., Giorgi, A., & Bischetti, G. B. (2019). How to renew soil bioengineering for slope stabilization: Some proposals. *Landscape and Ecological Engineering, 15*(1), 37–50. DOI: 10.1007/s11355-018-0359-9

Glavinovic, R., Vukic, L., & Peronja, I. (2023). Analysis of Management Models of Regional Ports in the Republic of Croatia. *Journal of Marine Science and Engineering, 11*(2), 332. DOI: 10.3390/jmse11020332

Glenn, O., Schwab, R.K., and Telcott, E. (1981) 'Soil and water Conservation engineering', Wiley publication.

Glina, B., Yetunde, F. M., Mendyk, L., & Piernik, A. (2024). Soil water repellency and its importance for the climate-smart sustainable management of fen peatland soils in Central Poland. *Geoderma Regional, 39*, e00867. DOI: 10.1016/j.geodrs.2024.e00867

Golzar, F., & Silveira, S. (2021). Impact of wastewater heat recovery in buildings on the performance of centralized energy recovery – A case study of Stockholm. *Applied Energy, 297*, 117141. DOI: 10.1016/j.apenergy.2021.117141

Grigg, N. (2023). *Water resource management: Principle methods and tools*. Wiley Publication.

Gude, V. G., Gadhamshetty, V., & Kandiah, R. (Eds.). (2020). *Sustainable Water: Resources, Management and Challenges*. Nova Science Publishers.

Guler, H. G., Arikawa, T., Oei, T., & Yalciner, A. C. (2015). Performance of rubble mound breakwaters under tsunami attack, a case study: Haydarpasa Port, Istanbul, Turkey. *Coastal Engineering, 104*, 43–53. DOI: 10.1016/j.coastaleng.2015.07.007

Gumz, J., & Fettermann, D. C. (2024). User's perspective in smart meter research: State-of-the-art and future trends. *Energy and Building, 308*, 114025. DOI: 10.1016/j.enbuild.2024.114025

Guo, J., Zhou, M., Wang, C., & Mei, L. (2012). The application of the model of coordinate S-transformation for stability analysis of datum points in high-precision GPS deformation monitoring networks. *Journal of Applied Geodesy, 6*(3–4), 143–148. DOI: 10.1515/jag-2012-0028

Haffner, M. E. A., & Hulse, K. (2019). A fresh look at contemporary perspectives on urban housing affordability. *International Journal of Urban Sciences, 25*(sup1), 59–79. https://doi.org/.DOI: 10.1080/12265934.2019.1687320

Hajibeigi, P., Pazhouhanfar, M., Grahn, P., & Nazif, H. (2023). Enhancing citizens' perceived restoration potential of green facades through specific architectural attributes. *Buildings (Basel, Switzerland), 13*(9), 2356. DOI: 10.3390/buildings13092356

Halicioglu, F. H., & Gurel, K. (2023). The perceived quality of green buildings: From a review of occupant satisfaction towards a conceptual framework. *International Journal of Quality &Amp. International Journal of Quality & Reliability Management, 41*(3), 757–777. DOI: 10.1108/IJQRM-10-2022-0307

Hamim, O. F., Hoque, M. S., McIlroy, R. C., Plant, K. L., & Stanton, N. A. (2020). A sociotechnical approach to accident analysis in a low-income setting: Using Accimaps to guide road safety recommendations in Bangladesh. *Safety Science, 124*, 104589. DOI: 10.1016/j.ssci.2019.104589

Hanzawa, M., Matsumoto, A., & Tanaka, H. (2012). STABILITY OF WAVE-DISSIPATING CONCRETE BLOCKS OF DETACHED BREAKWATERS AGAINST TSUNAMI. *Coastal Engineering Proceedings, 33*(33), 33. Advance online publication. DOI: 10.9753/icce.v33.structures.24

Hardjito, D., Wallah, S. E., Sumajouw, D. M., & Rangan, B. V. (2004). On the development of fly ash-based geopolymer concrete. *Materials Journal, 101*(6), 467–472.

Harris, S., Weinzettel, J., Bigano, A., & Källmén, A. (2020). Low carbon cities in 2050? GHG emissions of European cities using production-based and consumption-based emission accounting methods. *Journal of Cleaner Production, 248*, 119206. DOI: 10.1016/j.jclepro.2019.119206

Hassebo, A., & Tealab, M. (2023). Global models of smart cities and potential iot applications: A review. *IoT, 4*(3), 366–411. DOI: 10.3390/iot4030017

He, K., & Ye, J. (2021). Physical Modeling of the Stability of a Revetment Breakwater Built on Reclaimed Coral Calcareous Sand Foundation in the South China Sea—Regular Wave. *Applied Sciences (Basel, Switzerland), 11*(5), 5. Advance online publication. DOI: 10.3390/app11052325

Helser, J., Perumal, P., & Cappuyns, V. (2022). Valorizing (cleaned) sulfidic mine waste as a resource for construction materials. *Journal of Environmental Management, 319*, 115742. DOI: 10.1016/j.jenvman.2022.115742 PMID: 35849929

Henderson, J. V., & Turner, M. A. (2020). Urbanization in the developing world: Too early or too slow? *The Journal of Economic Perspectives*, *34*(3), 150–173. DOI: 10.1257/jep.34.3.150

Herath, N., Vaz-Serra, P., Hui, F. K. P., Mendis, P., & Aye, L. (2023). Risk Mitigation Measures in Green Building Projects: *An Investigation. In International Conference on Sustainable Built Environment* (pp. 277-289). Springer. DOI: 10.1007/978-981-99-3471-3_20

Holladay, M. (2023, May 26). *Interview with Ann Deminster*. From Green Building Advisor: https://www.greenbuildingadvisor.com/article/interview-with-ann-edminster

Hosseinzadeh-Bandbafha, H., Nizami, A. S., Kalogirou, S. A., Gupta, V. K., Park, Y. K., Fallahi, A., Sulaiman, A., Ranjbari, M., Rahnama, H., Aghbashlo, M., Peng, W., & Tabatabaei, M. (2022). Environmental life cycle assessment of biodiesel production from waste cooking oil: A systematic review. *Renewable & Sustainable Energy Reviews*, *161*, 112411. DOI: 10.1016/j.rser.2022.112411

Hsien, C., Low, J. S. C., Fuchen, S. C., & Han, T. W. (2019). Life cycle assessment of water supply in Singapore—A water-scarce urban city with multiple water sources. *Resources, Conservation and Recycling*, *151*, 104476. DOI: 10.1016/j.resconrec.2019.104476

Htet, A., Liana, S. R., Aung, T., & Bahumik, A. (2023). The Intersection of Facade Engineering and Building Information Modeling: Opportunities and Challenges. *Journal of Technology Innovations and Energy*, *2*(3), 94–110. DOI: 10.56556/jtie.v2i3.643

Htet, A., Liana, S. R., Aung, T., & Bhaumik, A. (2023). Economic and Environmental Impact Analysis of Green Building Materials in Facade Engineering. *International Journal of Membrane Science and Technology*, *10*(2), 2805–2814. DOI: 10.15379/ijmst.v10i2.2973

Huang-Lachmann, J. T., & Lovett, J. C. (2016). How cities prepare for climate change. *Cities (London, England)*, *54*, 36–44. DOI: 10.1016/j.cities.2015.11.001

Hur, D.-S., Kim, C.-H., Kim, D.-S., & Yoon, J.-S. (2008). Simulation of the nonlinear dynamic interactions between waves, a submerged breakwater and the seabed. *Ocean Engineering*, *35*(5), 511–522. DOI: 10.1016/j.oceaneng.2007.12.002

Hu, X., Li, Y., Sun, Y., Zhang, C., Ding, W., & Deng, J. (2022). Water permits trading framework for urban water demand management based on smart metering. *Journal of Environmental Management, 304*, 114208. DOI: 10.1016/j.jenvman.2021.114208 PMID: 34891052

Ibrahim, M. I. M. (2016). Estimating the sustainability returns of recycling construction waste from building projects. *Sustainable Cities and Society, 23*, 78–93. DOI: 10.1016/j.scs.2016.03.005

Igos, E., Benetto, E., Meyer, R., Baustert, P., & Othoniel, B. (2019). How to treat uncertainties in life cycle assessment studies? *The International Journal of Life Cycle Assessment, 24*(4), 794–807. DOI: 10.1007/s11367-018-1477-1

ILFI. (2022). *Certified Case Studies*. From International Living Future Institute: https://living-future.org/program/lbc/

Ilia, A., & O'Donnell, J. (2018). An Assessment of Two Models of Wave Propagation in an Estuary Protected by Breakwaters. *Journal of Marine Science and Engineering, 6*(4), 4. Advance online publication. DOI: 10.3390/jmse6040145

Instituto Nacional de Estatística (INE). (2021). Censos 2021. Available online: https://censos.ine.pt/xportal/xmain?xpid=CENSOS&xpgid=censos_quadro&QUADRO=1&selTab=tab1 (accessed on 7 Junho 2024).

Izvoltova, J., Villim, A., & Kozak, P. (2014). Determination of Geometrical Track Position by Robotic Total Station. *Procedia Engineering, 91*, 322–327. DOI: 10.1016/j.proeng.2014.12.068

Jalaei, F., Jalaei, F., & Mohammadi, S. (2020). An integrated BIM-LEED application to automate sustainable design assessment framework at the conceptual stage of building projects. *Sustainable Cities and Society, 53*, 101979. DOI: 10.1016/j.scs.2019.101979

Janjua, S., Sarker, P., & Biswas, W. (2019). Sustainability assessment of a residential building using a life cycle assessment approach. *Chemical Engineering Transactions, 72*, 19–24. DOI: 10.3303/CET1972004

Jankauskiene, D., Dorosevas, V., & Pukite, V. (2021). Monitoring and Analysis of Seaport Piers Structural Elements Displacements. MAPAN, 1-6.

Jankauskiene, D., Dorosevas, V., & Volkovas, V. (2018). The crack analysis in monitoring procedures of Seaport pier concrete constructions. *Journal of Measurements in Engineering, 6*(1), 1. Advance online publication. DOI: 10.21595/jme.2018.19763

Janker, J., & Mann, S. (2020). Understanding the social dimension of sustainability in agriculture: A critical review of sustainability assessment tools. *Environment, Development and Sustainability*, *22*(3), 1671–1691. DOI: 10.1007/s10668-018-0282-0

Jankovic, S. D., & Curovic, D. M. (2023). Strategic integration of artificial intelligence for sustainable businesses: Implications for data management and human user engagement in the digital era. *Sustainability (Basel)*, *15*(21), 15208. DOI: 10.3390/su152115208

Jayakody, S., Gallage, C., & Ramanujam, J. (2019). Performance characteristics of recycled concrete aggregate as an unbound pavement material. *Heliyon*, *5*(9), e02494. Advance online publication. DOI: 10.1016/j.heliyon.2019.e02494 PMID: 31687586

Jeng, D. S., Cha, D. H., Lin, Y. S., & Hu, P. S. (2001). Wave-induced pore pressure around a composite breakwater. *Ocean Engineering*, *28*(10), 1413–1435. DOI: 10.1016/S0029-8018(00)00059-7

Jesper, M., Schlosser, F., Pag, F., Walmsley, T. G., Schmitt, B., & Vajen, K. (2021). Large-scale heat pumps: Uptake and performance modelling of market-available devices. *Renewable & Sustainable Energy Reviews*, *137*, 110646. DOI: 10.1016/j.rser.2020.110646

Jiang, N. J., Tang, C. S., Hata, T., Courcelles, B., Dawoud, O., & Singh, D. N. (2020). Bio-mediated soil improvement: The way forward. *Soil Use and Management*, *36*(2), 185–188. DOI: 10.1111/sum.12571

Jimenez, A.-F., Cardenas, P.-F., & Jimenez, F. (2021). Smart water management approach for resource allocation in High-Scale irrigation systems. *Agricultural Water Management*, *256*, 107088. DOI: 10.1016/j.agwat.2021.107088

Jin, R., Yuan, H., & Chen, Q. (2019). Science mapping approach to assisting the review of construction and demolition waste management research published between 2009 and 2018. *Resources, Conservation and Recycling*, *140*, 175–188. DOI: 10.1016/j.resconrec.2018.09.029

Joensuu, T., Edelman, H., & Saari, A. (2020). Circular economy practices in the built environment. *Journal of Cleaner Production*, *276*, 124215. DOI: 10.1016/j.jclepro.2020.124215

Joseph, B. (2017) Environmental science and engineering, Mc Grew Hill publication.

Joyce, K., Koehler, M., & Vogel, M. (undated). *Trimble Engineering and Construction Group Westminster, Colorado, USA*.

Kaab, A., Sharifi, M., Mobli, H., Nabavi-Pelesaraei, A., & Chau, K. W. (2019). Combined life cycle assessment and artificial intelligence for prediction of output energy and environmental impacts of sugarcane production. *The Science of the Total Environment*, *664*, 1005–1019. DOI: 10.1016/j.scitotenv.2019.02.004 PMID: 30769303

Kakkos, E., Heisel, F., Hebel, D. E., & Hischier, R. (2020). Towards urban mining—Estimating the potential environmental benefits by applying an alternative construction practice. A case study from Switzerland. *Sustainability (Basel)*, *12*(12), 5041. DOI: 10.3390/su12125041

Kaloop, M., Beshr, A. A. A., & Elshiekh, M. Y. (2008). *Using Total Station for monitoring the deformation of high strength concrete beams*. https://www.semanticscholar.org/paper/Using-Total-Station-for-monitoring-the-deformation-Kaloop-Beshr/f8b73755dfb4fad206da019c8cfda66753e557a3

Kamal, M. A., & Ahmed, E. (2023). Analyzing energy efficient design strategies in high-rise buildings with reference to hvac system. *Architecture Engineering and Science*, *4*(3), 183. DOI: 10.32629/aes.v4i3.1275

Kanyama, M. N., Shava, F. B., Gamundani, A. M., & Hartmann, A. (2024). Anomalies identification in Smart Water Metering Networks: Fostering improved water efficiency. *Physics and Chemistry of the Earth Parts A/B/C*, *134*, 103592. DOI: 10.1016/j.pce.2024.103592

Kanyama, M. N., Shava, F. B., Gamundani, A. M., & Hartmann, A. (2024). Machine learning applications for anomaly detection in Smart Water Metering Networks: A systematic review. *Physics and Chemistry of the Earth Parts A/B/C*, *134*, 103558. DOI: 10.1016/j.pce.2024.103558

Kapustenko, P., Arsenyeva, O., Fedorenko, O., & Kusakov, S. (2022). Integration of low-grade heat from exhaust gases into energy system of the enterprise. *Clean Technologies and Environmental Policy*, *24*(1), 67–76. DOI: 10.1007/s10098-021-02082-3

Karami, Z., & Kashef, R. (2020). Smart transportation planning: Data, models, and algorithms. *Transportation Engineering*, *2*, 100013. DOI: 10.1016/j.treng.2020.100013

Kasiulis, E., Punys, P., & Kofoed, J. P. (2015). Assessment of theoretical near-shore wave power potential along the Lithuanian coast of the Baltic Sea. *Renewable & Sustainable Energy Reviews*, *41*, 134–142. DOI: 10.1016/j.rser.2014.08.044

Kavya, M., Mathew, A., Shekar, P. R., & P, S. (2023). Short term water demand forecast modelling using artificial intelligence for smart water management. *Sustainable Cities and Society*, *95*, 104610. DOI: 10.1016/j.scs.2023.104610

Kermanshachi, S., Safapour, E., Anderson, S., Goodrum, P., Taylor, T., & Sadatsafavi, H. (2018). Exploring current scoping practices used in the development of transportation infrastructure projects. CSCE 12th International Transportation Specialty.

Khachatryan, H., Suh, D. H., Xu, W., Useche, P., & Dukes, M. D. (2019). Towards sustainable water management: Preferences and willingness to pay for smart landscape irrigation technologies. *Land Use Policy*, *85*, 33–41. DOI: 10.1016/j.landusepol.2019.03.014

Khandelwal, H., Dhar, H., Thalla, A. K., & Kumar, S. (2019). Application of life cycle assessment in municipal solid waste management: A worldwide critical review. *Journal of Cleaner Production*, *209*, 630–654. DOI: 10.1016/j.jclepro.2018.10.233

Khan, M., Salahuddin, T., Khan, Q., Al Alwan, B., & Almesfer, M. (2022). Numerical study of binary mixture and thermal analysis near a solar radiative heated surface. *Solar Energy*, *231*, 262–269. DOI: 10.1016/j.solener.2021.11.003

Kibert, C. (2016). *Kibert, C. J. (2016). Sustainable construction: green building design and delivery*. John Wiley & Sons.

Kibert, C. J. (2016). *Sustainable construction: green building design and delivery*. John Wiley & Sons.

Kim, A. A., Sadatsafavi, H., Anderson, S. D., & Bishop, P. (2017). Preparing for the future of transportation construction: Strategies for state transportation agencies. *Journal of Management Engineering*, *33*(3), 04016045. DOI: 10.1061/(ASCE)ME.1943-5479.0000494

Kim, Y. U., Yang, S., Yun, B. Y., & Kim, S. (2022). Evaluation of energy consumption in apartment buildings with biochar and phase-change material aggregate-applied artificial stone finishing materials. *International Journal of Energy Research*, *46*(9), 12772–12786. DOI: 10.1002/er.8043

Kim, Y., Kim, D., Park, K., & Kim, D. (2014). A Study on Analysis of Industrial Injury Characteristics of Aging Workers in Agriculture. *Journal of Ergonomics Society of Korea*, *33*(6), 477–486. DOI: 10.5143/JESK.2014.33.6.477

Kivirenko, O., Kostenko, S., Sorokin, V., Tsegelnyk, Y., Plankovskyy, S., & Kombarov, V. (2023). Fiberglass pipeline continuous filament winding automation. *International Journal of Mechatronics and Applied Mechanics*, *2023*(14), 241–247. DOI: 10.17683/ijomam/issue14.28

Klang, A., Vikman, P. Å., & Brattebø, H. (2003). Sustainable management of demolition waste—An integrated model for the evaluation of environmental, economic and social aspects. *Resources, Conservation and Recycling*, *38*(4), 317–334. DOI: 10.1016/S0921-3449(02)00167-2

Klemes, J. J., Arsenyeva, O., Kapustenko, P., & Tovazhnyanskyy, L. (2015). *Compact Heat Exchangers for Energy Transfer Intensification: Low Grade Heat and Fouling Mitigation*. CRC Press. DOI: 10.1201/b18862

Kombarov, V., Kryzhyvets, Y., Biletskyi, I., Tsegelnyk, Y., Aksonov, Y., & Piddubna, L. (2021). *Numerical control of fiberglass pipe bends manufacturing. In 2021 IEEE 2nd KhPI Week on Advanced Technology (KhPIWeek)*. IEEE., DOI: 10.1109/KhPIWeek53812.2021.9570068

Koreta, O. (n.d.). *Maritime structures — Part 7: Guide to the design and construction of breakwaters*. Available at https://www.academia.edu/37123655/Maritime_structures_Part_7_Guide_to_the_design_and_construction_of_breakwaters

Korkmaz, S., Horman, M., & Riley, D. (2009). Key attributes of a longitudinal study of green project delivery. *In Construction Research Congress 2009: Building a Sustainable Future* (pp. 558-567.). ASCE.DOI: 10.1061/41020(339)57

Kovačič, B., Kamnik, R., Pustovgar, A., & Vatin, N. (2016). Analysis of Precision of Geodetic Instruments for Investigating Vertical Displacement of Structures. *Procedia Engineering*, *165*, 906–917. DOI: 10.1016/j.proeng.2016.11.800

Kretschmer, F., Simperler, L., & Ertl, T. (2016). Analysing wastewater temperature development in a sewer system as a basis for the evaluation of wastewater heat recovery potentials. *Energy and Building*, *128*, 639–648. DOI: 10.1016/j.enbuild.2016.07.024

Krisnawijaya, N. N. K., Tekinerdogan, B., Catal, C., & van der Tol, R. (2024). Reference architecture design for developing data management systems in smart farming. *Ecological Informatics*, *81*, 102613. DOI: 10.1016/j.ecoinf.2024.102613

Krogness, C., Britt, K., & Owen, M. (2020). Trenchless structural rehabilitation of non-circular pipe in congested utility corridor of commercial business district. In Pulido, J. F., & Poppe, M. (Eds.), *Pipelines 2020* (pp. 466–474). American Society of Civil Engineers., DOI: 10.1061/9780784483206.053

Kugler, T., Schittenhelm, C., Volkmer, S., Ryba, M., Moormann, C., Kurth, D., & Koenigsdorff, R. (2022). Sustainable heating and cooling management of urban quarters. *Sustainability (Basel)*, *14*(7), 4353. DOI: 10.3390/su14074353

Kumar, R., Singh, R., Agrawal, A., & A. (2024). Chapter 5 - Global scenario of water footprints in smart cities. *Current Directions in Water Scarcity Research*, Volume 8, Pages 63-68.

Kumari, W. G. P., & Ranjith, P. G. (2019). Sustainable development of enhanced geothermal systems based on geotechnical research–A review. *Earth-Science Reviews, 199*, 102955. DOI: 10.1016/j.earscirev.2019.102955

Kumar, P., Nigam, G. K., Sinha, M. K., & Singh, A. (Eds.). (2022). *Water Resources Management and Sustainability*. Springer Singapore. DOI: 10.1007/978-981-16-6573-8

Kuranchie, F. A., Shukla, S. K., & Habibi, D. (2013). Mine wastes in Western Australia and their suitability for embankment construction. In Geo-Congress 2013: Stability and Performance of Slopes and Embankments III (pp. 1443-1452). DOI: 10.1061/9780784412787.145

Kuranchie, F. A., Shukla, S. K., & Habibi, D. (2016). Utilisation of iron ore mine tailings for the production of geopolymer bricks. *International Journal of Mining, Reclamation and Environment, 30*(2), 92–114. DOI: 10.1080/17480930.2014.993834

Lai, T. L., Robbins, H., & Wei, C. Z. (1979). Strong consistency of least squares estimates in multiple regression II. *Journal of Multivariate Analysis, 9*(3), 343–361. DOI: 10.1016/0047-259X(79)90093-9

Lamrani, B., Johannes, K., & Kuznik, F. (2021). Phase change materials integrated into building walls: An updated review. *Renewable & Sustainable Energy Reviews, 140*, 110751. DOI: 10.1016/j.rser.2021.110751

Laureano, P. (2013) 'Water conservation techniques in traditional settlements' Copal publishing group publication.

Laurent, A., Weidema, B. P., Bare, J., Liao, X., Maia de Souza, D., Pizzol, M., Sala, S., Schreiber, H., Thonemann, N., & Verones, F. (2020). Methodological review and detailed guidance for the life cycle interpretation phase. *Journal of Industrial Ecology, 24*(5), 986–1003. DOI: 10.1111/jiec.13012 PMID: 33746505

Lebedeva, R., Skripkiūnas, G., & Vasiljeva, L. (2012). The effects of seawater on the durability of hydrotechnical (concrete) structures in the port of Klaipeda. *Engineering Structures and Technologies, 4*(4), 111–118. DOI: 10.3846/2029882X.2012.753745

Lee, D., & Hess, D. J. (2020). Regulations for on-road testing of connected and automated vehicles: Assessing the potential for global safety harmonization. *Transportation Research Part A, Policy and Practice, 136*, 85–98. DOI: 10.1016/j.tra.2020.03.026

Lee, Y., Kemp, P. A., & Reina, V. J. (2022). Drivers of housing (un)affordability in the advanced economies: A review and new evidence. *Housing Studies*, *37*(10), 1739–1752. DOI: 10.1080/02673037.2022.2123623

Lennon, A., & Berg, N. G. (2022). Alternative places for alternative people? A changing ecovillage discourse from Othered lifestyle to another rurality. *Journal of Rural Studies*, *95*, 302–315. DOI: 10.1016/j.jrurstud.2022.09.024

Lephondo, I., Telukdarie, A., Munien, I., Onkonkwo, U., & Vermeulen, A. (2024). The Outcomes of Smart Irrigation System using Machine Learning to minimize water usage within the Agriculture Sector. *Procedia Computer Science*, *237*, 525–532. DOI: 10.1016/j.procs.2024.05.136

Li, M., Zhou, S., Shen, S., Wang, J., Yang, Y., Wu, Y., Chen, F., & Lei, Y. (2024). Climate-smart irrigation strategy can mitigate agricultural water consumption while ensuring food security under a changing climate. *Agricultural Water Management*, *292*, 108663. DOI: 10.1016/j.agwat.2023.108663

Lima, L., Trindade, E., Alencar, L., Alencar, M., & Silva, L. (2021). Sustainability in the construction industry: A systematic review of the literature. *Journal of Cleaner Production*, *289*, 125730. DOI: 10.1016/j.jclepro.2020.125730

Li, S. (2024). Sustainable material selection in architectural design. *Advances in Social Science, Education and Humanities Research*, *4*(10), 38476. DOI: 10.2991/978-2-38476-222-4_2

Li, S., Lu, Y., Kua, H. W., & Chang, R. (2020). The economics of green buildings: A life cycle cost analysis of non-residential buildings in tropic climates. *Journal of Cleaner Production*, *252*, 119771. DOI: 10.1016/j.jclepro.2019.119771

Li, S., Yu, L., Jiang, W., Yu, H., & Wang, X. (2022). The recent progress China has made in green mine construction, Part I: Mining groundwater pollution and sustainable mining. *International Journal of Environmental Research and Public Health*, *19*(9), 5673. DOI: 10.3390/ijerph19095673 PMID: 35565067

Li, S., Zheng, X., & Zeng, Q. (2023). Can green finance drive the development of the green building industry?—Based on the evolutionary game theory. *Sustainability (Basel)*, *15*(17), 13134. DOI: 10.3390/su151713134

Liu, Z., & Guo, A. (2021). Application of green building materials and multi-objective energy-saving optimization design. *International Journal of Heat and Technology, 39(1). International Journal of Heat and Technology*, *39*(1), 299–308. DOI: 10.18280/ijht.390133

Li, W., Wang, F., Liu, T., Xue, Q., & Liu, N. (2023). Peer effects of digital innovation behavior: An external environment perspective. *Management Decision*, *61*(7), 2173–2200. DOI: 10.1108/MD-06-2022-0865

Li, Y., Song, H., Sang, P., Chen, P. H., & Liu, X. (2019). Review of Critical Success Factors (CSFs) for green building projects. *Building and Environment*, *158*, 182–191. DOI: 10.1016/j.buildenv.2019.05.020

Li, Z., Ding, Z., & Zhang, Y. (2004, May). Development of sustainable cementitious materials. In *Proceedings of international workshop on sustainable development and concrete technology*, Beijing, China (pp. 55-76).

López, C. D., Carpio, M., Martín-Morales, M., & Zamorano, M. (2019). A comparative analysis of sustainable building assessment methods. *Sustainable Cities and Society*, *49*, 101611. DOI: 10.1016/j.scs.2019.101611

Lousada, S., Alves, R., & Fernandes, M. (2024). The impact of the Madeira Airport expansion on tourism and sustainable development in the city of Machico. *XV International Tourism Congress*. 2024. DOI: https://doi.org/DOI: 10.2478/9788367405287-019

Lousada, S., & Castanho, R. A. (2022). The Role of Ports in Tourism: Porto Santo Harbour. *Water (Basel)*, *14*(19), 19. Advance online publication. DOI: 10.3390/w14193176

Lousada, S., & Castanho, R. A. (2023). Port Structures, Maritime Transport, and Tourism. *Water, 15*(22). *Water (Basel)*, *22*(22), 3898. Advance online publication. DOI: 10.3390/w15223898

Louviere, J. J., Hensher, D. A., & Swait, J. D. (2000). *Stated choice methods: analysis and applications*. Cambridge University Press. DOI: 10.1017/CBO9780511753831

Lu, C., Wang, S., Wang, K., Gao, Y., & Zhang, R. (2020). Uncovering the benefits of integrating industrial symbiosis and urban symbiosis targeting a resource-dependent city: A case study of Yongcheng, China. *Journal of Cleaner Production*, *255*, 120210. DOI: 10.1016/j.jclepro.2020.120210

Lü, G., Batty, M., Strobl, J., Lin, H., Zhu, A. X., & Chen, M. (2019). Reflections and speculations on the progress in Geographic Information Systems (GIS): A geographic perspective. *International Journal of Geographical Information Science*, *33*(2), 346–367. DOI: 10.1080/13658816.2018.1533136

Lu, K., Jiang, X., Yu, J., Tam, V. W., & Skitmore, M. (2021). Integration of life cycle assessment and life cycle cost using building information modeling: A critical review. *Journal of Cleaner Production*, *285*, 125438. DOI: 10.1016/j.jclepro.2020.125438

Luo, H., Liu, J., Fang, W., Love, P. E., Yu, Q., & Lu, Z. (2020). Real-time smart video surveillance to manage safety: A case study of a transport mega-project. *Advanced Engineering Informatics*, *45*, 101100. DOI: 10.1016/j.aei.2020.101100

Lu, Y., Chang, R., Shabunko, V., & Lay Yee, A. T. (2019). he implementation of building-integrated photovoltaics in Singapore: Drivers versus barriers. *Energy*, *168*, 400–408. DOI: 10.1016/j.energy.2018.11.099

Lv, Z., Kong, G., Liu, H., & Ng, C. W. (2020). Effects of soil type on axial and radial thermal responses of field-scale energy piles. *Journal of Geotechnical and Geoenvironmental Engineering*, *146*(10), 06020018. DOI: 10.1061/(ASCE)GT.1943-5606.0002367

Machiels, L., Paajanen, M., Jones, P. T., & Binnemans, K. (2022). Near-zero-waste recycling of low-grade sulphidic mining waste for critical-metal, mineral and construction raw-material production in a circular economy.

Mahdi, B. S., Sulaiman, N., Shehab, M. A., Shafie, S., Hizam, H., & Hassan, S. L. B. M. (2024). Optimization of operating cost and energy consumption in a smart grid. *IEEE Access : Practical Innovations, Open Solutions*, *12*, 18837–18850. DOI: 10.1109/ACCESS.2024.3354065

Mahmoud, A. S. (2022). Overview of Green Roof Technology as a Prospective Energy Preservation Technique in Arid Regions. *Engineering Technology and Applied Science Research*, *12*(4), 8982–8989. DOI: 10.48084/etasr.4919

Maida, M. J. (2015). Case Studies: An update on the SPD. Leeds City: Leeds City Council.

Maida, M. J. (2015). *Sustainable Construction in Practice: Performance of sustainable buildings in Leeds*. Leeds City.

Maida, Y., Sakata, H., Qu, Z., Maegawa, T., & Suzuki, H. (2021). Cyclic loading test of lightly reinforced concrete wall piers with slit dampers in RC frames. *Engineering Structures*, *236*, 112099. DOI: 10.1016/j.engstruct.2021.112099

Mallinson, D. J., Azevedo, L., Best, E., Robles, P., & Wang, J. (2023). The future of ai is in the states: The case of autonomous vehicle policies. *Business and Politics*, *26*(2), 180–199. DOI: 10.1017/bap.2023.19

Manandhar, R., Kim, J. H., & Kim, J. T. (2019). Environmental, social and economic sustainability of bamboo and bamboo-based construction materials in buildings. *Journal of Asian Architecture and Building Engineering*, *18*(2), 49–59. DOI: 10.1080/13467581.2019.1595629

Mannan, M., & Al-Ghamdi, S. G. (2020). Environmental impact of water-use in buildings: Latest developments from a life-cycle assessment perspective. *Journal of Environmental Management*, *261*, 110198. DOI: 10.1016/j.jenvman.2020.110198 PMID: 32148271

Martinez, A., DeJong, J., Akin, I., Aleali, A., Arson, C., Atkinson, J., Bandini, P., Baser, T., Borela, R., Boulanger, R., Burrall, M., Chen, Y., Collins, C., Cortes, D., Dai, S., DeJong, T., Del Dottore, E., Dorgan, K., Fragaszy, R., & Zheng, J. (2022). Bio-inspired geotechnical engineering: Principles, current work, opportunities and challenges. *Geotechnique*, *72*(8), 687–705. DOI: 10.1680/jgeot.20.P.170

Martins, N. P., Srivastava, S., Simão, F. V., Niu, H., Perumal, P., Snellings, R., Illikainen, M., Chambart, H., & Habert, G. (2021). Exploring the potential for utilization of medium and highly sulfidic mine tailings in construction materials: A review. *Sustainability (Basel)*, *13*(21), 12150. DOI: 10.3390/su132112150

Massaro, M., Secinaro, S., Dal Mas, F., Brescia, V., & Calandra, D. (2021). Industry 4.0 and circular economy: An exploratory analysis of academic and practitioners' perspectives. *Business Strategy and the Environment*, *30*(2), 1213–1231. DOI: 10.1002/bse.2680

Masson, V., Lemonsu, A., Hidalgo, J., & Voogt, J. (2020). Urban climates and climate change. *Annual Review of Environment and Resources*, *45*(1), 411–444. DOI: 10.1146/annurev-environ-012320-083623

Matarneh, S. T., Danso-Amoako, M., Al-Bizri, S., Gaterell, M., & Matarneh, R. (2019). Building information modeling for facilities management: A literature review and future research directions. *Journal of Building Engineering*, *24*, 100755. DOI: 10.1016/j.jobe.2019.100755

Mateus de Brito, J. (2008). Landslide of the slope adjacent to the peripheral retaining wall of the Machico hypermarket on Madeira Island.

Mathews, M. E., Shaji, A. E., Anand, N., Andrushia, A. D., Chin, S. C., & Lubloy, E. (2023). *IoT-based BIM integrated model for energy and water management in smart homes*. Intelligent Edge Computing for Cyber Physical Applications, Intelligent Data-Centric Systems. DOI: 10.1016/B978-0-323-99412-5.00009-5

Matsuya, I., Katamura, R., Sato, M., Iba, M., Kondo, H., Kanekawa, K., Takahashi, M., Hatada, T., Nitta, Y., Tanii, T., Shoji, S., Nishitani, A., & Ohdomari, I. (2010). Measuring relative-story displacement and local inclination angle using multiple position-sensitive detectors. *Sensors (Basel)*, *10*(11), 9687–9697. DOI: 10.3390/s101109687 PMID: 22163434

Mauser, W. (2009). *Water resources: efficient, sustainable and equitable use*. Haus Publishing. DOI: 10.2307/jj.10984417

Mays, L. (2006) Water resources sustainability, McGraw Hill publication.

Meenakshi, P. (2019) Elements of environmental and engineering, PHI learning publication.

Meerow, S., & Newell, J. P. (2021). Urban resilience for whom, what, when, where, and why? In *Geographic Perspectives on Urban Sustainability* (pp. 43–63). Routledge. DOI: 10.4324/9781003130185-3

Meerow, S., Newell, J. P., & Stults, M. (2016). Defining urban resilience: A review. *Landscape and Urban Planning*, *147*, 38–49. DOI: 10.1016/j.landurbplan.2015.11.011

Meglin, R., Kytzia, S., & Habert, G. (2022). Regional circular economy of building materials: Environmental and economic assessment combining material flow analysis, input-output analyses, and life cycle assessment. *Journal of Industrial Ecology*, *26*(2), 562–576. DOI: 10.1111/jiec.13205

Meyer, M. D. (2016). *Transportation planning handbook*. John Wiley & Sons. DOI: 10.1002/9781119174660

Mirdul, D. (2018) Water sustainability, Notion press publication.

Mi, Z., Guan, D., Liu, Z., Liu, J., Viguié, V., Fromer, N., & Wang, Y. (2019). Cities: The core of climate change mitigation. *Journal of Cleaner Production*, *207*, 582–589. DOI: 10.1016/j.jclepro.2018.10.034

Mizutani, N., & Mostafa, A. M. (1998). Nonlinear Wave-Induced Seabed Instability Around Coastal Structures. *Coastal Engineering Journal*, *40*(2), 131–160. DOI: 10.1142/S0578563498000091

Mocharla, I. R., Selvam, R., Govindaraj, V., & Muthu, M. (2022). Performance and life-cycle assessment of high-volume fly ash concrete mixes containing steel slag sand. *Construction & Building Materials*, *341*, 127814. DOI: 10.1016/j.conbuildmat.2022.127814

Moghaddam, F. B., Mir, J. M. F., Yanguas, A. B., Delgado, I. N., & Dominguez, E. R. (2020). Building orientation in green facade performance and its positive effects on urban landscape: Case study of an urban block in Barcelona. *Sustainability (Basel)*, *12*(21), 9273. DOI: 10.3390/su12219273

Mohajerani, A., Burnett, L., Smith, J. V., Kurmus, H., Milas, J., Arulrajah, A., Horpibulsuk, S., & Abdul Kadir, A. (2019). Nanoparticles in construction materials and other applications, and implications of nanoparticle use. *Materials (Basel), 12*(19), 3052. DOI: 10.3390/ma12193052 PMID: 31547011

Mohamed, W. W., Osman, N., & Abdullah, R. (2023). A review of bioengineering techniques for slope stability in Malaysia. *International Journal of Environmental Science and Technology*, *20*(3), 3467–3482. DOI: 10.1007/s13762-022-04235-3

Mohammad, H. D., Karri, R. R., Tyagi, I. and Scholz, M. (2023) Water, the environment and the sustainable development goals, Elsevier publication.

Moni, S. M., Mahmud, R., High, K., & Carbajales-Dale, M. (2020). Life cycle assessment of emerging technologies: A review. *Journal of Industrial Ecology*, *24*(1), 52–63. DOI: 10.1111/jiec.12965

Montefrio, M. J. F., Lee, X. R., & Lim, E. (2021). Aesthetic politics and community gardens in Singapore. *Urban Geography*, *42*(10), 1459–1479. DOI: 10.1080/02723638.2020.1788304

Montgomery, R., Schirmer Jr, H., & Hirsch, A. (2015). Improving environmental sustainability in road projects. Citation of Environment and natural resources global practice discussion paper, (2).

Morchid, A., Jebabra, R., Khalid, H. M., El Alami, R., Qjidaa, H., & Jamil, M. Q. (2024). IoT-based smart irrigation management system to enhance agricultural water security using embedded systems, telemetry data, and cloud computing. *Results in Engineering*, *23*, 102829. DOI: 10.1016/j.rineng.2024.102829

Moreau, F., Dauteuil, O., Bour, O., & Gavrilenko, P. (2006). GPS measurements of ground deformation induced by water level variations into a granitic aquifer (French Brittany). *Terra Nova*, *18*(1), 50–54.

Moreno-Sader, K., Meramo-Hurtado, S. I., & González-Delgado, A. D. (2020). Environmental sustainability analysis of chitosan microbeads production for pharmaceutical applications via computer-aided simulation, WAR and TRACI assessments. *Sustainable Chemistry and Pharmacy*, *15*, 100212. DOI: 10.1016/j.scp.2020.100212

Morishita, N., Heidenreich, M., Hemmers, R., Vankann, M., Sahakari, T., Vainio, T., & Österreicher, D. (2017). EU-GUGLE: A Sustainable Renovation for Smarter Cities from a Pilot Project. In Bisello, A. (Eds.), *Smart and Sustainable Planning for Cities and Regions* (pp. 353–382). Springer., DOI: 10.1007/978-3-319-44899-2_21

Mother, D. (2018) Civil engineering basics: Water, wastewater and storm water conveyance, Create space independent publishing platform.

Murthy, S. (2020) Water resources engineering: Principles and practice, New age international publication.

Murthy, C. S. (2002). *Water resources engineering: Principles and practice*. New Age International.

Murty, V.N. (2018) 'Land and water management engineering' Jain Brothers publication.

Mutani, G., & Todeschi, V. (2020). The effects of green roofs on outdoor thermal comfort, urban heat island mitigation and energy savings. *Atmosphere (Basel)*, *11*(2), 123. DOI: 10.3390/atmos11020123

Najed, F., Djunaedy, E., Utami, A. R. I., Sari, I. P., Rithwan, F., & Sofyan, S. (2023). The effect of baseline establishment on energy savings estimate in green building certification. *Journal of Physics: Conference Series*, *2596*(1), 012051. DOI: 10.1088/1742-6596/2596/1/012051

Nakhaei, M., Ahmadi, A., Gheibi, M., Chahkandi, B., Hajiaghaei-Keshteli, M., & Behzadian, K. (2023). A smart Sustainable decision Support system for water management oF power plants in water stress regions. *Expert Systems with Applications*, *230*, 120752. DOI: 10.1016/j.eswa.2023.120752

Nawaz, A., Su, X., Din, Q. M. U., Khalid, M. I., Bilal, M., & Shah, S. A. R. (2020). Identification of the h&s (Health and safety factors) involved in infrastructure projects in developing countries-a sequential mixed method approach of OLMT-project. *International Journal of Environmental Research and Public Health*, *17*(2), 635. DOI: 10.3390/ijerph17020635 PMID: 31963777

Newing, A., Hibbert, O., Van-Alwon, J., Ellaway, S., & Smith, A. (2023). Smart water metering as a non-invasive tool to infer dwelling type and occupancy – Implications for the collection of neighbourhood-level housing and tourism statistics. *Computers, Environment and Urban Systems*, *105*, 102028. DOI: 10.1016/j.compenvurbsys.2023.102028

Nguy n, T., & Waibel, M. (2023). Promoting urban health through the green building movement in vietnam: An intersectoral perspective. *Sustainability (Basel)*, *15*(13), 10296. DOI: 10.3390/su151310296

Niza, I. L., de Souza, M. P., da Luz, I. M., & Broday, E. E. (2024). Sick building syndrome and its impacts on health, well-being and productivity: A systematic literature review. *Indoor and Built Environment*, *33*(2), 218–236. DOI: 10.1177/1420326X231191079

Nuaklong, P., Sata, V., & Chindaprasirt, P. (2016). Influence of recycled aggregate on fly ash geopolymer concrete properties. *Journal of Cleaner Production, 112*, 2300–2307. DOI: 10.1016/j.jclepro.2015.10.109

Nutley, S. (2003). Transport and Social Exclusion: An Overview of the Literature. *Transport Policy, 10*(1), 1–11.

Nwodo, M. N., & Anumba, C. J. (2019). A review of life cycle assessment of buildings using a systematic approach. *Building and Environment, 162*, 106290. DOI: 10.1016/j.buildenv.2019.106290

Oberascher, M., Kinzel, C., Kastlunger, U., Kleidorfer, M., Zingerle, C., Rauch, W., & Sitzenfrei, R. (2021). Integrated urban water management with micro storages developed as an IoT-based solution – The smart rain barrel. *Environmental Modelling & Software, 139*, 105028. DOI: 10.1016/j.envsoft.2021.105028

Oberascher, M., Rauch, W., & Sitzenfrei, R. (2022). Towards a smart water city: A comprehensive review of applications, data requirements, and communication technologies for integrated management. *Sustainable Cities and Society, 76*, 103442. DOI: 10.1016/j.scs.2021.103442

Ohaedeghasi, C. I., Ezeokoli, F. O., Agu, N. N., Okolie, K. C., & Ugochukwu, S. C. (n.d.). rospects for improving building maintenance management in nigerian public universities: A case study of nnamdi azikiwe university. *Journal of Engineering Research and Reports, 21*(4), 53–59. DOI: 10.9734/jerr/2021/v21i417458

Ok, B., Sarici, T., Talaslioglu, T., & Yildiz, A. (2020). Geotechnical properties of recycled construction and demolition materials for filling applications. *Transportation Geotechnics, 24*, 100380. DOI: 10.1016/j.trgeo.2020.100380

Okorie, O., Obi, M., Russell, J., Charnley, F., & Salonitis, K. (2021). A triple bottom line examination of product cannibalisation and remanufacturing: A review and research agenda. *Sustainable Production and Consumption, 27*, 958–974. DOI: 10.1016/j.spc.2021.02.013

Osinubi, K. J., Eberemu, A. O., Ijimdiya, T. S., Yakubu, S. E., Gadzama, E. W., Sani, J. E., & Yohanna, P. (2020). Review of the use of microorganisms in geotechnical engineering applications. *SN Applied Sciences, 2*(2), 1–19. DOI: 10.1007/s42452-020-1974-2

Østergaard, P. A., Duic, N., Noorollahi, Y., & Kalogirou, S. (2022). Renewable energy for sustainable development. *Renewable Energy, 199*, 1145–1152. DOI: 10.1016/j.renene.2022.09.065

Paccoud, A., Hesse, M., Becker, T., & Gorczyńska, M. (2021). Land and the housing affordability crisis: Landowner and developer strategies in Luxembourg's facilitative planning context. *Housing Studies*, *0*, 1–18. DOI: 10.1080/02673037.2021.1950647

Pal, J. G. (2023) An introduction to water management techniques for professional engineers, independently published.

Panda, S. C. (2018). *Principles and practices of water management, M/S Agrobios*. Publication.

Pan, Y., & Zhang, L. (2021). Roles of artificial intelligence in construction engineering and management: A critical review and future trends. *Automation in Construction*, *122*, 103517. DOI: 10.1016/j.autcon.2020.103517

Papagiannis, F., Gazzola, P., Burak, O., & Pokutsa, I. (2018). Overhauls in water supply systems in Ukraine: A hydro-economic model of socially responsible planning and cost management. *Journal of Cleaner Production*, *183*, 358–369. DOI: 10.1016/j.jclepro.2018.02.156

Pappu, A., Saxena, M., & Asolekar, S. R. (2007). Solid wastes generation in India and their recycling potential in building materials. *Building and Environment*, *42*(6), 2311–2320. DOI: 10.1016/j.buildenv.2006.04.015

Patton, W., & McMahon, M. (2021). Career development and systems theory: Connecting theory and practice. In *Career Development and Systems Theory*. Brill. DOI: 10.1163/9789004466210

Paumgartten, P. (2023). The business case for high performance green buildings: Sustainability and its financial impact. *Journal of Facilities Management*, *2*(1), 26–34. DOI: 10.1108/14725960410808096

Peña, C., Civit, B., Gallego-Schmid, A., Druckman, A., Pires, A. C., Weidema, B., Mieras, E., Wang, F., Fava, J., Canals, L. M., Cordella, M., Arbuckle, P., Valdivia, S., Fallaha, S., & Motta, W. (2021). Using life cycle assessment to achieve a circular economy. *The International Journal of Life Cycle Assessment*, *26*(2), 215–220. DOI: 10.1007/s11367-020-01856-z

Pham, D. H., Kim, B., Lee, J., & Ahn, Y. (2020). An investigation of the selection of LEED version 4 credits for sustainable building projects. *Applied Sciences (Basel, Switzerland)*, *10*(20), 7081. DOI: 10.3390/app10207081

Phillipson, J., Gorton, M., Turner, R., Shucksmith, M., Aitken-McDermott, K., Areal, F., Cowie, P., Hubbard, C., Maioli, S., McAreavey, R., Souza-Monteiro, D., Newbery, R., Panzone, L., Rowe, F., & Shortall, S. (2020). The COVID-19 Pandemic and Its Implications for Rural Economies. *Sustainability (Basel)*, *2020*(12), 3973. DOI: 10.3390/su12103973

Phoon, K. K., & Zhang, W. (2023). Future of machine learning in geotechnics. Georisk. *Assessment and Management of Risk for Engineered Systems and Geohazards*, *17*(1), 7–22. DOI: 10.1080/17499518.2022.2087884

PIETRAM. (2016). Plano Integrado Estratégico de Transportes da Região Autónoma da Madeira (PIETRAM) – 2014/2020.

Pilehvar, S., Cao, V. D., Szczotok, A. M., Valentini, L., Salvioni, D., Magistri, M., Pamies, R., & Kjøniksen, A. L. (2017). Mechanical properties and microscale changes of geopolymer concrete and Portland cement concrete containing micro-encapsulated phase change materials. *Cement and Concrete Research*, *100*, 341–349. DOI: 10.1016/j.cemconres.2017.07.012

Piotrowska, B., Słyś, D., Kordana-Obuch, S., & Pochwat, K. (2020). Critical analysis of the current state of knowledge in the field of waste heat recovery in sewage systems. *Resources*, *9*(6), 72. DOI: 10.3390/resources9060072

Pliuhin, V., Tsegelnyk, Y., Plankovskyy, S., Aksonov, O., & Kombarov, V. (2023). Implementation of induction motor speed and torque control system with reduced order model in ANSYS Twin Builder. In Cioboată, D.D. (eds.) *International Conference on Reliable Systems Engineering (ICoRSE)–* 2023 (pp. 514–531). Springer. https://doi.org/DOI: 10.1007/978-3-031-40628-7_42

Pokhrel, S., Amiri, L., Poncet, S., Sasmito, A. P., & Ghoreishi-Madiseh, S. A. (2022). Renewable heating solutions for buildings; a techno-economic comparative study of sewage heat recovery and Solar Borehole Thermal Energy Storage System. *Energy and Building*, *259*, 111892. DOI: 10.1016/j.enbuild.2022.111892

Pomianowski, M. Z., Johra, H., Marszal-Pomianowska, A., & Zhang, C. (2020). Sustainable and energy-efficient domestic hot water systems: A review. *Renewable & Sustainable Energy Reviews*, *128*, 109900. DOI: 10.1016/j.rser.2020.109900

Ponce-Ortega, J. M., Nápoles-Rivera, F., Lira-Barragán, L. F., & Ramírez-Márquez, C. (2024). Sustainable Water Systems.

Prabhu, R. S., Anuradha, R., & Jude, A. B. (2022). Understanding the mining waste as resources in self-compacting concrete: A numerical study on sustainable construction. *Resources Policy*, *78*, 102806. DOI: 10.1016/j.resourpol.2022.102806

Pregnolato, M., Gunner, S., Voyagaki, E., De Risi, R., Carhart, N., Gavriel, G., Tully, P., Tryfonas, T., Macdonald, J., & Taylor, C. (2022). Towards Civil Engineering 4.0: Concept, workflow and application of Digital Twins for existing infrastructure. *Automation in Construction*, *141*, 104421. DOI: 10.1016/j.autcon.2022.104421

Provan, D. J., Woods, D. D., Dekker, S. W., & Rae, A. J. (2020). Safety II professionals: How resilience engineering can transform safety practice. *Reliability Engineering & System Safety*, *195*, 106740. DOI: 10.1016/j.ress.2019.106740

Prozuments, A., Borodinecs, A., Zaharovs, S., Banionis, K., Monstvilas, E., & Norvaišienė, R. (2023). Evaluating reduction in thermal energy consumption across renovated buildings in latvia and lithuania. *Buildings (Basel, Switzerland)*, *13*(8), 1916. DOI: 10.3390/buildings13081916

Puchol-Salort, P., O'Keeffe, J., van Reeuwijk, M., & Mijic, A. (2021). An urban planning sustainability framework: Systems approach to blue green urban design. *Sustainable Cities and Society*, *66*, 102677. DOI: 10.1016/j.scs.2020.102677

Punetha, P., Samanta, M., & Sarkar, S. (2019). Bioengineering as an effective and ecofriendly soil slope stabilization method: A review. Landslides: Theory, practice and modelling, 201-224. DOI: 10.1007/978-3-319-77377-3_10

Qin, L. (2023). Impacts of green rating systems on the economy and society. *IOP Conference Series. Earth and Environmental Science*, *1204*(1), 012006. DOI: 10.1088/1755-1315/1204/1/012006

Qin, N., & Tian, H. (2022). Operational analysis of a newly untreated sewage source heat pump with a plate heat exchanger. *Heat and Mass Transfer*, *58*(4), 683–693. DOI: 10.1007/s00231-021-03141-7

Radcliffe, J. C. (2019). History of water sensitive urban design/low impact development adoption in Australia and internationally. In *Approaches to water sensitive urban design* (pp. 1–24). Woodhead Publishing. DOI: 10.1016/B978-0-12-812843-5.00001-0

Rao, S., & Gayana, B. C. (2021). Experimental Study on Iron Ore Tailings as Aggregate in Development of Bricks and Concrete. *Research Transcripts in Materials*, *1*, 51–68.

Rauch, H. P., von der Thannen, M., Raymond, P., Mira, E., & Evette, A. (2022). Ecological challenges* for the use of soil and water bioengineering techniques in river and coastal engineering projects. *Ecological Engineering*, *176*, 106539. DOI: 10.1016/j.ecoleng.2021.106539

Raymond, A. J., DeJong, J. T., Kendall, A., Blackburn, J. T., & Deschamps, R. (2021). Life cycle sustainability assessment of geotechnical ground improvement methods. *Journal of Geotechnical and Geoenvironmental Engineering, 147*(12), 04021161. DOI: 10.1061/(ASCE)GT.1943-5606.0002646

Raymond, A. J., Tipton, J. R., Kendall, A., & DeJong, J. T. (2020). Review of impact categories and environmental indicators for life cycle assessment of geotechnical systems. *Journal of Industrial Ecology, 24*(3), 485–499. DOI: 10.1111/jiec.12946

Reddy, V. J., Hariram, N. P., Ghazali, M. F., & Kumarasamy, S. (2024). Pathway to sustainability: An overview of renewable energy integration in building systems. *Sustainability (Basel), 16*(2), 638. DOI: 10.3390/su16020638

Regona, M., Yigitcanlar, T., Hon, C. K. H., & Teo, M. (2023). Mapping two decades of ai in construction research: A scientometric analysis from the sustainability and construction phases lenses. *Buildings (Basel, Switzerland), 13*(9), 2346. DOI: 10.3390/buildings13092346

Rezaei, F., Bulle, C., & Lesage, P. (2019). Integrating building information modeling and life cycle assessment in the early and detailed building design stages. *Building and Environment, 153*, 158–167. DOI: 10.1016/j.buildenv.2019.01.034

Ribeiro, P. J. G., & Gonçalves, L. A. P. J. (2019). Urban resilience: A conceptual framework. *Sustainable Cities and Society, 50*, 101625. DOI: 10.1016/j.scs.2019.101625

Robinson, R. J., Jones, A. A., and Woo, M.K. (2007) Managing water resources in physical and social environment, IGU home of geography publication.

Rogers, M., & Enright, B. (2023). *Highway engineering*. John Wiley & Sons.

Roh, C.-G., & Kim, J. (2022). What Are More Efficient Transportation Services in a Rural Area? A Case Study in Yangsan City, South Korea. *International Journal of Environmental Research and Public Health, 2022*(19), 11263. DOI: 10.3390/ijerph191811263 PMID: 36141536

Roque, A. J., Paleologos, E. K., O'Kelly, B. C., Tang, A. M., Reddy, K. R., Vitone, C., Mohamed, A.-M. O., Koda, E., Goli, V. S. N. S., Vieira, C. S., Fei, X., Sollecito, F., Vaverková, M. D., Plötze, M., Petti, R., Podlasek, A., Puzrin, A. M., Cotecchia, F., Osiński, P., & Singh, D. N. (2021). Sustainable environmental geotechnics practices for a green economy. *Environmental Geotechnics, 9*(2), 68–84. DOI: 10.1680/jenge.21.00091

Rose, C. M., & Stegemann, J. A. (2018). From waste management to component management in the construction industry. *Sustainability (Basel), 10*(1), 229. DOI: 10.3390/su10010229

Rosenzweig, C., Solecki, W. D., Romero-Lankao, P., Mehrotra, S., Dhakal, S., & Ibrahim, S. A. (Eds.). (2018). *Climate change and cities: Second assessment report of the urban climate change research network.* Cambridge University Press. DOI: 10.1017/9781316563878

Roy, P.K., Roy, M.B. and Pal, S. (2021) Advances in water resources management for sustainable use, Springer publication.

Russo, G., Marone, G., & Di Girolamo, L. (2021). Hybrid energy piles as a smart and sustainable foundation. *Journal of Human, Earth, and Future*, *2*(3), 306–322. DOI: 10.28991/HEF-2021-02-03-010

Ruzaikin, V., & Lukashov, I. (2023). Pump assistant ground-source two-phase thermosiphon for the heating and cooling of households. In Nechyporuk, M. (Eds.), *Integrated Computer Technologies in Mechanical Engineering – 2022* (pp. 665–676). Springer., DOI: 10.1007/978-3-031-36201-9_55

Sabat, V., Shaikh, M., Kanap, M., Chaudhari, M., Suryawanshi, S., & Knadgouda, K. (2015). Use of iron ore tailings as a construction material. *International Journal of Conceptions on Mechanical and Civil Engineering*, *3*(2), 1–6.

Sadrizadeh, S., Yao, R., Yuan, F., Awbi, H., Bahnfleth, W., Bi, Y., Cao, G., Croitoru, C., de Dear, R., Haghighat, F., Kumar, P., Malayeri, M., Nasiri, F., Ruud, M., Sadeghian, P., Wargocki, P., Xiong, J., Yu, W., & Li, B. (2022). Indoor air quality and health in schools: A critical review for developing the roadmap for the future school environment. *Journal of Building Engineering*, *57*, 104908. DOI: 10.1016/j.jobe.2022.104908

Safari, K., & AzariJafari, H. (2021). Challenges and opportunities for integrating BIM and LCA: Methodological choices and framework development. *Sustainable Cities and Society*, *67*, 102728. DOI: 10.1016/j.scs.2021.102728

Sahoo, S., & Singh, S. P. (2022). Strength and durability properties of expansive soil treated with geopolymer and conventional stabilizers. *Construction & Building Materials*, *328*, 127078. DOI: 10.1016/j.conbuildmat.2022.127078

Saja, A. A., Goonetilleke, A., Teo, M., & Ziyath, A. M. (2019). A critical review of social resilience assessment frameworks in disaster management. *International Journal of Disaster Risk Reduction*, *35*, 101096. DOI: 10.1016/j.ijdrr.2019.101096

Salas-Zapata, W. A., & Ortiz-Muñoz, S. M. (2019). Analysis of meanings of the concept of sustainability. *Sustainable Development (Bradford)*, *27*(1), 153–161. DOI: 10.1002/sd.1885

Salehi, S., Arashpour, M., Kodikara, J., & Guppy, R. (2021). Sustainable pavement construction: A systematic literature review of environmental and economic analysis of recycled materials. *Journal of Cleaner Production*, *313*, 127936. DOI: 10.1016/j.jclepro.2021.127936

Salman, M. Y., & Hasar, H. (2023). Review on environmental aspects in smart city concept: Water, waste, air pollution and transportation smart applications using IoT techniques. *Sustainable Cities and Society*, *94*, 104567. DOI: 10.1016/j.scs.2023.104567

Sampene, A., Li, C., Agyeman, F., & Brenya, R. (2021). Analysis of the BRICS countries' pathways towards a low-carbon environment. *BRICS Journal of Economics*, *2*(4), 77–102. DOI: 10.38050/2712-7508-2021-4-4

Santos, J., Neves, M., & Silva, L. (2004). *Rubble-Mound Breakwater Inspection in Portugal.*, DOI: 10.1061/40733(147)21

Sarı Hassoun, S. E., & Ayad, H. (2020). Renewable Energy and Sustainable Development: Evidence from 17 OECD Countries/Renewable Energy and Sustainable Development: Evidence from 17 OECD Countries. *Uluslararası Ekonomi İşletme ve Politika Dergisi*, *4*(1), 41–60. DOI: 10.29216/ueip.653074

Sawada, Y., Miyake, M., Miyamoto, J., & Kawabata, T. (2015). Numerical Analysis on Stability of Caisson-Type Breakwaters under Tsunami-Induced Seepage. *Transportation Infrastructure Geotechnology*, *2*(3), 120–138. DOI: 10.1007/s40515-015-0022-y

Sawarkar, A. D., Shrimankar, D. D., Kumar, A., Kumar, A., Singh, E., Singh, L., Kumar, S., & Kumar, R. (2020). Commercial clustering of sustainable bamboo species in India. *Industrial Crops and Products*, *154*, 112693. DOI: 10.1016/j.indcrop.2020.112693

Sayed, E. T., Olabi, A. G., Alami, A. H., Radwan, A., Rezk, A., & Abdelkareem, M. A. (2023). Renewable energy and energy storage systems. *Energies*, *16*(3), 1415. DOI: 10.3390/en16031415

Sayegh, M. A., Jadwiszczak, P., Axcell, B. P., Niemierka, E., Bryś, K., & Jouhara, H. (2018). Heat pump placement, connection and operational modes in European district heating. *Energy and Building*, *166*, 122–144. DOI: 10.1016/j.enbuild.2018.02.006

Sayyadi, R., & Awasthi, A. (2018). A simulation-based optimisation approach for identifying key determinants for sustainable transportation planning. *International Journal of Systems Science: Operations & Logistics*, *5*(2), 161–174.

Scanlon, K., Lunde, J., & Whitehead, C. (2008). Mortgage Product Innovation in Advanced Economies: More Choice, More Risk. *European Journal of Housing Policy*, *8*(2), 109–131. DOI: 10.1080/14616710802037359

Schaefer, C., Stelter, A., Holl-Supra, S., Weber, S., & Niehaves, B. (2022). The Acceptance and Use Behavior of Shared Mobility Services in a Rural Municipality. *Smart Cities*, *2022*(5), 1229–1240. DOI: 10.3390/smartcities5040062

Schiller, P. L., & Kenworthy, J. (2017). *An introduction to sustainable transportation: Policy, planning and implementation*. Routledge. DOI: 10.4324/9781315644486

Schneider, D. (2006). *TERRESTRIAL LASER SCANNING FOR AREA BASED DEFORMATION ANALYSIS OF TOWERS AND WATER DAMNS*.

Sciara, G. C. (2017). Metropolitan transportation planning: Lessons from the past, institutions for the future. *Journal of the American Planning Association*, *83*(3), 262–276. DOI: 10.1080/01944363.2017.1322526

Sedayu, A., Setiono, A. R., Subaqin, A., & Gautama, A. G. (2020). Improving the performance of construction project using green building principles. *Asian Journal of Civil Engineering*, *21*(8), 1443–1452. DOI: 10.1007/s42107-020-00289-1

Segui, P., Safhi, A. E. M., Amrani, M., & Benzaazoua, M. (2023). Mining wastes as road construction material: A review. *Minerals (Basel)*, *13*(1), 90. DOI: 10.3390/min13010090

Selvakumar, A., & Tafuri, A. N. (2012). Rehabilitation of aging water infrastructure systems: Key challenges and issues. *Journal of Infrastructure Systems*, *18*(3), 202–209. DOI: 10.1061/(ASCE)IS.1943-555X.0000091

Sevela, P., Frenger, J., Schnieders, J., & Pfluger, R. (2022). Potential of WASTE WATER HEAT RECOVERY in reducing the EU's energy need. In *CLIMA 2022 Conference Proceedings* (pp. 2099–2105). Delft University of Technology. https://doi.org/DOI: 10.34641/clima.2022.439

Seyis, S. (2019). Pros and cons of using building information modeling in the AEC industry. *Journal of Construction Engineering and Management*, *145*(8), 04019046. DOI: 10.1061/(ASCE)CO.1943-7862.0001681

Shah, P., & Dahal, K. (2023). Problems associated with implementation of bioengineering in hill road construction in Nepal. *Journal of Engineering and Applied Sciences (Asian Research Publishing Network)*, *70*(1), 30. DOI: 10.1186/s44147-023-00198-5

Shailesh, K., Ramswaroop, R. Y., & Nidhi, C. (2024). Ai-driven energy management systems for smart buildings. *Power System Technology*, *48*(1), 322–337. DOI: 10.52783/pst.280

Sharifi, A. (2016). A critical review of selected tools for assessing community resilience. *Ecological Indicators*, *69*, 629–647. DOI: 10.1016/j.ecolind.2016.05.023

Sharma, R. (2024) water management in India: A legal approach, Universal academic books publishers and distributors.

Shen, C., Wang, M., Zhang, C., Dong, Y., Ni, L., & Jiang, Y. (2020). Experimental investigation on a novel sewage-resource-based system with functions of heat recovery and water purification. *Applied Thermal Engineering*, *165*, 114580. DOI: 10.1016/j.applthermaleng.2019.114580

Shergold, I., & Parkhurst, G. (2012). Understanding and Addressing Rural Transport Issues: The Role of Sustainable Transport Solutions. *Transport Reviews*, *32*(6), 745–763.

Shimon, C. (2024) Water management: Prioritizing justice and sustainability, Island press publication.

Shreekant, R. L., Aruna, M., & Vardhan, H. (2016). Utilisation of mine waste in the construction industry–a critical review. International journal of Earth sciences and engineering, 9(01), 182-195.

Shufrin, I., Pasternak, E., & Dyskin, A. (2023). Environmentally friendly smart construction review of recent developments and opportunities. *Applied Sciences (Basel, Switzerland)*, *13*(23), 12891. DOI: 10.3390/app132312891

Simão, F. V., Chambart, H., Vandemeulebroeke, L., Nielsen, P., Adrianto, L. R., Pfister, S., & Cappuyns, V. (2022). Mine waste as a sustainable resource for facing bricks. *Journal of Cleaner Production*, *368*, 133118. DOI: 10.1016/j.jclepro.2022.133118

Singh, G. (1990) 'Manual of soil and water conservation practices' Oxford and IBH publication.

Singh, A. (2023). The Current Scenario of Sustainable Planning and Development of the Indian Built Environment. In Cobîrzan, N., Muntean, R., & Felseghi, R. (Eds.), *Circular Economy Implementation for Sustainability in the Built Environment* (pp. 106–138). IGI Global., DOI: 10.4018/978-1-6684-8238-4.ch005

Singh, A. (2023). The Current Trends and Need of Infrastructure Projects in the Country India. In Popescu, C., Yu, P., & Wei, Y. (Eds.), *Achieving the Sustainable Development Goals Through Infrastructure Development* (pp. 235–261). IGI Global., DOI: 10.4018/979-8-3693-0794-6.ch010

Singh, A. (2023). The Present and Future Role of Artificial Intelligence in the Promotion of Sustainable Practices in the Construction Sector. In Kahraman, C., Sari, I. U., Oztaysi, B., Cebi, S., Cevik Onar, S., & Tolga, A. Ç. (Eds.), *Intelligent and Fuzzy Systems. INFUS 2023. Lecture Notes in Networks and Systems* (Vol. 759). Springer., DOI: 10.1007/978-3-031-39777-6_2

Singh, A. (2023). The Recent Trend of Artificial Neural Network in the Field of Civil Engineering. In Vasant, P. (Eds.), *Intelligent Computing and Optimization. ICO 2023. Lecture Notes in Networks and Systems* (Vol. 855). Springer., DOI: 10.1007/978-3-031-50158-6_32

Singh, A., Verdi, A., Haver, D., Sapkota, A., & Iradukunda, J. C. (2024). Using a soil moisture sensor-based smart controller for autonomous irrigation management of hybrid bermudagrass with recycled water in coastal Southern California. *Agricultural Water Management*, *299*, 108906. DOI: 10.1016/j.agwat.2024.108906

Singh, M., & Ahmed, S. (2021). IoT based smart water management systems: A systematic review. *Materials Today: Proceedings*, *46*(Part 11), 5211–5218. DOI: 10.1016/j.matpr.2020.08.588

Singh, P., Raj, A., & Yadav, B. (2022). Impacts of agriculture-based contaminants on groundwater quality. In *Sustainability of Water Resources: Impacts and Management* (pp. 249–261). Springer International Publishing. DOI: 10.1007/978-3-031-13467-8_16

Sivakumar Babu, G. L. (2024). Reliability and Risk Analysis in Geotechnical and Geoenvironmental Engineering. Indian Geotechnical Journal, 1-33. DOI: 10.1007/s40098-024-00909-6

Sliwa, T., Sapińska-Śliwa, A., Wysogląd, T., Kowalski, T., & Konopka, I. (2021). Strength Tests of Hardened Cement Slurries for Energy Piles, with the Addition of Graphite and Graphene, in Terms of Increasing the Heat Transfer Efficiency. *Energies*, *14*(4), 1190. DOI: 10.3390/en14041190

Slupheny, M. M. (2018) Introduction to environment management, PHI learning publication.

Song, J. (2023). The green building development and assessment method – energy and cost analysis in china. *Highlights in Science. Engineering and Technology, 29*, 149–154. DOI: 10.54097/hset.v29i.4534

Soni, A., Das, P. K., Hashmi, A. W., Yusuf, M., Kamyab, H., & Chelliapan, S. (2022). Challenges and opportunities of utilizing municipal solid waste as alternative building materials for sustainable development goals: A review. *Sustainable Chemistry and Pharmacy, 27*, 100706. DOI: 10.1016/j.scp.2022.100706

Soyombov, D. (2024). The role of policy and regulation in promoting green buildings. *World Journal of Advanced Research and Reviews, 22*(1), 139–150. DOI: 10.30574/wjarr.2024.22.1.1047

Spriet, J., McNabola, A., Neugebauer, G., Stoeglehner, G., Ertl, T., & Kretschmer, F. (2020). Spatial and temporal considerations in the performance of wastewater heat recovery systems. *Journal of Cleaner Production, 247*, 119583. DOI: 10.1016/j.jclepro.2019.119583

Standar, A., Kozera, A., & Satoła, Ł. (2021). The Importance of Local Investments Co-Financed by the European Union in the Field of Renewable Energy Sources in Rural Areas of Poland. *Energies, 2021*(14), 450. DOI: 10.3390/en14020450

Sukumar, N., Chopp, D. L., & Moran, B. (2003). Extended finite element method and fast marching method for three-dimensional fatigue crack propagation. *Engineering Fracture Mechanics, 70*(1), 29–48. DOI: 10.1016/S0013-7944(02)00032-2

Sumalatha, J. (2023) Environmental geotechnical engineering, Notion press publication.

Sundaresan, S., Kumar, K. S., Kumar, T. A., Ashok, V., & Julie, E. G. (2021). *Blockchain architecture for intelligent water management system in smart cities*. Blockchain for Smart Cities. DOI: 10.1016/B978-0-12-824446-3.00006-5

Suresh, R. (2005) 'Soil and water conservation engineering' Standard book publisher and distributor.

Surya, R., Ali Berawi, M., & Sari, M. (2024). Development of a green building assessment system at the green building planning stage in indonesia. *E3S Web of Conferences, 517*, 05017. .DOI: 10.1051/e3sconf/202451705017

Suryanarayanan, S., Shivaraju, H. P., & Jenkins, D. (2024). *Water management in developing countries and sustainable development*. Springer Publication. DOI: 10.1007/978-981-99-8639-2

Sushma, N., Suresh, H. N., Mohana Lakshmi, J., & Santhosh Kumar, K. B. (2024). Experimental investigation on wireless integrated smart system for energy and water resource management in Indian smart cities. *Results in Engineering*, *23*, 102687. DOI: 10.1016/j.rineng.2024.102687

Szabo, S., Apipoonanon, C., Pramanik, M., Leeson, K., & Singh, D. R. (2021). Perceptions of an ageing agricultural workforce and farmers' productivity strategies: Evidence from Prachinburi Province, Thailand. *Outlook on Agriculture*, *50*(3), 294–304. DOI: 10.1177/00307270211025053

Szymanska, E., & Koloszko-Chomentowska, Z. (2022). Sustainable Innovative Mobility Solutions Preferred by Inhabitants of Rural Areas—The Case of Lithuania and Poland. *Sustainability (Basel)*, *2022*(14), 6603. DOI: 10.3390/su14116603

Taherian, G., Hosseini Nourzad, S. H., & Neyestani, M. (2024). Customizing a sustainability evaluation framework for Infrastructure projects in developing countries: The case study of Iran. *Sustainable and Resilient Infrastructure*, *9*(2), 168–191. DOI: 10.1080/23789689.2023.2272462

Tahwia, A. M., El-Far, O., & Amin, M. (2022). Characteristics of sustainable high strength concrete incorporating eco-friendly materials. *Innovative Infrastructure Solutions*, *7*(1), 1–13. DOI: 10.1007/s41062-021-00609-7

Tang, Z., Li, W., Tam, V. W., & Xue, C. (2020). Advanced progress in recycling municipal and construction solid wastes for manufacturing sustainable construction materials. *Resources, Conservation and Recycling. X*, *6*, 100036. DOI: 10.1016/j.rcrx.2020.100036

Tasalloti, A., Chiaro, G., Murali, A., Banasiak, L., Palermo, A., & Granello, G. (2021). Recycling of end-of-life tires (ELTs) for sustainable geotechnical applications: A New Zealand perspective. *Applied Sciences (Basel, Switzerland)*, *11*(17), 7824. DOI: 10.3390/app11177824

Tasan, M., Ozkan, Y., Ozgur, A. O., & Ozpinar, A. (2024). Data plateau: A unified analytics platform with intuitive interfaces for real-time and ml-driven insights. *Orclever Proceedings of Research and Development*, *4*(1), 73–89. DOI: 10.56038/oprd.v4i1.457

Tatar, J., & Hamilton, H. R. (2016). Comparison of laboratory and field environmental conditioning on FRP-concrete bond durability. *Construction & Building Materials*, *122*, 525–536. DOI: 10.1016/j.conbuildmat.2016.06.074

Tavares, M., Perez-Sanchez, M., Carravetta, A., Coronado-Hernandez, O. E., Lopez-Jimenez, P. A., & Ramos, H. M. (2024). Smart Feasibility Optimization of Hybrid Renewable Water Supply Systems by Digital Twin Technologies: A Multicriteria Approach Applied to Isolated Cities. *Sustainable Cities and Society*, *115*, 105834. DOI: 10.1016/j.scs.2024.105834

Tayerani Charmchi, A. S., Ifaei, P., & Yoo, C. (2021). Smart supply-side management of optimal hydro reservoirs using the water/energy nexus concept: A hydropower pinch analysis. *Applied Energy*, *281*, 116136. DOI: 10.1016/j.apenergy.2020.116136

Taylor, K.-Y. (2019). *Race for Profit: How Banks and the Real Estate Industry Undermined Black Homeownership*. University of North Carolina Press. DOI: 10.5149/northcarolina/9781469653662.001.0001

Temizel-Sekeryan, S., Wu, F., & Hicks, A. L. (2021). Global scale life cycle environmental impacts of single-and multi-walled carbon nanotube synthesis processes. *The International Journal of Life Cycle Assessment*, *26*(4), 656–672. DOI: 10.1007/s11367-020-01862-1

Teng, F., Zhang, Y., Yan, W., Wang, X., & Zhang, K. (2023). Restoring Force Model of Precast Segmental Reinforced Concrete Piers after Seawater Freeze–Thaw Cycles. *Coatings*, *13*(1), 1. Advance online publication. DOI: 10.3390/coatings13010016

Ternel, C., Bouter, A., & Melgar, J. (2021). Life cycle assessment of mid-range passenger cars powered by liquid and gaseous biofuels: Comparison with greenhouse gas emissions of electric vehicles and forecast to 2030. *Transportation Research Part D, Transport and Environment*, *97*, 102897. DOI: 10.1016/j.trd.2021.102897

Thomson, H., Simcock, N., Bouzarovski, S., & Petrova, S. (2019). Energy poverty and indoor cooling: An overlooked issue in Europe. *Energy and Building*, *196*, 21–29. DOI: 10.1016/j.enbuild.2019.05.014

Tijan, E., Jović, M., Panjako, A., & Žgaljić, D. (2021). The Role of Port Authority in Port Governance and Port Community System Implementation. *Sustainability (Basel)*, *13*(5), 5. Advance online publication. DOI: 10.3390/su13052795

Tiwari, K., and Kuma, R. (2022) Advances in civil engineering for sustainable development- Geotechnical, transportation & water resources, Walnut publication.

Tiwari, N., Satyam, N., & Sharma, M. (2021). Micro-mechanical performance evaluation of expansive soil biotreated with indigenous bacteria using MICP method. *Scientific Reports*, *11*(1), 10324. DOI: 10.1038/s41598-021-89687-2 PMID: 33990644

Tocco, B., Bailey, A., & Davidova, S. (2013). Determinants to Leave Agriculture and Change Occupational Sector: Evidence from an Enlarged Eu. Factor Markets Working Paper, 32.

Tofallis, C. (2008). Least Squares Percentage Regression. *Journal of Modern Applied Statistical Methods; JMASM*, *7*(2), 526–534. DOI: 10.22237/jmasm/1225513020

Tokede, O., & Traverso, M. (2020). Implementing the guidelines for social life cycle assessment: Past, present, and future. *The International Journal of Life Cycle Assessment*, *25*(10), 1910–1929. DOI: 10.1007/s11367-020-01814-9

Toniolo, S., Tosato, R. C., Gambaro, F., & Ren, J. (2020). Life cycle thinking tools: Life cycle assessment, life cycle costing and social life cycle assessment. In *Life Cycle Sustainability Assessment for Decision-Making* (pp. 39–56). Elsevier., DOI: 10.1016/B978-0-12-818355-7.00003-8

Tovazhnyanskyy, L., Klemeš, J. J., Kapustenko, P., Arsenyeva, O., Perevertaylenko, O., & Arsenyev, P. (2020). Optimal design of welded plate heat exchanger for ammonia synthesis column: An experimental study with mathematical optimisation. *Energies*, *13*(11), 2847. DOI: 10.3390/en13112847

Tran, C.-T., Lin, C., & Tsai, C.-C. (2023). Bragg Resonance of Water Waves by Multiple Permeable Thin Barriers over Periodic Breakwaters. *Water (Basel)*, *15*(3), 3. Advance online publication. DOI: 10.3390/w15030495

Trubia, S., Severino, A., Curto, S., Arena, F., & Pau, G. (2020). Smart roads: An overview of what future mobility will look like. *Infrastructures*, *5*(12), 107. DOI: 10.3390/infrastructures5120107

Tsai, C.-P., Chen, H.-B., & Lee, F.-C. (2006). Wave transformation over submerged permeable breakwater on porous bottom. *Ocean Engineering*, *33*(11), 1623–1643. DOI: 10.1016/j.oceaneng.2005.09.006

Uddin, M. J., & Jeong, Y.-K. (2021). Urban river pollution in Bangladesh during last 40 years: Potential public health and ecological risk, present policy, and future prospects toward smart water management. *Heliyon*, *7*(2), e06107. DOI: 10.1016/j.heliyon.2021.e06107 PMID: 33659727

Umoh, A. A., Nwasike, C. N., Tula, O. A., Adekoya, O. O., & Gidiagba, J. O. (2024). A review of smart green building technologies: Investigating the integration and impact of ai and iot in sustainable building designs. *Computer Science & IT Research Journal*, *5*(1), 141–165. DOI: 10.51594/csitrj.v5i1.715

UNDESA. (2018). World urbanization prospects: The 2018 revision. *Retrieved*, (August), 26.

Usatenko, D. V., & Biletskyi, I. V. (2023). *Block of sewage collector* (UA Patent 154296). Ukrainian National Office for Intellectual Property and Innovations. https://iprop-ua.com/inv/ew1g50e8/ [in Ukrainian]

USGBC. (2024). *Green Building*. From U.S. Green Building Council: https://www.usgbc.org/articles

Vadruccio, R., Salvadori, G., & Tumino, A. (2024). Smart metering and Internet of Things for efficient water management. *Procedia Computer Science*, *239*, 1498–1505. DOI: 10.1016/j.procs.2024.06.324

van Der Werf, H. M., Knudsen, M. T., & Cederberg, C. (2020). Towards better representation of organic agriculture in life cycle assessment. *Nature Sustainability*, *3*(6), 419–425. DOI: 10.1038/s41893-020-0489-6

Vega-Araujo, D., Martinez-Arguelles, G., & Santos, J. (2020). Comparative life cycle assessment of warm mix asphalt with recycled concrete aggregates: A Colombian case study. *Procedia CIRP*, *90*, 285–290. DOI: 10.1016/j.procir.2020.02.126

Velásquez, J. R., Schwartz, M., Phipps, L. M., Restrepo-Baena, O. J., Lucena, J., & Smits, K. M. (2022). A review of the environmental and health implications of recycling mine tailings for construction purposes in artisanal and small-scale mining communities. *The Extractive Industries and Society*, *9*, 101019. DOI: 10.1016/j.exis.2021.101019

Verhagen, T. J., Sauer, M. L., van der Voet, E., & Sprecher, B. (2021). Matching demolition and construction material flows, an urban mining case study. *Sustainability (Basel)*, *13*(2), 653. DOI: 10.3390/su13020653

Verma, I., & Taegen, J. (2019). Access to Services in Rural Areas from the Point of View of Older Population—A Case Study in Finland. *International Journal of Environmental Research and Public Health*, *2019*(16), 4854. DOI: 10.3390/ijerph16234854 PMID: 31810348

Visintin, P., Ali, M. M., Albitar, M., & Lucas, W. (2017). Shear behaviour of geopolymer concrete beams without stirrups. *Construction & Building Materials*, *148*, 10–21. DOI: 10.1016/j.conbuildmat.2017.05.010

Vitale Brovarone, E., & Cotella, G. (2020). Improving Rural Accessibility: A Multilayer Approach. *Sustainability (Basel)*, *2020*(12), 2876. DOI: 10.3390/su12072876

Vystavna, Y., Schmidt, S. I., Diadin, D., Rossi, P. M., Vergeles, Y., Erostate, M., Yermakovych, I., Yakovlev, V., Knöller, K., & Vadillo, I. (2019). Multi-tracing of recharge seasonality and contamination in groundwater: A tool for urban water resource management. *Water Research*, *161*, 413–422. DOI: 10.1016/j.watres.2019.06.028 PMID: 31226539

Wang, B., Arsenyeva, O., Zeng, M., Klemeš, J. J., & Varbanov, P. S. (2022). An advanced Grid Diagram for heat exchanger network retrofit with detailed plate heat exchanger design. *Energy*, *248*, 123485. DOI: 10.1016/j.energy.2022.123485

Wang, X., Wei, H., Luo, N., Luo, H., Zhou, X., Qin, B., Mei, Y., & Zhang, Y. (2024). Smart Materials and Micro/Nanoarchitectonics for Water Harvesting: From Fundamental Mechanism to Advanced Application. *Composites. Part A, Applied Science and Manufacturing*, *184*, 108241. DOI: 10.1016/j.compositesa.2024.108241

Wang, Z., Luther, M. B., Amirkhani, M., Liu, C., & Horan, P. (2021). State of the art on heat pumps for residential buildings. *Buildings (Basel, Switzerland)*, *11*(8), 350. DOI: 10.3390/buildings11080350

Wehbi, Z., Taher, R., Faraj, J., Ramadan, M., Castelain, C., & Khaled, M. (2022). A short review of recent studies on wastewater heat recovery systems: Types and applications. *Energy Reports*, *8*, 896–907. DOI: 10.1016/j.egyr.2022.07.104

Wei, B., & Yang, S. (2021). Polymer composites with expanded graphite network with superior thermal conductivity and electromagnetic interference shielding performance. *Chemical Engineering Journal*, *404*, 126437. DOI: 10.1016/j.cej.2020.126437

Weiland, S., Hickmann, T., Lederer, M., Marquardt, J., & Schwindenhammer, S. (2021). The 2030 agenda for sustainable development: Transformative change through the sustainable development goals? *Politics and Governance*, *9*(1), 90–95. DOI: 10.17645/pag.v9i1.4191

Weisse, R., Dailidienė, I., Hünicke, B., Kahma, K., Madsen, K., Omstedt, A., Parnell, K., Schöne, T., Soomere, T., Zhang, W., & Zorita, E. (2021). Sea level dynamics and coastal erosion in the Baltic Sea region. *Earth System Dynamics: ESD*, *12*(3), 871–898. DOI: 10.5194/esd-12-871-2021

Weisser, W. W., Hensel, M., Barath, S., Culshaw, V., Grobman, Y. J., Hauck, T., Joschinski, J., Ludwig, F., Mimet, A., Perini, K., Roccotiello, E., Schloter, M., Shwartz, A., Hensel, D. S., & Vogler, V. (2022). Creating ecologically sound buildings by integrating ecology, architecture and computational design. *People and Nature*, *5*(1), 4–20. DOI: 10.1002/pan3.10411

Wilson, A., Tewdwr-Jones, M., & Comber, R. (2019). Urban planning, public participation and digital technology: App development as a method of generating citizen involvement in local planning processes. *Environment and Planning. B, Urban Analytics and City Science*, *46*(2), 286–302. DOI: 10.1177/2399808317712515

Wilson, M. P., & Worrall, F. (2021). The heat recovery potential of 'wastewater': A national analysis of sewage effluent discharge temperatures. *Environmental Science. Water Research & Technology*, *7*(10), 1760–1777. DOI: 10.1039/D1EW00411E

Wolfsgruber, N., Tanda, A., Archodoulaki, V. M., & Burgstaller, C. (2023). Influence of filler type and content on thermal conductivity and mechanical properties of thermoplastic compounds. *Polymer Engineering and Science*, *63*(4), 1094–1105. DOI: 10.1002/pen.26266

Wu, D., Wang, H., & Seidu, R. (2020). Smart data driven quality prediction for urban water source management. *Future Generation Computer Systems*, *107*, 418–432. DOI: 10.1016/j.future.2020.02.022

Wurbs, R.A. and James, W.P. (2015) Water resources engineering, Pearson publication.

Wu, S. R., Fan, P., & Chen, J. (2015). Incorporating culture into sustainable development: A cultural sustainability index framework for green buildings. *Sustainable Development (Bradford)*, *24*(1), 64–76. DOI: 10.1002/sd.1608

Wu, Z. Y., Chew, A., Meng, X., Cai, J., Pok, J., Kalfarisi, R., Lai, K. C., Hew, S. F., & Wong, J. J. (2023). High Fidelity Digital Twin-Based Anomaly Detection and Localization for Smart Water Grid Operation Management. *Sustainable Cities and Society*, *91*, 104446. DOI: 10.1016/j.scs.2023.104446

Wu, Z., Mao, Z., & Shen, W. (2021). Integrating Multiple Datasets and Machine Learning Algorithms for Satellite-Based Bathymetry in Seaports. *Remote Sensing (Basel)*, *13*(21), 21. Advance online publication. DOI: 10.3390/rs13214328

Yang, Q., Pu, L., & Huang, S. (2024). Review of urbanization-associated farmland research in china: A sustainability perspective. *Land (Basel)*, *13*(4), 534. DOI: 10.3390/land13040534

Yellishetty, M., Karpe, V., Reddy, E. H., Subhash, K. N., & Ranjith, P. G. (2008). Reuse of iron ore mineral wastes in civil engineering constructions: A case study. *Resources, Conservation and Recycling*, *52*(11), 1283–1289. DOI: 10.1016/j.resconrec.2008.07.007

Yi, H., Xu, G., Cheng, H., Wang, J., Wan, Y., & Chen, H. (2012). An overview of utilization of steel slag. *Procedia Environmental Sciences*, *16*, 791–801. DOI: 10.1016/j.proenv.2012.10.108

Yin, S., Liu, J., & Zeng-lin, H. (2022). Relationship between urban morphology and land surface temperature a case study of nanjing city. *PLoS One*, *17*(2), e0260205. DOI: 10.1371/journal.pone.0260205 PMID: 35139084

You, W. & Zhong, L. (2023). Exploring hydrogen-based energy storage systems for canadian residential buildings: an energy evaluation methodology. *E3S Web of Conferences, 396*, 04001. .DOI: 10.1051/e3sconf/202339604001

Youssef, Z., Alshuwaikhat, H., & Reza, I. (2021). Modeling the Modal Shift towards a More Sustainable Transport by Stated Preference in Riyadh, Saudi Arabia. *Sustainability (Basel)*, *2021*(13), 337. DOI: 10.3390/su13010337

Yudelson, J. (2009). *Green Building Trends: Europe* (1st ed.). Island Press.

Zaman, A., & Hedayetullah, M. (2022). *Sustainable Water Resource Development and Management*. Apple Academic Press. DOI: 10.1201/9781003180494

Zannin, J., Ferrari, A., Larrey-Lassalle, P., & Laloui, L. (2022). Early-stage thermal performance design of thermo-active walls implemented in underground infrastructures. *Geomechanics for Energy and the Environment*, *30*, 100218. DOI: 10.1016/j.gete.2020.100218

Zantalis, F., Koulouras, G., Karabetsos, S., & Kandris, D. (2019). A Review of Machine Learning and IoT in Smart Transportation. *Future Internet*, *2019*(11), 94. DOI: 10.3390/fi11040094

Zarte, M., Pechmann, A., & Nunes, I. L. (2019). Decision support systems for sustainable manufacturing surrounding the product and production life cycle–A literature review. *Journal of Cleaner Production*, *219*, 336–349. DOI: 10.1016/j.jclepro.2019.02.092

Zhang, A., Zhong, R. Y., Farooque, M., Kang, K., & Venkatesh, V. G. (2020). Blockchain-based life cycle assessment: An implementation framework and system architecture. *Resources, Conservation and Recycling*, *152*, 104512. DOI: 10.1016/j.resconrec.2019.104512

Zhang, F., Chen, H., Li, X., Li, H., Lv, T., Zhang, W., & Yang, Y. (2017). Experimental study of the mechanical behavior of FRP-reinforced concrete canvas panels. *Composite Structures*, *176*, 608–616. DOI: 10.1016/j.compstruct.2017.05.072

Zhang, J., Fujiwara, A., & Thein, S. (2008). Capturing Travelers' Stated Mode Choice Preferences Under Influence of Income in Yangon City, Myanmar. *Journal of Transportation Systems Engineering and Information Technology*, *2008*(8), 49–62. DOI: 10.1016/S1570-6672(08)60032-X

Zhang, J., Liu, Y., Zhu, W., Pei, Y., Yin, Y., Qin, Y., Li, X., & Guiver, M. D. (2021). Self-adjusting anode catalyst layer for smart water management in anion exchange membrane fuel cells. *Cell Reports. Physical Science*, *2*(3), 100377. DOI: 10.1016/j.xcrp.2021.100377

Zhang, K., Li, Q., Zhang, J., Shi, H., Yu, J., Guo, X., & Du, Y. (2022). Simulation and Analysis of Back Siltation in a Navigation Channel Using MIKE 21. *Journal of Ocean University of China*, *21*(4), 893–902. DOI: 10.1007/s11802-022-5052-9

Zhang, W. (2023). Smart building skin design with dynamic climate adaptability of smart cities based on artificial intelligence. *Advances in Multimedia*, *2023*, 1–12. DOI: 10.1155/2023/2859195

Zhang, Y., Wang, H., Gao, W., Wang, F., Zhou, N., Kammen, D. M., & Ying, X. (2019). A survey of the status and challenges of green building development in various countries. *Sustainability (Basel)*, *11*(19), 5385. DOI: 10.3390/su11195385

Zhao, L. (2019). Fast Decentralized Data Analytics in IoT Wireless Networks. *IEEE Access*, *PP*, 1–1. DOI: 10.1109/ACCESS.2019.2960494

Zhao, X., Zuo, J., Wu, G., & Huang, C. (2019). A bibliometric review of green building research 2000–2016. *Architectural Science Review*, *62*(1), 74–88. DOI: 10.1080/00038628.2018.1485548

Zhong, W., Schröder, T., & Bekkering, J. (2022). Biophilic design in architecture and its contributions to health, well-being, and sustainability: A critical review. *Frontiers of Architectural Research*, *11*(1), 114–141. DOI: 10.1016/j.foar.2021.07.006

Zhu, H., Wang, T., Wang, Y., & Li, V. C. (2021). Trenchless rehabilitation for concrete pipelines of water infrastructure: A review from the structural perspective. *Cement and Concrete Composites*, *123*, 104193. DOI: 10.1016/j.cemconcomp.2021.104193

Zhu, H., Zhang, D., Goh, H. H., Wang, S., Ahmad, T., Mao, D., Liu, T., Zhao, H., & Wu, T. (2023). Future data center energy-conservation and emission-reduction technologies in the context of smart and low-carbon city construction. *Sustainable Cities and Society*, *89*, 104322. DOI: 10.1016/j.scs.2022.104322

Živković, M., & Ivezić, D. (2022). Utilizing sewage wastewater heat in district heating systems in Serbia: Effects on sustainability. *Clean Technologies and Environmental Policy*, *24*(2), 579–593. DOI: 10.1007/s10098-021-02063-6

Zohuri, B. (2017). Forced convection heat transfer. In *Thermal-Hydraulic Analysis of Nuclear Reactors* (pp. 323–345). Springer., DOI: 10.1007/978-3-319-53829-7_9

About the Contributors

Sérgio António Neves Lousada holds an International PhD. in Civil Engineering – Hydraulics, at the University of Minho (UMinho), Portugal and an International PhD. in Sustainable Territorial Development at the University of Extremadura (UEx), Spain. He teaches at the University of Madeira (UMa), field of Hydraulics, Environment and Water Resources and Construction. Furthermore, he collaborates with the Research Group on Environment and Spatial Planning (MAOT) - University of Extremadura (UEx), Badajoz, Spain; VALORIZA - Research Center for the Enhancement of Endogenous Resources, Polytechnic Institute of Portalegre (IPP), Portugal; CITUR-Madeira - Centre for Tourism Research, Development and Innovation, Madeira, Portugal; RISCO - Civil Engineering Department of University of Aveiro, Aveiro, Portugal; OSEAN – Outermost Regions Sustainable Ecosystem for Entrepreneurship and Innovation, Madeira, Portugal. Moreover, he holds and International Interuniversity Master's Degree in Ports and Coasts Engineering at the University of Las Palmas de Gran Canaria (ULPGC) and at the University of La Laguna (ULL).

***

Raul Manuel Costa Alves holds a position as a Técnico Superior at the Câmara Municipal de Machico, in the Subunit of Municipal Works, with a complexity level of 3. His responsibilities involve providing consultative services, conducting studies, programming, evaluating and applying technical and/or scientific methods and processes that support and prepare decision-making. He prepares opinions, projects, and activities related to defining municipal policies in the area of the organizational unit. He also conceives and carries out projects and plans for private works, preparing, organizing, and supervising their construction, maintenance, and repair. He performs calculations to ensure the strength and stability of the considered works, taking into account factors such as the nature of the construction materials to be used. He prepares the program and coordinates operations as the work progresses, and is also responsible for supervising maintenance and repair work on existing constructions. In addition, he oversees and provides technical direction for construction projects and conducts technical inspections.

Olga Arsenyeva is a Professor at the Department of Automation and Computer-Integrated Technologies of the O.M. Beketov National University of Urban Economy in Kharkiv (Ukraine). She received her Ph.D. degree in Chemical Engineering from the National Technical University "Kharkiv Polytechnic Institute" (Ukraine) in 2005 and a D.Sc. degree in Chemical Engineering from National Technical University "Kharkiv Polytechnic Institute" (Ukraine) in 2014. In 2017 she had an Alexander von Humboldt fellowship for experienced researchers at the Chair of Fluid Process Engineering of Paderborn University (Germany). Since 2019 she has been a Ukrainian Delegate of the Computer Aided Process Engineering Working Party in the European Federation of Chemical Engineering (EFCE). She is a reviewer of international scientific journals: Applied Thermal Engineering (Elsevier), Energy (Elsevier), Journal of Cleaner Production (Elsevier), and editorial-board member of Energy Sources, Part A: Recovery, Utilization, and Environmental Effects (Taylor and Francis). She was an invited professor to the Brno University of Technology (Czech Republic) and the Pázmány Péter Catholic University, Institute for Process Systems Engineering and Sustainability (Hungary). Her research activities mainly focus on Automation and Process Integration; Heat and Mass Transfer; Plate Heat Exchanger Design; Optimization of Heat Exchanger Networks; and Energy-saving Technologies.

Ihor Biletskyi is a Rector at the O.M. Beketov National University of Urban Economy in Kharkiv (Ukraine). He received his Ph.D. degree in Project Management from O.M. Beketov National University of Urban Economy in Kharkiv (Ukraine) in 2016 and a D.Sc. degree in Economics from Odesa National University of Technology (Ukraine) in 2023. He is an expert of the Ministry of Education and Science of Ukraine for examining research and development projects of the thematic direct "Energy and energy efficiency", and a Member of the Expert Council of the Sikorsky Challenge Startup Competition. His research activities mainly focus on Energy Efficiency, BIM Technologies, Smart Services, and Project Management.

José Cabezas has a Ph.D. in Ecology and has extensive research experience in sustainable management, conservation, and recovery of natural ecosystems (water, soil, fauna, flora, etc.), land use planning, environmental impact, remote sensing and GIS and spatial data infrastructure (SDI), aspects that are fundamentally oriented towards the conservation of transboundary Mediterranean ecosystems and their sustainability. His projects have focused on the development of geographic models of environmental and territorial management for rural areas with low population density for better and more coherent management of the territory, with a special focus on cross border cooperation.

Jige Sandipan is working as assistant professor in botany from 25 years and also five year experience in subject horticultural science in under graduate college. He has visited Dubai, Indonesia, Thailand, Singapore, Malaysia, Germany, France, Netherlands, Switzerland and Belgium attend conferences. He has also published more than thirty book chapter and research paper at national and international level. Also work on different academic committees at college and university level. He also work as research paper and chapter reviewer.

Petro Kapustenko is a key Researcher at SPIL – Sustainable Process Integration Laboratory, NETME Centre, FME, Brno University of Technology – VUT Brno, Czech Republic. From 1994 to 2022 he was the head of the Centre for Energy Saving Process Integration at National Technical University Kharkiv Polytechnic Institute (NTU KhPI), Kharkiv, Ukraine, founded with the support of DPI UMIST under the British Council Know How Link project. In 2000, he became a professor at NTU KhPI. In 1991 he cofounded Sodrugestvo-T engineering company and became its chief executive officer. The company has been involved in several projects funded by the EC and the World Bank aimed at the renovation of district heating networks and industry using process integration methods. They implemented advanced energy-saving equipment like CHP units, compact heat exchangers, heat pumps, and modern district heating substations. Main fields of interest: energy saving and efficiency, process Integration, resource reuse and Circular Economy (CE), Computer-integrated engineering, compact heat exchangers, Total Site, renewables, waste to energy and waste heat utilization, GHG, and water footprint reduction.

Sergiy Plankovskyy is Head of the School of Energy, Information and Transport Infrastructure at the O.M. Beketov National University of Urban Economy in Kharkiv (Ukraine). He received his Dr.Sc. degree in Aerospace Engineering from the National Aerospace University "Kharkiv Aviation Institute" (Ukraine), where he later served as Dean of the Faculty of Aircraft Engineering. He has more than 35 years of international experience in research and development, with a focus on processing with intense energy fluxes, including plasma, laser, and detonation technologies. He is the author of over 200 refereed publications and has successfully supervised 11 Ph.D. students. His research interests include innovative manufacturing technologies for the digital industry and advanced high-energy processing methods. He is also a member of the Expert Council on Mechanical Engineering of the National Agency for Higher Education Quality Assurance (NAQA), Ukraine.

Sandeep Poddar, Deputy Vice Chancellor (Research & Innovation), Member of Board of Studies, Lincoln University College, Malaysia, has been graduated from University of Calcutta in 1993 with Honours in Zoology, he has obtained Post Graduate Diploma in Dietetics from All India Institute of Hygiene and Public Health 1995, Master of Science in Zoology with specialization in Biochemical Genetics from Dayalbagh Educational Institute 1998 with distinction. He has completed PhD in Zoology from Vivekananda Institute of Medical Sciences on Cytotoxicity in 2004. After completing PhD he pursued Post Doctoral Research in different projects on Hemoglobinopathies and Oral Cancer mutation. He has published several research papers, filed patent and organized international conferences, and also edited books in Malaysia, Australia and India. Dr Sandeep is Fellow member of Indian Academic Research Association, and World Researchers Associations and Life members of Different associations including Indian Science Congress Association. Dr Sandeep is founder Assistant Secretary of Dr Tarak Nath Podder Memorial Foundation, Kolkata, India

Aditya Singh is an Independent Researcher, Editorial Review Board Member in 6 American International Journals (IJSSMET, IJSDS, IJPAEI, IJESGT, IJDSST, IJRLEDM), Reviewer in 8 American, 4 Singaporean & 1 Indonesian International Journals (IJUPSC, IJCVIP, IRMJ, JTA, IJAIBM, IJCEWM, JITR, IJSSSP, JDSIS, AAES, JCCE, SWT, & IJEECS) these 19 International Journals include some Scopus & WoS indexed ones, Mentor (Research) at Academy Innova World, Kalipur, India, & Ex-Alumni Mentor (LPU) at present. He had completed his regular B. Tech degree in Civil Engineering from Lovely Professional University, India, in 2020. He is currently affiliated with Amrita Vishwa Vidyapeetham, Coimbatore, India. He has 30 publications under his name, in which he is a single author in 21 of them, whereas in 6 of them he is the main author. Most of his publications are Indexed in Scopus. He had also worked as a Reviewer/Editorial Review Board Member for 153 times in the evaluation of manuscripts so far in Conferences, Research Books and Journals of International stature. His global presence as a professional could also be seen with 25 Paper Presentations in 22 Conferences in 9 countries (Thailand, Sri Lanka, Egypt, India, Cambodia, Bangladesh, Turkey, United Kingdom, & Germany); as well as his position as Program Committee Member, and Scientific Committee Member in International Conferences, and Reviewer in total 8 International Conferences (including countries Thailand, Cambodia, USA, India, and UK), and Keynote Speaker in Faculty Development Programme, and so on. He had also won 2 International Awards((Thailand) & (Cambodia)) as a Reviewer in 2023. He had worked on 52 positions so far in his professional career, including positions in Journals, Conferences, Research Books of International stature.

Subramanian Yavana Rani, Professor of MBA in the Decision Science area. Since receiving her MBA and Ph.D. in Management, she has devoted her career to Research and Teaching. She also has a Master's degree in Structural Engineering. She is the University rank holder in MBA. She has 16 years of teaching experience. Dr. Rani has received Young Scientist Fellowship Award from Tamilnadu State Council for Science and Technology (TNSCST). She has completed her fellowship from Indian Institute of Science (IISc), Bangalore. She is the life time member in Indian Society for Technical Education (ISTE). Dr. Yavana Rani has done two research projects sponsored by Indian Council for Social Science and Research (ICSSR). She is the UBA coordinator for the project under Unnat Bharat Abhiyan Scheme, Govt of India. She has published more than 20 research articles in Scopus indexed and Springer journals. She has presented papers in National and International conferences in India and abroad. She is the editor for many reputed journals. She has delivered key note address in countries like Egypt, Malaysia. Prof. Rani holds many administrative roles such as Dean Student Affairs, Director Coordination for various activities. She has been Project Director for Entrepreneurship Cell (EDC) sponsored by Department of Science and Technology (DST). She also coordinated many programmes. Professor has got funding for conducting conferences, International visit from various funding agencies like DST and ICSSR.

Natalia Teliura is Head of the Research Department at the O.M. Beketov National University of Urban Economy in Kharkiv (Ukraine). She received her Ph.D. degree in Environmental Safety from the Ukrainian Scientific and Research Institute of Ecological Problems (Ukraine) in 2019. She has more than 10 years of international experience in research and development. She is an expert of the National Agency for Higher Education Quality Assurance (NAQA), Ukraine, and a Member of the Expert Council (Environmental and Energy Security, "Green" Energy) of the Sikorsky Challenge Startup Competition. Her research activities mainly focus on Innovative Solutions for Environmental Protection, Environmental, and Energy Security.

Yevgen Tsegelnyk is a Deputy Head (Research) at the School of Energy, Information and Transport Infrastructure of the O.M. Beketov National University of Urban Economy in Kharkiv (Ukraine). He received his Ph.D. degree in Aerospace Engineering from the National Aerospace University "Kharkiv Aviation Institute" (Ukraine) in 2010. He has more than 15 years of international experience in research and development. He had an international traineeship as a Post-Doctorate at the

Czech Technical University in Prague (Czech Republic) in 2014. His research interests cover innovation technologies, including the methods of manufacturing for the digital industry, processing with intense energy fluxes (technologies using plasma, laser, and detonation sources), additive and subtractive manufacturing technologies, and applied computing techniques.

Index

H

I

J

L

M

P

R

S

T

U

www.ingramcontent.com/pod-product-compliance
Lightning Source LLC
LaVergne TN
LVHW080612300926
849162LV00005B/1136